AI Advancements in Internet of Things, Smart Healthcare, and Intelligent Devices

Muneer Khan
Cadre Tech Labs, USA

Shantanu Awasthi
Missouri Southern State University, USA

Mohd Sakib
UPES University, Dehradun, India

Mohd Vaseem Khan
Bansal Institute of Engineering and Technology, India

IGI Global
Scientific Publishing
Publishing Tomorrow's Research Today

Vice President of Editorial	Melissa Wagner
Director of Acquisitions	Mikaela Felty
Director of Book Development	Jocelynn Hessler
Production Manager	Mike Brehm
Cover Design	Jose Rosado

Published in the United States of America by
IGI Global Scientific Publishing
701 East Chocolate Avenue
Hershey, PA, 17033, USA
Tel: 717-533-8845 | Fax: 717-533-7115
Website: https://www.igi-global.com E-mail: cust@igi-global.com

Library of Congress Cataloging-in-Publication Data

Names: Khan, Muneer, 1996- editor | Awasthi, Shantanu, 1989- editor | Sakib, Mohd, 1993- editor | Khan, Mohammad Vaseem, 1980- editor
Title: AI advancements in Internet of Things, smart healthcare, and intelligent devices / edited by Muneer Khan, Shantanu Awasthi, Mohd Sakib, Mohd Vaseem Khan.
Description: Hershey, PA : IGI Global Scientific Publishing, [2026] | Includes bibliographical references and index. | Summary: "This book presents a comprehensive exploration of the powerful convergence between Artificial Intelligence (AI), Machine Learning (ML), and the Internet of Things (IoT), with a strong emphasis on their integration into intelligent devices"-- Provided by publisher.
Identifiers: LCCN 2025030692 (print) | LCCN 2025030693 (ebook) | ISBN 9798337357270 hardcover | ISBN 9798337357287 paperback | ISBN 9798337357294 ebook
Subjects: LCSH: Internet of things--Industrial applications | Artificial intelligence--Industrial applications
Classification: LCC TK5105.8857 .A398 2026 (print) | LCC TK5105.8857 (ebook)
LC record available at https://lccn.loc.gov/2025030692
LC ebook record available at https://lccn.loc.gov/2025030693

British Cataloguing in Publication Data
A Cataloguing in Publication record for this book is available from the British Library.

Dedicated with heartfelt gratitude to the beloved teaching and research community, whose tireless pursuit of knowledge continues to inspire generations.

Editorial Advisory Board

Sachin Gupta, *MAIT, Delhi, India*
Vijay H. *Kalmani Rajarambapu Institute of Technology, India*
Hammad, *Aligarh Muslim University, India*
Nazia Hasa, *Teerthanker Mahaveer University, India*
Balqees Talal Hasan, *IT College /Ninevah University, Iraq*
Nazia Hasan, *Teerthanker Mahaveer University, India*
Akhtar Husain, *MJPRU Rohilkhand University, India*
Uvesh Husain, *Mazoon College, Oman*
Jayesh Kumar Chandravadan Jani, *Dharmsinh Desai University, India*
Vijay S Jariwala, *Sardar Patel University, India*
Md. Saquib Jawed, *Bennett University, India*
S. Jeyakumar, *Garden City University Bangalore, India*
Rishabh Jha, *Lambton College, Canada*
Sudhanshu Kumar Jha, *University of Allahabad, India*
Yugal Jindle, *Meta Platforms Inc., USA*
Sriprasadh K., *SRM Institute of Science and Technology, India*
Murshid Kamal, *Aligarh Muslim University, India*
Amol Kasture, *Ajeenkya DY Patil University, India*
Ihtiram Raza Khan, *Jamia Hamdard, India*
Mohd Zain Khan, *Aligarh Muslim University, India*
Mudassir Khan, *King Khalid University, Saudi Arabia*
Wasim Khan, *KL University Vijayawada, India*
Ajay Kumar, *NIIT (China) Limited, China*
M. Vinoth Kumar, *Vel Tech Rangarajan Sagunthala R&D Institute of Science and Technology, India*
Rajeev Kumar, *Infrastructure University Kuala Lumpur (IUKL), Malaysia*
Sanjeev Kumar, *G.L. Bajaj Institute of Technology and Management, India*
Fazal Manan, *North Dakota State University, USA*
Oleksandra Mandych, *State Biotechnological University, Ukraine*
Marina Jirotka, *University of Oxford, UK*
Nimish Mathure, *Loras College, USA*
Kamran Ali Memon, *North Dakota State University, USA*
Janelli M. Mendez, *LORMA Colleges, Philippines*
Mingling, *University of Texas, USA*
Nandita Mishra, *Chetana's Institute of Management and Research, India*
Pratyush Mishra, *Wake Forest University, USA*
Siddhartha Mishra, *Samsung Research (IIT Dhanbad), India*
Sukriti Mittal, *Amazon, USA*
Asif Moiz, *Madanapalle Institute of Technology and Science, India*
Pratik Mungekar, *Savitribai Phule Pune University, India*

Table of Contents

Detailed Table of Contents

Computer Vision (CV) deals with the specific study in which Input applied to the system is an Image while output obtained are some kind of description or information or code. Image compression is an important subset process of computer vision or machine vision in machine learning, as they handle large image datasets for their applications in various fields of image processing, medical image analysis, data analysis, pattern recognition, image classification, detection and several other such applications. To significantly reduce storage space capacity and transmission bandwidth requirements, compressed images are used. Many machine vision and computer vision models deal with large amount of training datasets which can be stored and used for deep learning (DL) model training only by image compression. Compressed images can be processed faster leading to faster computer vision tasks and improved model accuracy while maintaining the original quality.

The integration of artificial intelligence (AI) into healthcare represents one of the most transformative technological advances in modern medicine. As healthcare systems worldwide grapple with rising costs, aging populations, and increasing data volumes, AI emerges as a powerful solution to enhance diagnostic accuracy, streamline operations, and improve patient outcomes. This comprehensive analysis explores the current applications of AI in healthcare, examines the challenges facing implementation, and outlines future directions for this rapidly evolving field.

The Internet of Things highlights the forthcoming detection, computing, and connections trends. The Internet of Things has transformed the field of health care. Growing numbers of chronic patients as well as an ageing population provide significant challenges with respect to current healthcare and medical services, demanding the incorporation of medical care alongside modern information and communication technologies (ICT) in order to enhance treatments. As a result, the breakthroughs studied in the present research will capitalize on healthcare systems powered by the Internet of things.A systematic bibliometric review is conducted by using the data of 4950 articles derived from the scopus database. The present study has used VOS viewer and R Biblioshiny package for the bibliometric analysis. The results pointed out that the Internet of Things can help policymakers improve societal health and economy.

The early discovery and accurate classification of brain tumors are crucial for improved recovery and prognosis due to their high lethality. The study introduces an ensemble model utilizing transfer learning, incorporating EfficientNetV2B3 and ResNetRS100 to identify four categories of brain tumors: meningioma, glioma, pituitary and No tumor. Grad-CAM is used to localize important MRI regions while feature importance analysis explains each model's contribution to the final decision, enhancing trust and interpretability. The proposed ensemble model had an accuracy of 98.95%, precision of 98.60%, recall of 98.80% and F1-score of 98.70% demonstrating superior performance compared to alternative approaches on a publicly accessible dataset. This innovative approach to prediction generation utilizes transfer learning, explainable artificial intelligence and ensemble modeling. It produces intelligible predictions that are efficacious for healthcare decision-making.

Urban traffic congestion affects economies, the environment, and quality of life. Traditional systems with fixed timing or localized sensors lack adaptability for dynamic conditions. This chapter examines how Artificial Intelligence (AI) and the Internet of Things (IoT) facilitate real-time, data-driven traffic control, thereby reducing congestion and enhancing traffic flow. It examines AI techniques, including machine learning, graph models, and multi-agent reinforcement learning, applied to adaptive signal timing, traffic forecasting, incident detection, and autonomous response. Case studies, such as Pittsburgh's SURTRAC, Alibaba's City Brain, and Google's Green Light, demonstrate measurable gains in travel time, emissions, and throughput. A layered IoT-AI framework is proposed for data collection, decision-making, actuation, and continuous feedback, alongside discussions of key challenges, including privacy, connected vehicle integration, scalability, edge computing, and federated learning to unlock the full potential of intelligent traffic systems.

The Internet of Things (IoT) is a key technology for addressing major challenges through its network of interconnected objects with sensors, electronics, and software. It forms a crucial foundation for smart city development by enabling efficient information and communication management. This chapter explores IoT-enabled energy management in smart cities, focusing on the integration of Plug-in Electric Vehicles (PEVs) and Wind Turbines (WTs) as renewable energy sources. It highlights how smart homes and neighborhoods can collaborate for peak shaving and valley filling, aligning energy demand with WT generation and PEV storage capabilities. A case study demonstrates the design of an energy management system that monitors total energy use while smoothing the daily power curve through WT and PEV integration, thus enhancing grid stability. Simulation results validate the proposed control strategy under various load scenarios, showing its effectiveness in flattening load profiles, reducing load variance, and preventing blackouts in smart grids.

The (IIoT) transformative force is modern driving an unprecedented wave of digitalization and connectivity across sectors of physical machinery with enabled actuators, advanced data analytics, resulting in monitor, exchange, analyze systems, humans, IIoT empowers industries to achieve greater operational efficiency, predictive maintenance, improved product quality, and enhanced safety. Its impact is particularly visible in manufacturing, energy, transportation, healthcare, and supply chain management—industries that historically relied on manual operations and reactive maintenance are now leveraging IIoT to shift toward data-driven, proactive models. One of the most significant contributions of IIoT is in operational efficiency. Through connected devices and smart sensors, organizations gain granular, real-time visibility into every aspect of their operations. Machines can autonomously report performance metrics, detect anomalies, and suggest optimizations, leading to reduced downtime and better resource utilization.

This article introduces a signal-resilient surveillance platform for monitoring extreme environments. Its tele-operated robotic vehicle operates in perilous locations inaccessible to humans, featuring a critical radar-enabled system for real-time obstacle detection and directional mapping. This data is displayed on a simple red/green interface, enabling rapid operator interpretation under stress. A closed IP connection ensures secure, reliable data transfer, maintaining constant communication despite structural or electromagnetic interference. Key use cases include rescue missions in confined spaces like borewells and hazard detection in underground mines. Experiments confirmed the system maintains signal integrity and real-time data feedback in harsh conditions. By prioritizing signal resilience and operational simplicity, this platform overcomes critical limitations of existing technologies, paving the way for future autonomous systems in disaster management and environmental hazard detection.

Foreword

In the present era of rapid technological transformation, the convergence of Artificial Intelligence (AI), Machine Learning (ML), and the Internet of Things (IoT) has become a defining force in reshaping industries, societies, and human lives. From healthcare and transportation to agriculture and smart cities, these interconnected technologies are revolutionizing the way data is collected, processed, and applied. At the heart of this revolution lies the urgent need for comprehensive scholarly resources that not only explain the theoretical underpinnings of AI and IoT but also provide actionable insights into their practical integration.

AI Advancements in Internet of Things and Intelligent Devices responds to this need with remarkable depth and clarity. This book presents an exceptional collection of chapters that address foundational principles, emerging frameworks, and cutting-edge applications of AI in IoT ecosystems. It successfully bridges the gap between research and practice, offering readers both rigorous academic contributions and forward-looking perspectives on real-world implementations.

The volume is thoughtfully organized into four parts. The first part, *Foundations and Security in IoT*, lays the groundwork for understanding hybrid scheduling frameworks, blockchain-enabled resilience, adversarial defenses, quantum-resistant cryptography, and advanced intrusion detection. This section demonstrates that any meaningful discussion of IoT must begin with trust, privacy, and robust security mechanisms.

The second part, *Enabling Platforms and Frameworks*, highlights innovative approaches to software testing, cloud-based visualization, secure bioinformatics-driven models, and efficient architectures such as LoRaWAN. These chapters illustrate how scalable platforms form the backbone of modern IoT solutions.

Healthcare and wearable technologies form the focus of the third part, *AI in Healthcare and Wearable Devices*. From explainable AI models for brain tumour classification to AI-driven wearables for remote monitoring, these chapters showcase how intelligent systems are reshaping patient care and medical research, ultimately improving human well-being.

The final part, *Smart Environments and Applications*, broadens the horizon by examining intelligent traffic systems, smart energy management, resilient surveillance, industrial automation, and agricultural innovation. These contributions demonstrate the immense potential of AI-powered IoT to address challenges across diverse sectors and environments.

What makes this book particularly significant is not only its comprehensive scope but also its interdisciplinary orientation. The editors Mr. Muneer Khan, Dr. Shantanu Awasthi, Dr. Mohd Sakib, and Mohammad Vaseem Khan have brought together a distinguished group of contributors whose expertise spans computer science, engineering, healthcare, and applied AI. Their collective effort ensures that this book is not just a compilation of chapters, but a coherent and authoritative resource that will benefit students, researchers, practitioners, and policymakers alike.

I am confident that this book will serve as both a reference and an inspiration. It will guide scholars in their research, equip practitioners with practical insights, and encourage future innovations at the intersection of AI and IoT. I commend the editors and contributors for their outstanding work and wholeheartedly recommend this book to the global academic and professional community.

Durgesh Pant
Department of Science and Technology, Govt of Uttarakhand, India

Preface

One of the most important factors fuelling the next stage of digital transformation is the integration of Artificial Intelligence (AI) and the Internet of Things (IoT). Intelligent devices, empowered by advanced computational capabilities, are reshaping how societies operate, how industries function, and how individuals interact with technology. Despite significant progress, there remains a pressing need for scholarly resources that capture the diversity of research and practice at this intersection while also addressing the challenges of scalability, security, and real-world application. This book is our effort to respond to that need.

AI Advancements in Internet of Things and Intelligent Devices brings together contributions from leading experts and researchers across academia and industry. The chapters collectively highlight cutting-edge developments, novel frameworks, and emerging applications that illustrate the promise of AI in advancing IoT ecosystems. In particular, the book covers critical domains such as security and privacy, enabling platforms, healthcare and wearable technologies, and smart environments—domains that represent both the greatest opportunities and the most urgent challenges in this field.

CHAPTER OVERVIEW

Chapter 1: Un-Supervised Machine Learning for Computer Vision Image Compression

This chapter explores the critical role of image compression in computer vision tasks. Through unsupervised machine learning, the authors demonstrate how compression improves processing speed, reduces storage requirements, and maintains image quality—enabling efficient training of deep learning models.

Chapter 2: Artificial Intelligence in Healthcare – Applications, Challenges, and Future Directions

This chapter presents a thorough analysis of AI's role in healthcare, covering current applications such as diagnostics and operational optimization, while also addressing barriers like data privacy, ethical concerns, and implementation complexity.

Chapter 3: A Bibliometric Analysis of IoT-Based Applications in Smart Healthcare Systems

Through a comprehensive bibliometric review of nearly 5,000 articles, the authors map the research landscape of IoT in healthcare. Their analysis reveals key trends, research clusters, and future directions that can inform policymakers and researchers alike.

Chapter 4: An Explainable AI-Based Transfer Learning Ensemble Model for Brain Tumor Classification

Leveraging ensemble learning and explainable AI, this chapter presents a highly accurate model for classifying brain tumors. It emphasizes transparency and interpretability using Grad-CAM visualizations and feature attribution techniques to support clinical decision-making.

Chapter 5: A Smart Farmland for Crop Protection Using Sensors and IoT

Focusing on precision agriculture, this chapter introduces a dual-layered system integrating soil moisture and fire detection sensors. The solution enables real-time irrigation control and fire risk mitigation, thereby enhancing crop safety and resource efficiency.

Chapter 6: Enhancing Intelligence and Interoperability in Smart Homes through Sensor-Actuator Integration and Decentralized Innovation

This chapter offers a novel framework for smart homes that combines edge computing, cloud services, and decentralized governance to achieve seamless interoperability, user autonomy, and adaptive functionality.

Chapter 7: Intelligent Urban Traffic Management Using AI and IoT

Addressing one of the most pressing urban issues, this chapter outlines a comprehensive AI-IoT blueprint for adaptive traffic systems. Case studies and technical architectures illustrate how intelligent transport systems can reduce congestion and emissions while improving commuter experience.

Chapter 8: Smart City Energy Management via IoT, Wind Power, and Electric Vehicles

By integrating renewable energy sources like wind turbines and plug-in electric vehicles, this chapter showcases how IoT-enabled energy systems can flatten power demand curves, improve grid reliability, and support sustainable urban living.

Chapter 9: Industrial IoT Automation, Predictive Maintenance, and Smart Manufacturing

This chapter delves into how IIoT technologies enable real-time monitoring, predictive analytics, and autonomous decision-making across manufacturing and other industrial domains—driving efficiency, quality, and safety.

Chapter 10: AI-Driven IoT Surveillance Framework for Signal-Resilient Environmental Monitoring in Extreme Conditions

The final chapter presents a robust AI-IoT surveillance platform for disaster-prone or inaccessible environments. With a focus on signal resilience and usability, the system supports real-time monitoring in underground mines and rescue missions, offering a glimpse into the future of autonomous environmental intelligence.

If you are a graduate student, researcher, practitioner, or policymaker interested in learning more about the interplay between AI and the Internet of Things, then this book is for you. Even while the book dives into different problems and their answers in each chapter, it presents a unified picture of digital ecosystems that are smart, safe, and long-lasting.

We would like to express our deep appreciation to all the contributing authors for their dedication and scholarly excellence. Their collective expertise has shaped this book into a comprehensive and valuable resource. We also extend our gratitude to the reviewers and editorial team at IGI Global for their support and guidance throughout the publication process.

We hope that this volume will inspire further research, foster collaboration, and serve as a foundation for innovation in the rapidly evolving domains of AI and IoT.

Acknowledgment

I would like to express my heartfelt gratitude to all those who have supported me throughout the process of writing this book, *AI Advancements in Internet of Things and Intelligent Devices*, to be published by IGI Global, Hershey, Pennsylvania, USA.

First and foremost, I dedicate this work to my beloved parents, whose love, guidance, and unwavering encouragement have been the foundation of my journey. Their constant belief in me has been my greatest strength.

I am deeply thankful to my supervisors, Prof. Tamanna Siddiqui and Prof. Suhel Mustajab, for their invaluable mentorship, insightful guidance, and continuous support. Their encouragement and wisdom have been instrumental in shaping both this book and my academic path.

I would also like to acknowledge the contributions of all those who reviewed this work, offering their valuable time, constructive feedback, and thoughtful suggestions to enhance its quality.

To everyone who stood by me, inspired me, and believed in my vision, I extend my deepest gratitude. This book is as much yours as it is mine.

Mohd Sakib

Chapter 1
Un-Supervised Machine Learning for Computer Vision Image compression

Asif Moiz
https://orcid.org/0009-0001-5843-910X
National Institute of Technology, Patna, India

Bambam Kumar
National Institute of Technology, Patna, India

ABSTRACT

Computer Vision (CV) deals with the specific study in which Input applied to the system is an Image while output obtained are some kind of description or information or code. Image compression is an important subset process of computer vision or machine vision in machine learning, as they handle large image datasets for their applications in various fields of image processing, medical image analysis, data analysis, pattern recognition, image classification, detection and several other such applications. To significantly reduce storage space capacity and transmission bandwidth requirements, compressed images are used. Many machine vision and computer vision models deal with large amount of training datasets which can be stored and used for deep learning (DL) model training only by image compression. Compressed images can be processed faster leading to faster computer vision tasks and improved model accuracy while maintaining the original quality.

DOI: 10.4018/979-8-3373-5727-0.ch001

INTRODUCTION

Computer vision Image compression is the process of reducing the size of a digital image file by minimizing the amount of data required to represent it. The goal is to make images easier to store, transmit and to train the Machine Learning model, while preserving as much visual quality as possible *(Subramanya, 2001*). Compression techniques achieve this by eliminating redundancy and irrelevant information from the image data and efficient training of a model is possible *(R.C. Gonzalez & Woods, 2018*). The question arises why Image Compression is Needed. The answer to this can be explained as Computer vision digital images can be very large in size, especially high-resolution ones. Without using compression methodology, the storing and sharing of these images would consume excessive storage space and bandwidth *(Skodras et al., 2001).* Digital image compression makes it feasible to handle large volumes of image data efficiently, whether it's for websites, for training the model, medical archives, mobile apps, or digital photography *(Kumar et al., 2025; Patel, Suthar, & Thakar, 2019).*

The need for storing and sending visual data has increased dramatically in the current digital environment. Because of their high resolution and colour richness, images sometimes consume a significant amount of storage space. They may be challenging to effectively manage across a variety of digital platforms as a result. By reducing the file size without appreciably sacrificing visual quality, image compression techniques assist in addressing this issue (*Skodras et al., 2001; Wang et al., 2022;sakib et al, 2025*).

Optimizing storage is a primary motivation for image compression. Large amounts of picture data are frequently handled in applications like multimedia, satellite data processing, and medical imaging, which can swiftly deplete available storage capacity *(Fischer, Fleckenstein, Herglotz, & Kaup, 2021; Dax, Nagarajan, Li, & Werner, 2023; Zhang et al, 2022)* . Compression lowers costs and improves data management by enabling more images to be stored in the same amount of memory.

Image compression not only helps save storage but also speeds up data transfer and uses less bandwidth. This is particularly helpful in places where fast loading times and less data use enhance the user experience, such as websites, social media, and mobile apps *(Sujitha et al., 2021; Jung, Jin, & Kim, 2024).* Compressed images are essential for effective communication in real-time scenarios, such as video conversations or remote medical consultations *(Lu et al., 2024; Fang et al., 2025).* What remains is a condensed version of the image that captures the most relevant features. This streamlined representation makes it easier for algorithms to detect patterns and relationships, as they no longer have to process unnecessary details.

Another key benefit lies in the way compression naturally filters out noise. In particular, lossy compression techniques smooth out small, insignificant variations that might otherwise distract algorithms or skew results. As a result, models that rely on clean data distributions—such as clustering algorithms or manifold learning approaches—can perform better. In fact, compressed images often exhibit clearer groupings and structure, aiding in the discovery of natural clusters.

It's also worth noting that many unsupervised learning architectures, such as auto-encoders and variational auto-encoders (VAEs), are inherently built around the concept of compression. These models learn to represent images using a compressed, latent form. Rather than treating compression as a separate pre-processing step, they make it part of the model's learning objective. The compressed latent space captures only the most essential features of the data, which can then be used for tasks such as anomaly detection, image synthesis, or unsupervised categorization.

Computer Vision Image compression carries vital role in typically improving system performance. Smaller image sizes reduce processing time and accelerate file access, retrieval, and manipulation. This is particularly beneficial in resource-constrained environments, such as embedded systems or portable devices.

In summary, image compression is essential for efficient storage, faster transmission, cost reduction, and overall system optimization, making it a foundational technique in all fields that handle digital images and rely on digital images for storing information, sharing data for the purpose of training the model or in other advanced fields like medical image and Satellite image processing fields.

Computer Vision (CV) mechanism handles large image datasets for their applications in various fields of image processing, medical image analysis, data analysis, pattern recognition, image classification, detection and several other such applications. Although image compression is not the core-task of computer vision but then also it weighs essential aspect of computer vision especially for data storage or data base management system, data transmission and model training as one of the most crucial artificial intelligence (AI) work. To significantly reduce storage space capacity and transmission bandwidth requirements, compressed images are used. Many machine vision and computer vision models deal with large amount of training datasets which can be stored and used for deep learning (DL) model training only by image compression. Compressed images can be processed faster leading to faster computer vision tasks and improved model accuracy while maintaining the original quality. Image compression techniques are roughly classified into two groups based on whether the compressed image may be utilized precisely to recreate the original image. These are the lossy technique and lossless technique approaches. There are many compression techniques which need to be explored and discussed as we can never assume an AI world without image recognition, Pattern recognition, dataset storage, Data base management system, Training to a model, Validation and Testing of a model, Face recognition, Biometric system imaging, Medical imaging especially in ultra sound imaging, Magnetic Resonance imaging (or MRI) Computed Tomography (or CT) scan, X-ray images or any other computer vision and image processing. Meanwhile, it is always admired to have a conceptual knowledge of these domains. We also need to focus more on the idea of increasing storage capacity of our database which can only be possible by compression technology advancements. Numerous firms are working day and night on compression technique upgradations and development, but many more steps to go and achieve more improvement. Here we deal the basics of computer vision Image compression techniques in context of Un-Supervised Machine Learning domain.

How Image Compression Works

(Here, the general Steps involved are discussed pointwise)-

Transformation: It is the method to convert original spatial domain of an image to another domain, the image is often converted to a different domain (e.g., frequency domain using Discrete Cosine Transform (DCT), Discrete Fourier Transform (DFT) or wavelet Transform) using different mathematical tools of transformation, where redundant or less significant information becomes more apparent. It makes compression more efficient with compact representation and it reduces Redundancy and image is represented with fewer samples or coefficients. In the Reception side, this compressed encoded data is decoded and inverse transform is used to reconstruct the received image signal as in original form *(Healy & Mitchell, 1981)*.

Quantization: It is the compression process of decreasing the number of distinct colours or intensity levels of an image. By this way, automatically the number of data required to represent also decreases. Values in this case are approximated to reduce precision, especially in lossy methods *(Johnston, 2018; Wang et al., 2022)*. This step contributes most to size reduction.

Encoding: This process involves transformation of an image data into digital format with maximum compact form making it easier to reduce the file size, store and manage the data, keeping original information preserved. The data is compressed using coding techniques that represent frequent patterns with fewer bits (e.g., entropy coding). It enables transmission of image signals faster over network which is now very essential especially for web pages, social media signalling, online communication and many more *(Chen et al., 2021; Xie, Cheng, & Chen, 2021)*.

Figure 1. Basic Structure of Lossy Image compression and Lossless Image compression showing various stages

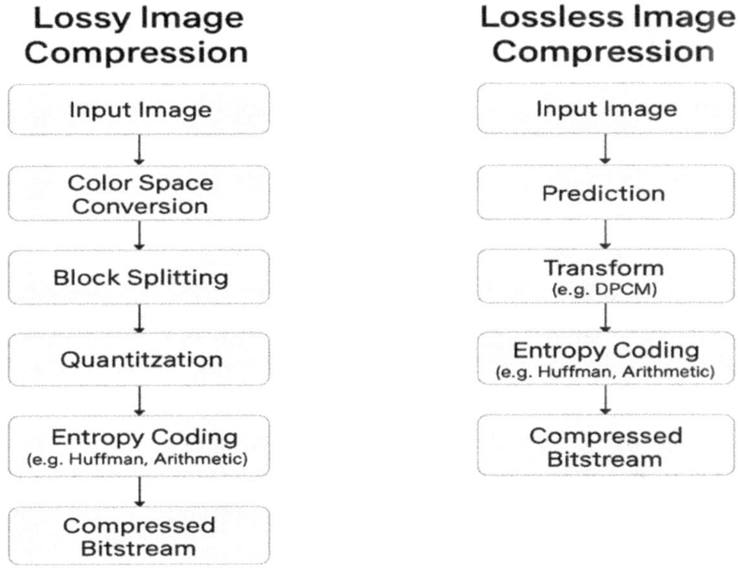

Benefits of Image Compression

- **Reduced Storage Requirements**: It enables more images to be stored on devices or servers. It helps large dataset files to consume less space so as to save the bandwidth requirements *(Fang et al., 2025; Kumar et al., 2025).*
- **Faster Transmission**: Due to this, Smaller files are uploaded and downloaded so transmission of data is more quickly executed over networks *(Lu et al., 2024; Fischer et al., 2021).*
- **Lower Costs**: This is achieved as it saves bandwidth and storage expenses in large-scale systems like cloud platforms or social media apps *(Sujitha et al., 2021; Sakib & Mustajab, 2024).* Image compression helps in reducing the size of digital image files by eliminating repetitive or non-essential information. This reduction makes storage and transmission more efficient, especially in bandwidth-limited environments. Compressed images are faster to load, which improves performance in web and mobile applications. In fields like healthcare and remote sensing, compression enables quicker data handling while maintaining necessary image clarity *(Dax et al., 2023; Patel et al., 2019*). It also supports cost-effective data management by minimizing memory and transmission needs.

Applications of Image Compression

- **Web and mobile media sharing** – It is used in world wide web and internetworking, for data storage, data sharing, model training and mobile media sharing. This is especially critical for users on slower connections or older devices. In mobile media sharing, particularly on apps like WhatsApp, Instagram, or Facebook, image compression ensures that users can send and receive pictures without consuming excessive mobile data. Compressed images strike a balance between maintaining acceptable visual quality and minimizing file size, which is crucial for users in areas with limited or expensive data plans. Lower storage requirements, enhanced or improved sharing speeds, Real time media processing where Applications like Snapchat, Instagram Stories, or live video broadcasting apps often need to compress images and video frames on-the-fly before uploading them. Image compression ensures that even real-time media streams remain smooth and uninterrupted.
- **Optimization for Content Delivery Networks (CDNs)**- Many web services use CDNs to distribute images globally. Compressing these images helps reduce latency and speeds up content delivery, especially on mobile networks, by ensuring that smaller image files are cached and transferred quickly.
- **Improved Model Training in Mobile AI Applications-** For mobile applications that use machine learning models (e.g., photo classification or augmented reality filters), compressed images are often used as input. Compressing the input images reduces pre-processing time and allows models to run faster on limited hardware, such as smartphones or tablets.
- **Digital photography** – For the use of information storage and information sharing in the context of various social, personal, economical and medical purposes, Digital photography is an essential medium which plays important role in computer vision as well as human vision.
- **Medical imaging archives** – Medical imaging archives deal with large volumes of data generated from modalities like CT Scan (Computed Tomography scan), Medical Imaging Resonance (MRI), Positron Emission Tomography (PET), X-rays, and ultrasound imaging systems (*Skodras et al., 2001; Sheena & Nair, 2024*). These images are typically stored in DICOM (Digital Imaging and Communications in Medicine) format and require long-term storage for diagnostic, legal, and research purposes. Due to the high resolution and 3D/4D nature of many scans, image compression becomes essential to manage storage, transmission, and processing Health issues related problems needs deep reporting and thorough check-ups depending on the condition of the patient and advice of the doctor or medical counsellor. So, Image compression is indispensable in medical imaging archives, ensuring efficient stor-

age, faster access, secure transmission, and scalable AI applications. The key lies in choosing the right compression method based on the clinical use-case balancing storage efficiency with diagnostic integrity.

- **Remote sensing and satellite image transmission** - Remote sensing systems and earth observation satellites capture vast amounts of high-resolution imagery used for agriculture, urban planning, climate monitoring, disaster management, and defence *(Dax et al., 2023; Wang et al., 2024)*. These images are often large and multidimensional (multi-spectral or hyper-spectral), making storage and transmission highly challenging.

Computer Vision Image compression plays a vital role in optimizing the vital complicated tasks by reducing data size without compromising critical information.

- **Video streaming** - As video is a sequence of compressed images, Video streaming has become an essential part of modern life, from entertainment platforms like YouTube and Netflix and other such platforms to live video conferencing, telemedicine, and surveillance *(Ahmmed, Paul, Murshed, & Taubman, 2021; Toderici, 2017)*. A video is essentially a sequence of images (frames) shown rapidly to create motion. Due to the massive data size of video, image compression is at the heart of video streaming, enabling efficient storage, transmission, and real-time playback. A video stream contains Frames per second (FPS) – usually 24 to 60, with Each frame is equal to an image (e.g., 1080p = 2 million pixels per frame). Without compression, video files are gigabytes in size, even for a few minutes. So, Video Compression is most essential tool to reduce large space consumption. Thus, video compression algorithms use image compression techniques (for intra-frame compression) and additional strategies to reduce redundancy across frames (inter-frame compression). Since a video is fundamentally a sequence of images, image compression directly governs video compression efficiency. It reduces spatial and temporal redundancy, enabling real-time streaming, efficient bandwidth usage, scalable quality, and improved user experience. As internet demand grows and 4K/8K content becomes widespread, advanced image compression will remain a core enabler of next-generation video streaming technologies.

Role of Image Compression in Key Areas

1. **PACS (Picture Archiving and Communication System):**
   - Reduces data storage costs.
   - Enhances speed of image retrieval.

- Supports integration with EHR (Electronic Health Records).

2. **Tele-radiology and Telemedicine:**
    - Faster upload/download of compressed images across remote networks.
    - Helps doctors diagnose without physical media.
3. **Data Backup and Archival Systems:**
    - Enables long-term retention of imaging data using affordable cloud or magnetic storage.
    - Compression standards ensure compatibility and regulatory compliance (e.g., HIPAA, FDA, IHE).
4. **Medical AI and ML Pipelines:**
    - Pre-processing via lossy or lossless compression reduces training time.
    - Compressed datasets are easier to distribute across data centres.
5. **Clinical Considerations:**
    - Compression Ratio vs. Diagnostic Quality must be balanced.
    - Studies show that visually lossless JPEG2000 can compress images up to 5–10 times without affecting radiological interpretation.
    - Radiologist acceptance and compression standards (DICOM Part 10) are crucial for legal and clinical trust.

Emerging Trends

- Deep learning-based compression (Auto-encoders, GANs) for smart and adaptive image storage.
- Region-of-interest (ROI) compression: High-resolution for diagnostic zones, compressed elsewhere.
- Cloud-native compression formats for large-scale medical repositories.

Two Compression techniques used widely are:

1. Intra-frame Compression (within a single frame):
    - It treats each frame as a separate image.
    - It uses image compression techniques like JPEG, Discrete Cosine Transform (DCT), Wavelet compression.
    - It removes spatial redundancy (e.g., smooth areas, repetitive textures).
2. Inter-frame Compression (between multiple frames):
    - It Exploits temporal redundancy (unchanged objects between frames).
    - It only stores changes (motion vectors and difference frames).
    - Common techniques used in this type are MPEG, H.264, HEVC (H.265), AV1.

Types of Image Compression

There are two broad categories of image compression:

1. Lossless or Reversible Computer vision compression approach-

Lossless or Reversible data compression technologies enable complete retrieval of the input image from the compressed (or encoded) image. In this type, the original image can be perfectly reconstructed from the compressed version. No information is lost. These are commonly referred to as noiseless because they add no noise in the signal. This is commonly referred to an entropy coding since it employs statistics or decomposition processes to reduce or eliminate redundancy. Only some of the frequent applications, such as medical diagnostic imaging analysis, employ lossless approaches of compression methods. So, we can say that it is useful where accuracy is critical, such as in technical drawings or medical scans of any modality like (MRI, CT Scan, X-Ray or Ultra Sonography or any other form of tests being done) for analysis of health check-up reports of a patient. Popular techniques used for these include encoding methods like Run-Length Encoding (RLE), Huffman Coding, Lempel-Ziv-Welch (LZW) and PNG format etc.

Run-Length Encoding (RLE) –

Run-Length Encoding (RLE) is one of the simplest and most widely used lossless image compression techniques. It is particularly effective in compressing images that contain large regions of uniform intensity or repeated patterns. The fundamental idea behind RLE is to represent consecutive occurrences of the same data value with a single value and a count. Instead of storing each pixel individually, the algorithm stores the pixel value once along with the number of times it is repeated in a continuous sequence, known as a "run."

For example, a sequence such as AAAAABBBCC would be encoded as 5A3B2C.

In the context of computer vision image compression, RLE is commonly applied to Binary Digit Map images (also called bitmap images which is a computer programming terminology and is a digital image file format, strict meaning is a map of bits or mapped array of bits), where pixels with the same colour or grayscale value appear in consecutive horizontal or vertical runs. A Bitmap represents a memory organization mainly used for storing digital images. This is especially useful in images like line drawings, icons, or binary images, where long runs of the same colour occur frequently.

Run Length Encoding offers several advantages which can be explained as:

- **Simplicity** or **Simple to implement**: The encoding and decoding processes are straightforward and computationally efficient. Run-Length Encoding is based on a simple principle: it stores repeated values as a single entry followed by a count. RLE is known for its straightforward method of compressing data. Both the compression (encoding) and decompression (decoding) steps are easy to understand and implement, requiring minimal processing power and no complicated algorithms. This makes it suitable for systems with limited computational resources.

However, Disadvantages of Run-Length Encoding (RLE) which can be explained as:

- **Ineffective for complex or detailed images:** For images like natural photographs, where pixel values vary frequently and unpredictably, RLE is not effective. It may even increase the size of the file rather than reduce it, as there are fewer long runs of repeating values to compress.

  While RLE may not be ideal for compressing all types of images—especially those with high complexity—it remains valuable in specific contexts. Its low processing requirements and ability to preserve original image data make it a practical choice for applications where speed and simplicity are more important than achieving the highest compression ratio. RLE also serves as a fundamental concept in many hybrid or advanced compression techniques.

  Huffman Encoding –

Huffman Encoding is a widely recognized method for compressing data without any loss of information. Introduced by David A. Huffman in 1952, this technique plays a crucial role in the compression of digital media such as images, audio files, and text.

At the heart of Huffman encoding lies the idea of using shorter binary codes for more common symbols and longer codes for less frequent ones. This strategy minimizes the overall number of bits needed to represent the original data, leading to efficient compression.

The process starts by analysing the frequency of each unique symbol in the dataset. These frequencies are then used to construct a Huffman Tree—a binary tree where each leaf node represents a symbol, and the path from the root to that leaf determines its code. Moving left in the tree adds a '0' to the code, while moving right adds a '1'. The result is a set of prefix-free codes, meaning no code is a prefix of another, ensuring unambiguous decoding.

Each symbol is treated as a leaf node with its frequency as the weight.

The two nodes with the smallest frequencies are combined to form a new internal node, whose weight is the sum of the two. This process is repeated until a single tree remains. Huffman Encoding stands out not only for its simplicity and effectiveness but also because it provides the optimal solution for symbol-based prefix coding under a given set of frequencies.

Each symbol's unique code is created by moving through the Huffman tree—starting at the root and following a specific path down to the symbol's leaf. A binary digit is added at each step, typically using '0' for a left turn and '1' for a right. This method ensures that no code is a prefix of another, making the encoding both efficient and unambiguous.

Because no code-word is a prefix of another, instantaneous decoding is possible, which makes Huffman coding particularly efficient for real-time or streaming applications.

In image compression, Huffman coding is often used as a final stage after transformation and quantization, such as in JPEG compression. The encoding process takes advantage of how often each quantized coefficient appears, allowing it to represent them more compactly and efficiently.

Advantages of Huffman Encoding:

Preserves original data with near-optimal efficiency: Huffman Encoding belongs to the class of lossless compression techniques, which means the data can be fully recovered after decompression with no loss of quality or information. It performs especially well when the likelihood of each symbol is known ahead of time. By giving shorter codes to symbols that occur more frequently and longer ones to those that appear less often, it reduces the overall average number of bits needed—making the compression very efficient for symbol-level encoding.

Limitations of Huffman Encoding:

- **Depends heavily on accurate symbol frequency analysis**: The effectiveness of Huffman coding hinges on knowing how often each symbol appears in the data. If the frequency estimates are off — especially in unpredictable or real-time environments — the compression won't be as efficient. The resulting code structure may fail to represent the data optimally, potentially reducing the benefit of compression.

Fixed Huffman codes can be suboptimal if the symbol distribution shifts.

- **Not ideal for data with shifting patterns:** Huffman Encoding assumes that symbol frequencies stay relatively consistent. In scenarios like live data streams or content with frequent variations, this assumption breaks down.

Since the code assignments don't adjust to changing patterns, the compression becomes less efficient over time.

Even with these drawbacks, Huffman Encoding continues to be a foundational concept in the world of data compression. Its straightforward logic and reliable performance in stable data environments make it a valuable tool, especially when used alongside other techniques in hybrid systems.

Lempel-Ziv-Welch (LZW) Compression -

LZW is a lossless data compression algorithm that falls under the category of dictionary-based techniques. Introduced by Terry Welch in 1984, it builds upon the earlier LZ78 method developed by Abraham Lempel and Jacob Ziv. Over time, LZW gained wide acceptance for its balance of speed and effectiveness and became the underlying compression mechanism for several popular file formats, including GIF, TIFF, and PDF.

How LZW Compression Works:

LZW compression operates by building a dictionary of symbol sequences as it processes the data. Both the encoder and decoder work in sync, allowing efficient compression without any loss of information *(Abd-Alzhra & Al-Tamimi, 2022)*.. Here's how the process unfolds:

1. **Starting Dictionary:** The compression begins with a predefined dictionary that includes all basic single-character entries — for example, all 256 ASCII characters.
2. **Reading Input and Matching Sequences:** The algorithm reads the input data one character at a time. It keeps checking if the current sequence, when combined with the next character, matches an existing entry in the dictionary.
3. **Adding New Entries:** When a combination of characters isn't found in the dictionary, it gets added as a new entry, and the code for the longest previously known match is written to the output.
4. **Output of Codes:** Instead of transmitting the actual characters or sequences, the encoder sends out the numeric codes corresponding to dictionary entries. This significantly reduces the amount of data being transferred.

On the decoding side, the process is mirrored. The decoder starts with the same base dictionary and builds it up in the same order by following the sequence of received codes. Because both sides construct the dictionary in real-time and identically, there's no need to share it explicitly — ensuring a compact yet accurate restoration of the original data.

Advantages of LZW compression:

- **Automatically adapts to the data:** One of the key strengths of LZW is that it doesn't rely on any pre-calculated symbol frequencies. Instead, it builds its dictionary dynamically while scanning the input. This makes it flexible and capable of adjusting to patterns as they emerge—ideal for compressing data where the structure isn't known in advance or changes over time.
- **Great for data with repeated patterns:** LZW shines when dealing with input that includes recurring sequences or repeated segments. As it detects and stores these patterns in its dictionary, it can substitute long strings of data with short codes.
- **Broad usage and software support:** LZW has been widely adopted across various formats and tools, thanks to its effectiveness and simplicity.

- **Limitations or demerits of LZW Compression:**

- **Less effective on high-entropy data**: If data is highly random or lacks repetition, compression gains are minimal.
- **Fixed dictionary size**: In some implementations, dictionary growth is capped, which may limit performance on very large files.

LZW stands as a powerful technique in lossless compression, combining simplicity, speed, and broad applicability, especially in systems where fast, reversible compression is required without prior statistical modelling.

PNG (Portable Network Graphics) format-

PNG (Portable Network Graphics) is a widely used lossless image compression format designed for efficient storage and transmission of raster graphics *(Skodras et al., 2001; Johnston, 2018)*. It was developed as an open and patent-free alternative to the proprietary GIF format and officially published as an ISO standard (ISO/IEC 15948) in 2004.

The PNG format supports lossless compression, which means that image data is preserved exactly during encoding and decoding, with no degradation in quality. This makes PNG especially suitable for images that require high fidelity, such as line art, textual graphics, logos, and images with transparency.

Key Features of PNG (Portable Network Graphics)

1. **Lossless Compression Using DEFLATE:**

PNG employs the DEFLATE compression method, which blends LZ77 (a dictionary-based approach) with Huffman coding. This hybrid model removes redundancy in image data while ensuring that no visual detail is lost, making PNG an excellent choice when image quality must remain intact.

2. **Transparency Through Alpha Channel Support:**

One standout feature of PNG is its support for varying levels of transparency through an 8-bit alpha channel. This means each pixel can have 256 levels of opacity, allowing images to blend smoothly with any background—especially useful in web and UI design.

3. **Broad Colour Format Compatibility:**

PNG handles multiple colour models including grayscale, indexed colour, and truecolor (24-bit RGB). For use cases requiring higher precision, such as medical or scientific imaging, it also supports 48-bit RGB and 16-bit grayscale images, delivering richer colour depth and detail.

4. **Built-in Error Detection**:

To enhance reliability during data transfer, PNG files include Cyclic Redundancy Checks (CRC). This mechanism helps detect errors in the file stream, improving robustness and preventing corrupted images from being displayed.

5. **Adam7 Interlacing for Progressive Display:**

PNG supports Adam7 interlacing, a method that displays a coarse preview of the image quickly, then refines it as more data is loaded. This is particularly beneficial on slow connections, giving users a preview while the rest of the image downloads.

6. **No Native Support for Animation:**

Unlike GIF files, standard PNGs don't support animation. However, a variant called APNG (Animated PNG) was developed to fill this gap, offering animation capabilities while retaining PNG's high-quality compression and transparency features.

Advantages of PNG

- **Free and Open Format:** PNG is an open standard, meaning it's not restricted by patents or licensing fees. This has contributed to its widespread adoption across platforms and software.
- **Efficient Lossless Compression:** While maintaining image quality, PNG compresses data efficiently—especially for images with large areas of flat colour or simple patterns.

- **Advanced Transparency and Colour Support:** PNG offers rich colour depth along with full alpha channel transparency, allowing smooth blending and accurate colour representation in high-quality graphics.

Limitations of PNG

Larger File Sizes for Photographic Content: When dealing with complex or photographic images, PNG files tend to be larger than those compressed using lossy formats like JPEG, which can make storage and transmission less efficient.

No Built-in Animation (Standard PNG): Unlike GIFs, standard PNG does not support animation. While APNG (Animated PNG) exists as an extension, it's not universally supported across all platforms and browsers.

Here is the comparison of all the above types in a Tabular form in the following table:

Table 1. Image Compression techniques comparison

Parameter	Run-Length Encoding (RLE)	Huffman Coding	Lempel-Ziv-Welch (LZW)	PNG Format
Type	Statistical (simple redundancy-based)	Statistical (entropy-based)	Dictionary-based	Hybrid (DEFLATE: LZ77 + Huffman)
Compression Ratio	Low to moderate (data-dependent)	Moderate	Moderate to high	High (for lossless methods)
Data Suitability	Repetitive data (e.g., icons, fax)	General-purpose	Text and image files with patterns	Complex images with transparency
Speed (Compression/ Decompression)	Very fast	Fast	Moderate	Moderate
Memory Requirement	Minimal	Low	Moderate	Moderate
Adaptability	Not adaptive	Static or semi-adaptive	Adaptive	Adaptive (based on image content)
Support for Colour Depth	Limited	N/A (used in combination)	Limited to moderate	High (supports 8, 24, and 48-bit)
Transparency Support	No	No	No	Yes (alpha channel supported)
Usage in File Formats	TIFF, PCX	JPEG (entropy coding), PNG	GIF, TIFF	PNG

continued on following page

Table 1. Continued

Parameter	Run-Length Encoding (RLE)	Huffman Coding	Lempel-Ziv-Welch (LZW)	PNG Format
Strengths	Simple, fast, easy to implement	Optimal for known symbol frequency	Excellent for repetitive sequences	Widely supported, efficient, transparent
Limitations	Ineffective on complex or noisy images	Requires frequency analysis	Less efficient on non-repetitive data	Larger files than lossy formats

LOSSLESS IMAGE COMPRESSION: THEORETICAL EXPLANATION

Lossless image compression refers to a set of techniques that reduce the size of image files while preserving every detail of the original data. After decompression, the image is restored exactly as it was—down to the last pixel—ensuring that no information is lost in the process *(Chen et al., 2021; Wang et al., 2022).*

This level of precision is vital in areas where data integrity is critical, such as medical imaging, scientific visualization, and archival of legal or historical documents. Even the slightest alteration in such contexts could compromise reliability or interpretation.

Lossless compression techniques generally focus on:

1. **Statistical Redundancy**: Many pixel values in an image may repeat or show predictable patterns. These repetitions can be represented using fewer bits through various encoding schemes.
2. **Spatial Redundancy**: Neighbouring pixels often have similar values, especially in smooth regions of an image. Predictive models can estimate pixel values based on neighbours, storing only the difference (which is often small).

Lossless compression methods make heavy use of statistical redundancy. Instead of storing exact values for every pixel, they:

- Use **prediction models** to estimate pixel values based on their neighbours and store only the small differences (called residuals).
- Apply **entropy coding** to represent frequently occurring pixel values or patterns with shorter binary codes.

For example:

- **Huffman coding** assigns shorter codes to more frequent pixel values and longer codes to rare ones.
- **Run-Length Encoding (RLE)** replaces sequences of the same value with a count and the value itself (e.g., "10 white pixels" instead of storing "white" 10 times).

Benefits of Exploiting Statistical Redundancy

- **Reduces File Size**: Fewer bits are needed to represent the image.
- **No Information Loss**: Since the redundancy is just repetition, it can be compacted without changing any actual data.
- **Faster Transmission**: Smaller size means quicker upload/download and less storage usage.

Example
Imagine a row of pixels with values like:
120, 120, 120, 120, 120, 120
Instead of saving all six values, a compression algorithm can store
(Value: 120, Count: 6)
This compact form represents the same data more efficiently.

So, we can conclude that Statistical redundancy is a key factor that allows images to be compressed without losing quality. By recognizing and removing repetitive or predictable patterns, image compression algorithms can store the same visual information using fewer bits, improving efficiency in storage and communication.

Typical Workflow of a Lossless Image Compression

Figure 2. Structure of Lossless Image compression showing various stages

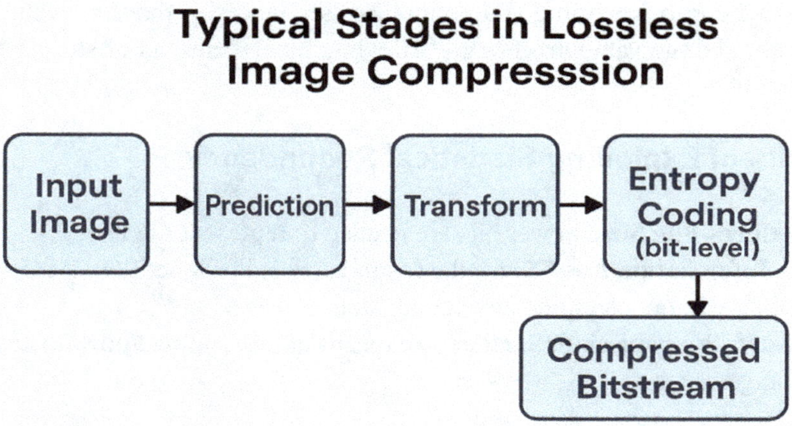

The above figure shows the structure of a lossless image compression algorithm which typically involves the following steps:

1. **Modelling or Prediction**: This phase attempts to estimate the value of a pixel based on previous pixels. For instance, if a region is smooth, the next pixel might be similar to its neighbours. Predictive coding stores the difference between the actual and predicted value, often resulting in smaller values that can be encoded efficiently.
2. **Transformation (optional)**: Some methods use reversible transforms (like the Burrows-Wheeler Transform or integer wavelet transforms) to reorganize data in a way that makes it easier to compress.
3. **Entropy Encoding**: This final stage involves encoding the data in a compact form. Popular methods include:
   - o **Huffman Coding**: Frequently occurring symbols are represented with shorter codes.
   - o **Arithmetic Coding**: Encodes the entire image as a single number between 0 and 1, based on symbol probabilities.
   - o **Run-Length Encoding (RLE)**: Replaces sequences of the same value with a count and a single value, which is useful in uniform areas.

Popular Lossless Compression Formats

- **PNG (Portable Network Graphics)**: Widely used on the web, PNG uses DEFLATE compression (a combination of LZ77 and Huffman coding) for lossless compression.
- **GIF (Graphics Interchange Format)**: Uses Lempel-Ziv-Welch (LZW) compression. Though it supports only 256 colours, it preserves image data exactly.
- **TIFF (Tagged Image File Format)**: Often used in publishing and medical imaging, it supports both lossless and lossy compression, including LZW and ZIP.

Advantages of Lossless Image Compression:

- **Perfect Reconstruction**: The original image can be restored bit-for-bit.
- **Data Integrity**: Essential for applications where even minor changes can have significant consequences.
- **No Visual Degradation**: Unlike lossy methods, image quality remains unchanged regardless of compression level.

Limitations of a lossless image compression:

- **Lower Compression Ratio**: Compared to lossy techniques, the reduction in file size is typically smaller.
- **More Computational Resources**: Some advanced lossless algorithms may be slower or require more memory.

Irreversible or Lossy Computer Vision Compression Approach

Lossy approaches also called Irreversible compression, is a technique that reduces the size of the image data by permanently eliminating some of the information. This process achieves significant size reduction while maintaining acceptable visual quality. It is widely used in applications such as digital photography, web image delivery, and mobile devices, where storage space and bandwidth are limited. Lossy compression contrasts with lossless methods, which retain all original data. The goal is to strike a balance between file size and visual fidelity. These techniques provide significantly greater compression ratio Because the quality of the recreated images is good enough for numerous applications used in data transfer, lossy approaches are commonly used. Using this procedure, the decompressed image is reasonably same as the original image but not completely identical. In the steps of prediction, change, and deconstruction is totally reversible. Quantization steps causes information loss. However, the entropy coding that occurs after the quantization process is considered lossy. Decoding step is then carried out in reverse., compressed data is first decoded using entropy decoding to get quantized data. The rebuilt image is then created by de-quantization then applying an inverse transformation. There are several parameters like Compression ratio, signal-to-noise ratio, and coding and decoding speed which make essential factors in determining the performance of lossy compression algorithms.

Lossy Image Compression: Theoretical Approach

Lossy image compression refers to a class of techniques that intentionally discard certain data from an image to reduce file size. The discarded data typically represent information that is less perceptible to the human visual system (HVS), making this form of compression particularly effective for natural images, such as photographs. Unlike lossless compression, where the original image can be perfectly reconstructed, lossy methods achieve higher compression ratios by accepting some degree of degradation in visual quality.

This method permanently removes certain data that are considered less important to human perception. Although some detail is lost, the resulting image still looks acceptable to the viewer. This approach achieves much higher compression rates and is used for everyday photos and online media.

General Workflow of Lossy Compression

The typical pipeline of a lossy compression algorithm includes the following stages:

Lossy image compression algorithms aim to significantly reduce file size by discarding information that has minimal impact on visual quality. This process generally follows a structured sequence of steps:

1. **Transformation:**

The first step involves converting the image from its original (spatial) domain into a frequency domain using mathematical transforms like the Discrete Cosine Transform (DCT) or Wavelet Transform. These transforms help reorganize the image data so that most of its energy is concentrated in a small number of frequency coefficients. This makes it easier to identify which parts carry essential visual information and which can be simplified or discarded.

2. Quantization:

Once the transformation is done, the frequency coefficients are quantized—a step where loss of information actually occurs. In this process, groups of similar values are rounded or mapped to single representative values. High-frequency components, which often correspond to fine details or noise, are given less precision or removed altogether. This significantly reduces the amount of data to be stored or transmitted.

3. **Encoding:**

After quantization, the remaining data is encoded using entropy coding techniques like Huffman coding or arithmetic coding. These methods compress the quantized values based on their frequency of occurrence, allowing the final output to be stored or transmitted in a much more compact form.

Each of these stages plays a critical role in balancing compression efficiency with acceptable image quality loss. This pipeline is commonly used in formats like JPEG, where reducing storage needs is more important than perfect reconstruction.

Figure 3. Structure of Irreversible or Lossy image compression showing various stages

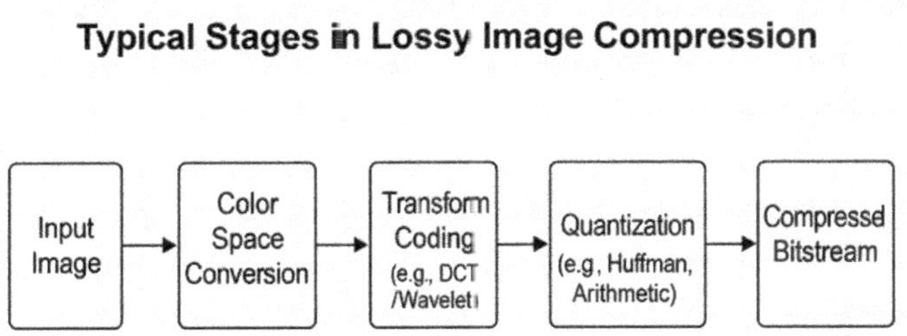

Examples of Lossy Compression Algorithms

- **JPEG format**: One of the most widely used formats for lossy compression, JPEG employs block-wise DCT, followed by quantization and Huffman encoding.
- **WebP** (in lossy mode): Developed by Google, this format supports both lossy and lossless modes and uses predictive coding and transform coding for compression.
- **HEIC (High-Efficiency Image Coding) format**: Based on the HEVC video standard, HEIC achieves higher compression ratios than JPEG at similar or better quality.

Classification of Lossy Compression Techniques:

1. **Transform-Based Techniques:** Transform-based methods play a central role in reducing image redundancy by converting spatial image data into a frequency domain. This process makes it easier to identify and compress parts of the image that are less visually important. Some commonly used transforms include:
   Discrete Cosine Transform (DCT):

DCT is the foundation of the JPEG compression standard. It works by dividing the image into small blocks (typically 8×8 pixels) and converting each block into a set of frequency components. Most of the visual information is concentrated in the low-frequency components, while the higher frequencies—often representing fine details or noise—can be more heavily compressed. This makes DCT highly effective for compacting energy and enabling efficient quantization.

Discrete Wavelet Transform (DWT):

Used in JPEG 2000, DWT offers a multi-resolution analysis of the image. Unlike DCT, it does not operate on fixed-size blocks but analyzes the entire image at various scales. This eliminates visible block boundaries, known as blocking artifacts, and provides better image quality at high compression ratios. Its hierarchical representation also supports progressive image transmission and scalable coding.

Karhunen–Loeve Transform (KLT):

KLT is considered the most efficient transform for energy compaction in theory, as it de-correlates data optimally based on the image's statistical properties. However, it is rarely used in practical systems because it requires prior knowledge of the image statistics and involves high computational cost. As a result, KLT is mostly limited to academic or simulation studies *(Cheng, Sun, Takeuchi, & Katto, 2020)*..

Fourier Transform (FT) is also an effective way of transform for periodic images but rarely used in modern image compression applications. So we mainly discuss the above three types of Transform based techniques, which we can summarize in Table below with following basic points:

Table 2. Transform based technique comparison

Feature	Discrete Cosine Transform (DCT)	Discrete Wavelet Transform (DWT)	Karhunen–Loève Transform (KLT)
Application	Widely used in JPEG	Core of JPEG 2000	Used in theoretical models, not in practical systems
Transformation Approach	Operates on fixed-size blocks (typically 8×8)	Analyzes the image at multiple resolutions	Uses eigenvectors based on image covariance
Energy Compaction	High energy compaction in low-frequency coefficients	Good energy compaction across multiple scales	Optimal energy compaction for any dataset
Visual Quality	May cause blocking artifacts at high compression	Avoids blocking artifacts; provides smoother images	Excellent, but computationally intensive
Compression Performance	Efficient for images with gradual intensity changes	Performs well on both smooth and detailed regions	Theoretically best but not scalable
Progressive Transmission	Not natively supported	Yes, supports progressive decoding	Not practical for real-time or streaming applications
Computational Complexity	Low; fast and hardware-friendly	Moderate; more complex than DCT	High; requires prior statistical analysis
Suitability for Real-Time Use	Excellent; used in real-time applications	Good; suitable for modern scalable codecs	Poor; not used in real-time systems

So the above Tabular form shows the basic comparison between the three important Transformation types with respect to different Parameters mentioned in the Table above., and can be summarized in other words as below:

DCT is a practical and efficient transform that has stood the test of time due to its simplicity and effectiveness, especially in block-based coding systems like JPEG.

DWT offers improved quality and scalability by working across multiple resolution levels, making it well-suited for advanced image formats like JPEG 2000.

KLT provides the theoretical gold standard for energy compaction but is limited to research applications due to its heavy computational demands.

2. **Predictive Coding Techniques**

Differential Pulse Code Modulation (DPCM) encodes the difference between predicted and actual pixel values, suitable for smooth images *(Taspınar, 2023).*

Lossy Predictive Models employ statistical or neural network-based prediction to reduce redundancy.

3. **Vector Quantization (VQ)**

Vector Quantization divides images into blocks and matches them to a predefined codebook. Fast but may introduce blocking artefacts.

4. **Subband Coding**

Subband Coding decomposes image into multiple frequency bands using filter banks, each quantized separately.

5. **Fractal Compression**

Fractal Compression exploits self-similarity in images, offering high compression but with slow encoding time.

6. **Neural Network-Based Compression**

Auto-encoder-based Compression uses deep learning models to learn compact representations of images.

GAN-based Compression utilizes generative adversarial networks to improve perceptual image quality.

Transformer-based Compression leverages attention mechanisms for more flexible and efficient compression *(Ballé et al., 2018; Minnen, Ballé, & Toderici, 2018; Wang et al., 2022; Chen et al., 2021; Xie et al., 2021; He et al., 2021;Akbari et al,2019*).

7. **Compressive Sensing**

Compressive Sensing samples fewer coefficients based on sparsity principles, ideal for certain low-sampling applications *(Pant et al., 2019; Sujitha et al., 2021)*.

8. **Dictionary Learning and Sparse Coding**

Sparse Coding learns a dictionary of basic elements and represents image patches sparsely for efficient encoding.

9. **Graph-Based Compression**

Graph-Based Methods represents images as graphs and applies transforms for compression, useful for irregular data.

10. **Region-Based or Segmentation-Based Compression**

Region-Based Compression applies stronger compression to less important areas and preserves quality in regions of interest *(Khan & Tripathi, 2024).*

11. **Hybrid Methods**

Hybrid approaches combine multiple methods (e.g., DCT + AI) to enhance performance and minimize artefacts.

There are several types of compression techniques which make use of sending compressed data for transmission and for training the model. We can discuss some of them like:

Coding Transformation act or popularly called Transformation coding is nothing but a data compression which deals with mathematical transformation reducing the redundancy in image data allowing natural data like audio voice signals or photo-graphic pictures to have efficient storing capacity and transmission. This has several advantages like it reduces redundancies, reduces bandwidth and has brilliant capacities to form images with fewer colours. JPEG is a widely used example of such type of transform coding.

Transform method - Transform algorithms are commonly utilized in lossy image compression systems because they outperform decorrelation. However, most transformations, such as discrete cosine transform (popularly called as DCT) and discrete Fourier transform (popularly known as DFT), involve real- or complex-valued transform coefficients that must be quantized for coding, making them difficult to utilize in the loss-less signal encoding. The discrete Walsh-Hadamard transform (or also called as WHT) is an specific transform algorithm that can be used directly for lossless signal encoding process. Because the WHT coefficients are binary fractions, there is no need for quantification.

Vector quantization (VQ) is another computer vision data compression technique which represents set of data points or vectors with the help of smaller set of representative vectors called code words. So it is also called code-word compression technology. In this computer vision technology, grouping of similar input vectors is done to form a "cluster" where each of them is assigned a "code-word" that is referred as its representative. The group of all such code words which are representing the data is called Codebook. A training set is also applied to generate the codebook of related representative vectors or code words. Clustering of identical input vectors are done and each cluster is assigned a representative code-word. Then in coding scheme section of transmitter, each and every input vector is compressed by finding the near-most code word. The address or index of the matched code word is transmitted and not the entire vector itself as such. Then in the decoding scheme section of receiver, same code word is retrieved from corresponding code book with the help of transmitted index. Thus by this original data is reconstructed in the receiver section.

Trade-Offs and Considerations

While lossy compression significantly reduces file size, it comes with certain trade-offs:

- **Quality Degradation**: Depending on the compression level, visible artifacts such as blurring, blocking, or ringing may appear.
- **Irreversibility**: Unlike lossless methods, original data cannot be fully recovered.
- **Application Specificity**: Lossy methods are well-suited for natural images but not ideal for images requiring high fidelity (e.g., medical imaging or legal documents).

Fundamental Concept

At its core, lossy image compression exploits two main characteristics:

1. **Psycho-visual Redundancy**: The human eye is less sensitive to high-frequency components and minor colour variations. Compression algorithms leverage this by reducing the precision of visually less important elements.

Psycho-Visual Redundancy: Concept and Explanation

Psycho-visual redundancy refers to the parts of visual information in an image that can be removed or simplified without significantly affecting human perception. This concept is grounded in the way the human visual system (HVS) interprets visual data. Since the human eye and brain are more sensitive to certain types of information and less responsive to others, not all visual details are equally important from a perception standpoint.

Understanding the Human Visual System

The human eye is particularly good at noticing:

- **Luminance (brightness)** changes rather than colour details
- **Low-frequency information** (large, smooth areas or gradual transitions) over high-frequency details (sharp edges, fine textures)
- Changes in central vision more than peripheral vision

Because of these perceptual tendencies, some image details—especially subtle colour variations, fine textures, or high-frequency components—may not be noticeable when slightly altered or removed.

Application in Image Compression

In lossy image compression, psycho-visual redundancy is exploited to reduce file size while keeping the image visually acceptable. For example:

- In JPEG compression, after converting the image into frequency components using the Discrete Cosine Transform (DCT), higher-frequency coefficients (which the human eye is less sensitive to) are aggressively quantized or discarded.
- Similarly, chrominance data (colour information) can be stored at a lower resolution than luminance, since the eye is more sensitive to brightness differences than to colour variations.

By eliminating or simplifying parts of the image that are perceptually unimportant, compression algorithms can significantly reduce data while maintaining acceptable visual quality.

Real-Life Example

Consider a photograph with a clear blue sky. The sky may have tiny variations in colour and intensity, but the human eye perceives it as a smooth gradient. Psycho-visual redundancy allows compression algorithms to treat large areas of similar colour as uniform, discarding the fine variations that are unlikely to be noticed.

So, we can conclude that Psycho-visual redundancy highlights the gap between what is captured by a digital sensor and what is actually needed for visual interpretation. By focusing on perceptually important features and ignoring those less noticeable to the human eye, image compression methods can achieve better efficiency without compromising the viewing experience *(Sun, He, & Chen, 2021; Sakib & Siddiqui, 2023)*.

2. **Statistical Redundancy**: Natural images often contain repetitive patterns or correlations among pixels. These redundancies can be represented more efficiently using transform-based or predictive models.

Statistical Redundancy: Concept and Explanation

Statistical redundancy in image data refers to the presence of repeated or predictable patterns that do not carry new or unique information. In simple terms, it means that many parts of an image contain similar or correlated pixel values, and this repetition can be compressed without any loss of meaningful content.

What Causes Statistical Redundancy in Images

Digital images are composed of pixels arranged in a grid. In natural images, especially those of real-world scenes, neighbouring pixels often have similar colours and intensities. This leads to high correlation among adjacent pixels. For example:

- In a photo of the sky, large regions may have the same shade of blue.
- In a portrait, areas of skin might have gradual tone changes, producing similar values.

Because of this repetition or predictability, storing each pixel value individually becomes inefficient. Compression techniques identify and encode these recurring patterns more compact.

EVALUATION METRICS

To assess the effectiveness of lossy image compression, several metrics are commonly used:

- Compression Ratio: Measures how much the image has been compressed.
- Peak Signal-to-Noise Ratio (PSNR): Evaluates the quality of the reconstructed image.
- Structural Similarity Index (SSIM): Compares structural information between original and compressed images.
- Bitrate vs. Quality Trade-off: Assesses efficiency in various compression levels.

Some image Compressions comes under Lossless as well as Lossy techniques both:

Decorrelation Methods - Almost all forms of signals have correlation between samples, which has redundant information that can be eliminated using reversible decorrelation techniques. Decorrelation, sometimes known as "whitening," can be accomplished by several variety of approaches.

CONCLUSION

Image compression is a fundamental technique in digital image processing that balances file size with visual quality. By intelligently removing redundant or imperceptible information, compression algorithms make image storage and communication more practical and efficient across countless real-world applications.

Lossy image compression is a powerful technique for reducing image storage and transmission requirements by leveraging the limitations of human perception and inherent data redundancy. Although it sacrifices some fidelity, it remains the dominant method for photographic and multimedia content where storage efficiency and bandwidth savings are crucial.

Lossless image compression is a vital tool for preserving image fidelity while optimizing storage and transmission. By leveraging the natural structure and repetition within image data, these techniques enable efficient encoding without compromising content. Although they may not achieve the dramatic size reductions seen in lossy methods, their precision and reliability make them indispensable in many domains.

So, the Types of Image Compression discussed so far in this chapter can be summarised as follows:

1. **Lossless Digital Image Compression:**
    - o Preserves all original pixel data.
    - o Used where diagnostic accuracy is critical (e.g., radiology, pathology).
    - o Common techniques: Run-Length Encoding (RLE), Huffman Coding, Lempel-Ziv-Welch (LZW), JPEG-LS.
    - o DICOM-compliant.
2. **Lossy Digital Image Compression:**
    - o Slight reduction in image quality but massive savings in storage.
    - o Used in non-diagnostic or secondary-use scenarios (e.g., teaching, research, post-treatment reviews).
    - o Techniques: JPEG, JPEG2000, HEVC/H.265.
    - o Must be carefully applied — only where image degradation does not affect clinical outcomes

REFERENCES

Abd-Alzhra, A. S., & Al- Tamimi, M. S. H. (2022). Image Compression Using Deep Learning: Methods and Techniques. *Iraqi Journal of Science*, 1299–1312. DOI: 10.24996/ijs.2022.63.3.34

Ahmmed, A., Paul, M., Murshed, M., & Taubman, D. (2021, September). Human-machine collaborative video coding through cuboidal partitioning. In *Proceedings of the IEEE International Conference on Image Processing (ICIP)* DOI: 10.1109/ICIP42928.2021.9506150

Akbari, M., Liang, J., & Han, J. (2019). DSSLIC: Deep semantic segmentation-based layered image compression. In *Proceedings of the IEEE International Conference on Acoustics, Speech and Signal Processing (ICASSP)* (pp. 2042–2046) DOI: 10.1109/ICASSP.2019.8683541

Ballé, J., Minnen, D., Singh, S., Hwang, S. J., & Johnston, N. (2018). Variational image compression with a scale hyperprior. In *Proceedings of the 6th International Conference on Learning Representations (ICLR)*

Chen, T., Liu, H., Ma, Z., Shen, Q., Cao, X., & Wang, Y. (2021). End-to-end learnt image compression via non-local attention optimization and improved context modeling. *IEEE Transactions on Image Processing*, *30*, 3179–3191. DOI: 10.1109/TIP.2021.3058615 PMID: 33606630

Cheng, Z., Sun, H., Takeuchi, M., & Katto, J. (2020). Learned image compression with discretized Gaussian mixture likelihoods and attention modules. In *Proceedings of the IEEE/CVF Conference on Computer Vision and Pattern Recognition (CVPR)* (pp. 7939–7948). DOI: 10.1109/CVPR42600.2020.00796

Dax, G., Nagarajan, S., Li, H., & Werner, M. (2023). Compression supports spatial deep learning. *IEEE Journal of Selected Topics in Applied Earth Observations and Remote Sensing*, *16*(5), 702–713. DOI: 10.1109/JSTARS.2022.3226563

Fang, X., Wu, X., Li, F., Duan, Y., & Tao, X. (2025). Group image compression for dual use of machine and human vision. *IEEE Transactions on Circuits and Systems for Video Technology*, *35*(3), 2820–2831. DOI: 10.1109/TCSVT.2024.3486558

Fischer, K., Fleckenstein, F., Herglotz, C., & Kaup, A. (2021, June). Saliency-driven versatile video coding for neural object detection. In *Proceedings of the IEEE International Conference on Acoustics, Speech and Signal Processing (ICASSP)*. DOI: 10.1109/ICASSP39728.2021.9415048

Gonzalez, R. C., & Woods, R. E. (2018). *Digital Image Processing*. Pearson.

He, D., Zheng, Y., Sun, B., Wang, Y., & Qin, H. (2021). Checkerboard context model for efficient learned image compression. In *Proceedings of the IEEE/CVF Conference on Computer Vision and Pattern Recognition (CVPR)* (pp. 14771–14780). DOI: 10.1109/CVPR46437.2021.01453

. Healy, D.; Mitchell, O. (1981). *"Digital Video Bandwidth Compression Using Block Truncation Coding"*. IEEE Transactions on Communications. 29 (12): 1809. Bibcode:1981 ITCom..29.1809H. DOI: 10.1109/TCOM.1981.1094938

Johnston, N. (2018). Improved lossy image compression with priming and spatially adaptive bit rates for recurrent networks. In *Proceedings of the IEEE/CVF Conference on Computer Vision and Pattern Recognition (CVPR)* (pp. 4385–4393) DOI: 10.1109/CVPR.2018.00461

Jung, S., Jin, S., & Kim, H. (2024). A novel side-channel archive framework using deep learning-based leakage compression. *IEEE Access : Practical Innovations, Open Solutions*, *12*, 105326–105336. DOI: 10.1109/ACCESS.2024.3416199

Khan, M., & Tripathi, A. (2024, December). Digital Patient Eye. *IEEE Transactions on Circuits and Systems for Video Technology*, *34*(12).

Kumar, H. S., Moiz, A., Kalshetti, K. V., Jain, V., & Kumar, K. (2025, April–May). *Image compression and enhancement using deep learning algorithm*. In *Proceedings of the 2025 International Conference on Emerging Smart Computing and Informatics (ESCI 2025)*. https://doi.org/DOI: 10.1109/ESCI63694.2025.10988240

Lu, G., Ge, X., Zhong, T., Hu, Q., & Geng, J. (2024). Preprocessing enhanced image compression for machine vision. *IEEE Transactions on Circuits and Systems for Video Technology*, *34*(12), 13556–13568. DOI: 10.1109/TCSVT.2024.3441049

Minnen, D., Ballé, J., & Toderici, G. D. (2018). Joint autoregressive and hierarchical priors for learned image compression. In *Advances in Neural Information Processing Systems (NeurIPS)* (pp. 10771–10780).

Pant, D., Bhatt, A., Khan, M., Nautiyal, O. P., & Adhikari, P. (2019). Automated IoT based smart water quality assessment system. In *Proceedings of the 8th International Conference on System Modeling and Advancement in Research Trends (SMART 2019)* DOI: 10.1109/SMART46866.2019.9117271

Patel, M. I., Suthar, S., & Thakar, J. (2019, May 17). Survey on image compression using machine learning and deep learning. In *Proceedings of the 2019 International Conference on Intelligent Computing and Control Systems (ICCS 2019)* (pp. 1103–1105). DOI: 10.1109/ICCS45141.2019.9065473

Sakib, M., & Mustajab, S. (2024). Enhanced multi-variate time series prediction through statistical-deep learning integration: The VAR-stacked LSTM model. *SN Computer Science*, *5*(5), 573. DOI: 10.1007/s42979-024-02950-x

Sakib, M., & Mustajab, S. (2025, July 4). CEEMDAN-TCN-Auto LSTM: A Triple-Force ensemble using cross-stitch networks for short-term load forecasting. *Arabian Journal for Science and Engineering*. Advance online publication. DOI: 10.1007/s13369-025-10398-1

Sakib, M., & Siddiqui, T. (2023). Multi-network-based ensemble deep learning model to forecast Ross River virus outbreak in Australia. *International Journal of Pattern Recognition and Artificial Intelligence*, *37*(10), 2352015. Advance online publication. DOI: 10.1142/S0218001423520158

Sheena, K. S., & Nair, M. S. (2024). GenCoder: A novel convolutional neural network based autoencoder for genomic sequence data compression. *IEEE/ACM Transactions on Computational Biology and Bioinformatics*, *21*(3), 405–415. DOI: 10.1109/TCBB.2024.3366240 PMID: 38358865

Skodras, A., Christopoulos, C., & Ebrahimi, T. (2001). The JPEG 2000 still image compression standard. *IEEE Signal Processing Magazine*, *18*(5), 36–58. DOI: 10.1109/79.952804

Subramanya, A. (2001). Image compression technique. *IEEE Potentials*, *20*(1), 19–23. DOI: 10.1109/45.913206

Sujitha, B., Parvathy, V. S., Lydia, E. L., Rani, P., Polkowski, Z., & Shankar, K. (2021). Optimal deep learning based image compression technique for data transmission on industrial Internet of things applications. *Transactions on Emerging Telecommunications Technologies*, *32*(7), e3976. DOI: 10.1002/ett.3976

Sun, S., He, T., & Chen, Z. (2021). Semantic structured image coding framework for multiple intelligent applications. *IEEE Transactions on Circuits and Systems for Video Technology*, *31*(9), 3631–3642. DOI: 10.1109/TCSVT.2020.3042517

Taspınar, Y. S. (2023). Light weight convolutional neural network and low-dimensional images transformation approach for classification of thermal images. *Case Studies in Thermal Engineering*, *41*, 102670. DOI: 10.1016/j.csite.2022.102670

Toderici, G. (2017). Full resolution image compression with recurrent neural networks. In *Proceedings of the IEEE Conference on Computer Vision and Pattern Recognition (CVPR)* (pp. 5435–5443) DOI: 10.1109/CVPR.2017.577

Wang, F., Shen, L., Teng, Q., & Tian, Z. (2024). DSCIC: Deep screen content image compression. *IEEE Transactions on Circuits and Systems for Video Technology, 34*(11), 11965–11979. DOI: 10.1109/TCSVT.2024.3419575

Wang, S., Wang, Z., Wang, S., & Ye, Y. (2022). Deep image compression towards machine vision: A unified optimization framework. *IEEE Transactions on Circuits and Systems for Video Technology, 33*(6), 2979–2989. DOI: 10.1109/TCSVT.2022.3230843

Xie, Y., Cheng, K. L., & Chen, Q. (2021). Enhanced invertible encoding for learned image compression. In *Proceedings of the 29th ACM International Conference on Multimedia* (pp. 162–170). DOI: 10.1145/3474085.3475213

Zhang, H. (2022). ResNest: Split-attention networks. In *Proceedings of the IEEE/CVF Conference on Computer Vision and Pattern Recognition Workshops (CVPRW)* (pp. 2735–2745).

Chapter 2
Artificial Intelligence in Healthcare:
Applications, Challenges, and Future Directions

Feroz Ahmad
https://orcid.org/0000-0002-0191-2891
Suki AI, USA

ABSTRACT

The integration of artificial intelligence (AI) into healthcare represents one of the most transformative technological advances in modern medicine. As healthcare systems worldwide grapple with rising costs, aging populations, and increasing data volumes, AI emerges as a powerful solution to enhance diagnostic accuracy, streamline operations, and improve patient outcomes. This comprehensive analysis explores the current applications of AI in healthcare, examines the challenges facing implementation, and outlines future directions for this rapidly evolving field.

INTRODUCTION

Artificial Intelligence (AI) has augmented healthcare for more than sixty years, evolving from first knowledge-based systems to today's data-driven deep learning. The foundational research in clinical decision-making established statistical and probabilistic approaches for diagnosis (Ledley and Lusted 1959). During the 1970s, computer-aided diagnosis had significant clinical efficacy, especially in the triage of severe abdominal pain (Dombal et al. 1972). Rule-based expert systems like MYCIN for infectious disease treatment (Shortliffe 1976) and INTERNIST-1 for internal medicine (Miller, Pople, and Myers 1982) investigated the application

DOI: 10.4018/979-8-3373-5727-0.ch002

of codified medical knowledge to enhance clinician reasoning at the bedside. Although exhibiting impressive performance in some areas, adoption was limited due to challenges in knowledge acquisition, maintenance demands, and restricted generalizability beyond selected use cases.

The incorporation of AI into healthcare is propelled by several converging themes.

1. The swift expansion of medical data from electronic health records (EHRs), wearable technology, and imaging techniques.
2. Progress in machine learning (ML) and deep learning (DL) facilitating large-scale automated pattern recognition.
3. Escalating demand for economical, tailored, and preventative healthcare solutions.

However, a seamless integration of AI in healthcare requires solving critical issues, including data privacy, bias-free implementation, interpretability, and clinical validation. The multidisciplinary nature of AI-powered healthcare requires collaboration among clinicians, engineers, and policymakers in order to solve the aforementioned challenges before AI can have a deeper level impact in healthcare.

This chapter provides an in-depth understanding of artificial intelligence and machine learning in healthcare from multiple perspectives: (a) its applications in clinical and operational settings, (b) the practical hurdles to its success, and (c) future directions where the current field is headed.

ML Model Deployment Lifecycle

Artificial intelligence powered by machine learning models undergo significantly different development cycles than conventional software. Deployment of a machine learning model follows a structured two-phase lifecycle that consists of the development of robust, clinically validated models and then its subsequent serving into production healthcare systems.

The first phase is known as the model development phase, which focuses on the creation of precise, dependable, and clinically relevant machine learning models via a systematic approach. This phase begins with engineers defining the problem thoroughly, listing the important and edge case scenarios. The next step requires the implementer to acquire relevant data by gathering already existing datasets in the open domain or through proprietary means. In specific scenarios this step might require the creation of datasets to suit the tailored needs of the specific problem statement. Once the data-gathering step is achieved, several iterations of modeling, optimization, and evaluation follow with the objective to perform optimally on a partition of the aforementioned dataset, also known as the test set. This phase prioritizes validation and quality assurance steps to guarantee that the models fulfill the required standards of healthcare applications, where precision and dependability directly affect patient outcomes.

Followed by the second phase, the model serving phase, which concerns the utilization of the developed model from the first phase in real-world deployments. This phase deals with challenges of serving the models in operational healthcare systems. The model pipelines must utilize real-time clinical data with high accuracy with proper data validation to prevent any wrong model predictions. The model serving system also needs to handle abrupt load fluctuations with techniques like auto-scaling to dynamically scale the compute to spin up several instances of the model that can handle more requests in a unit time. Additionally, the model serving phase incorporates exhaustive utilization of monitoring systems to ensure model performance, clinical safety, and compute scaling.

Model Development Phase

The model development phase consists of five sequential stages that ensure the creation of clinically viable machine learning models:

1. **Problem Formulation:** The stepping stones of model development start with a precise description of the clinical problem being targeted. For example, if the intention is to design a system for pneumonia detection, the problem formulation step would require implementers to describe the inputs and outputs to the model, such as a chest X-ray image with a given resolution as input and a positive or negative class as outputs.
2. **Data Collection:** This phase prioritizes data quality, representativeness of the population, and following healthcare privacy rules, while also ensuring enough samples are acquired for adequate model training. Another important aspect of the data collection stage is to ensure data is gathered from several sources, including electronic health records, research samples, wearable devices, clinical databases, and medical imaging systems (high and low quality).
3. **Modeling:** This stage requires translating the problem statement into a systematic set of machine learning modeling, in which the implementer chooses an algorithm that fits the requirements based on several factors. Criteria include the nature of the problem, the amount and characteristics of data, and performance and operational constraints, which influence the modelling. In clinical problem statements, a heavy emphasis on model interpretability and explainability is critical for successful adoption of the final solution. This step concludes with the selection of the appropriate machine learning framework (e.g., TensorFlow, PyTorch, Apache Spark MLlib, Scikit-learn, etc.).

4. **Model Training and Optimization:** Following the selection of a model architecture and machine learning framework, the model goes through model training and optimization. The objective in this stage is for the model to learn patterns from the data and store the learned information as parameters of the model. The overall definition or mechanics of the model can be changed via some levers called hyperparameters. This stage focuses on tuning these hyperparameters to learn patterns from the data most effectively. Other important steps in this stage also include strategies around feature extraction, model compression, and quantization for efficient model serving to reduce compute costs and latency.
5. **Evaluation-Driven Model Quality Iteration:** The optimized model is put to methodological validation over relevant clinical metrics, cross-validation methods, and external validation datasets. This iterative process over steps 3, 4, and 5 occurs until models perform at an acceptable quality, efficiency, and clinical value to justify deployment consideration.

Model Serving Phase

Once the model development completes, it goes through the following model serving phase that deals with the following stages:

1. **Data Pipeline:** In the serving stage the model utilizes the input data pertinent to a patient or a user that is interfaced to the model using resilient data pipelines. The pipelines are responsible for preprocessing and feature extraction, leading to a data domain that the model expects. Another aspect of these pipelines deals with dealing with the large volume and velocity of the clinical data such that the most recent and accurate data is seen by the model during the inference step. This encompasses interaction with current healthcare information systems, real-time data validation, and streamlined data transformation pipelines specifically designed for clinical workflows.
2. **Optimized Scaled Serving:** Creating inference systems that are both efficient and scalable to handle the high traffic needs of healthcare apps is crucial for usability. This is achieved by modularizing the software and machine learning software components into containerized microservices such that each component can be load balanced and scaled individually as per the incoming requests' requirements. GPU acceleration along with distributed computing architectures are utilized for high reliability and low latency guarantees to support a wide range of load circumstances.
3. **Monitoring:** The final stage in serving emphasizes continuous checks on model performance, data drift, system health, and clinical outcomes to determine whether remediation steps are required. This is achieved through notifications

in real time for performance drops, automated triggers for retraining models, and ongoing evaluations of clinical effects to ensure clinical safety.

The aforementioned methodological sequence ensures high-standard model development and a reliable deployment of the model into the real world such that the expected performance from the research setting translates well into the practical settings. The following section provides a peek into how the model deployment life cycle actually integrates well in some real-life operational settings in the healthcare practice.

APPLICATIONS OF AI IN HEALTHCARE

Integration of artificial intelligence into healthcare systems is a big step forward in modern medicine and personal healthcare technology. It has transformed how healthcare providers diagnose, treat, and monitor patients. AI applications in healthcare are very diverse, ranging from image interpretation systems that can detect subtle pathological changes that the human eye cannot detect to sophisticated predictive analytics platforms that can precisely predict patient outcomes for a clinical treatment.

This technological revolution spans various medical domains, each offering distinct challenges and opportunities for AI integration.

The various clinical applications can be categorized into four primary domains: (a) clinical decision support systems that enhance physician reasoning, (b) medical imaging and diagnostics, which improve interpretation accuracy over imaging test reports such as CT scans, (c) personalized medicine approaches that customize treatments to the individual patient level, and (d) remote monitoring through telehealth services that provide care beyond conventional approaches.

Each domain utilizes specific AI methodologies, ranging from traditional machine learning algorithms like ensemble methods and probabilistic models to advanced deep learning architectures such as convolutional and recurrent neural networks and sophisticated natural language processing (NLP).

The clinical impact of these applications fundamentally addresses critical healthcare challenges such as improving diagnostic accuracy, personalizing treatment, reducing costs, and enhancing accessibility.

Clinical Decision Support Systems (CDSS)

Clinical Decision Support Systems are advanced technological frameworks that integrate AI algorithms with established clinical workflows, acting as intelligent assistants to enhance healthcare professionals' ability to execute evidence-based decisions while preserving clinical autonomy. CDSS serve as collaborators in the diagnostic and clinical processes, where the primary task of such a system is to analyze patient data. In this process the expectation of the system is to identify patterns that were learned by the machine learning model and flag them to the healthcare providers. Given that the models can be trained on significantly greater diversity and volume of datasets than what a human can ever be trained on, such systems can sometimes detect imperceptible signals to identify scenarios.

Modern clinical decision support system architectures use a mix of different AI algorithms and methods to enhance the accuracy of diagnoses and the effectiveness of treatments. Supervised learning models, like logistic regression for predicting binary outcomes, random forests for handling non-linear relationships and feature interactions, and gradient boosting machines like XGBoost for better predictive performance, are frequently used with traditional rule-based reasoning systems to develop CDSS for healthcare settings. This integration allows systems to look at patient data from electronic health records (EHRs), lab findings, and clinical measurements and provide outcomes with a high degree of accuracy (Sendak et al. 2020).

Handling unstructured clinical text remains a critical and nontrivial task within modern Clinical Decision Support Systems (CDSS). Recent transformer-based encoders, such as BERT (Bidirectional Encoder Representations from Transformers) and the domain-specific BioBERT, have revolutionized the text understanding capabilities in clinical settings. These models are now utilized to extract clinically salient entities, relations, and contextual cues from physician notes, discharge summaries, and broader clinical narratives (Lee et al. 2020). Additionally, these models can normalize medical terminology and abbreviations to effectively unwrap the layered language typical of healthcare documentation.

Uncertainty quantification is essential in clinical decision-making to inform clinicians about the confidence of the decision by the system. With Bayesian networks and probabilistic graphical models, modeling the inherent uncertainty in patient outcomes and clinical predictions is feasible; thus, clinicians can be exposed to confidence scores. Probabilistic approaches allow Clinical Decision Support Systems (CDSS) to deliver not only binary predictions but also detailed probabilistic risk scores that indicate the confidence levels linked to various diagnostic or prognostic evaluations (Lucas, van der Gaag, and Abu-Hanna 2004). This probabilistic framework enables clinicians to comprehend the reliability of AI-generated recommendations and integrate this uncertainty into their clinical reasoning process.

Some real-world examples show how these technologies can be used in clinical settings. IBM Watson for Oncology incorporated different AI methods working together to solve for common clinical workflows. It uses advanced NLP algorithms to systematically look at large amounts of medical literature, clinical recommendations, and historical case records. It additionally employs supervised classification algorithms to evaluate and prioritize treatment alternatives based on the unique needs of each patient and the principles of evidence-based medicine (Zhou et al. 2018). AI-powered triage systems use advanced gradient boosting techniques like XGBoost to look at a patient's health record data and rank cases based on how likely they are to get worse quickly. This is crucial in emergency care situations where quick and accurate triage choices are essential (Rajkomar, Dean, and Kohane 2019).

Illustrative CDSS end-to-end workflow

To make these abstractions more concrete, let us consider a representative (but simplified) hospital deployment that predicts ward patient deterioration every hour and surfaces the risk to clinicians inside the electronic health record (EHR).

1. **Data source:** Chart-event tables, vital-sign trends, and laboratory measurements are streamed from the *MIMIC-IV* research database, whose hourly structure mirrors that of many commercial EHRs (Johnson et al. 2023).
2. **Feature engineering:** The open-source tsfresh library automatically derives ~900 statistical and frequency-domain descriptors from 24-hour sliding windows of each signal.
3. **Model building:** A gradient-boosting classifier (XGBoost, early-stopped on log-loss) is trained to discriminate patients who will require an ICU transfer within the next 6 h, outperforming logistic-regression baselines reported in recent hemodynamic-risk studies (J. Xu et al. 2024).
4. **Explainability:** SHAP value analysis accompanies every prediction, typically highlighting rising lactate and falling mean arterial pressure as dominant drivers—information clinicians now expect for regulatory and trust reasons (Ponce-Bobadilla, Schmitt, and Valderrama 2024).
5. **Deployment:** The trained model is packaged as a Docker micro-service, exposed through a RESTful FHIR (Fast Healthcare Interoperability Resources) endpoint—a modern healthcare data exchange standard that enables seamless communication between different healthcare systems—and orchestrated via Kubernetes so that new EHR events trigger real-time inference and alert generation (Eapen, Sartipi, and Archer 2020). The service returns structured alerts that appear inside the clinician's EHR workspace, thereby completing the data-model-decision loop.

Figure 1. End-to-end pipeline for hourly ward-patient deterioration prediction

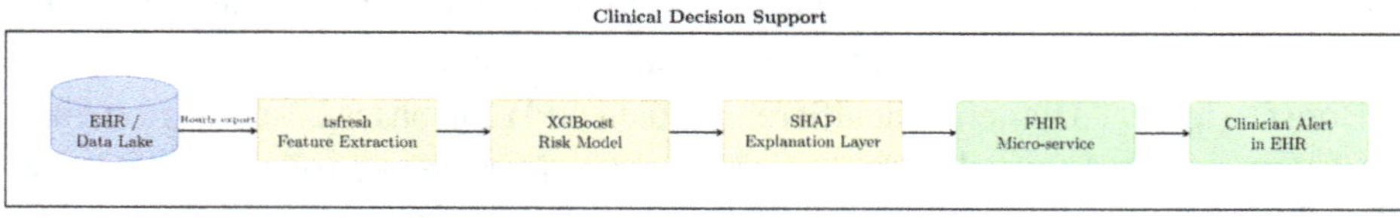

Prospective Evaluation

Hospitals that implement this kind of pipeline should follow recently published best-practice frameworks for clinical AI roll-out (Sendak et al. 2020). Those frameworks recommend continuously logging outcome metrics—such as alert lead-time, positive-predictive value, clinician-response latency, and ward code-blue frequency—to a dedicated monitoring database. The collection of this data facilitates systematic post-deployment auditing, fairness assessments, and regular model recalibration, thus ensuring the maintenance of the system's utility and safety within the local patient population.

Medical Imaging and Diagnostics

Medical imaging exemplifies a significant and clinically impactful application of deep learning in healthcare, radically transforming the domains of diagnostic radiology and pathology. This domain has seen major advances in automated image analysis, resulting in AI systems that perform diagnostic analysis comparable to or surpassing that of seasoned human specialists in a particular imaging task.

The clinical importance of such progress is considerable, as these technologies address critical issues in healthcare delivery, such as ensuring diagnostic consistency, improving speed, and expanding accessibility, particularly in settings with limited resources and scarce specialist availability.

Convolutional neural networks (CNNs) form the backbone of modern medical image analysis, capitalizing on their strength in learning hierarchical features directly from raw pixels. CNNs support reliable automation for detection, classification, and segmentation of disease across major imaging modalities such as X-rays, computed tomography (CT), magnetic resonance imaging (MRI), ultrasound, and histopathology (Esteva et al. 2017). The strength is attributed to CNNs' ability to extract subtle textures and spatial cues that clinicians may reasonably miss. Thereby delivering consistent accuracy without the drifts associated with fatigue or subjective variation of human evaluation, which makes these models particularly suitable for demanding clinical workflows.

Over the years, CNN architectures have advanced into more refined and task-oriented designs, each suited for different challenges of medical image processing. ResNet (Residual Networks) (Kaiming He et al., 2016) helps to overcome the vanishing gradient problem by using skip connections, while DenseNet (Densely Connected Networks) (Huang et al., 2017) improves the transfer of information between layers through dense links. EfficientNet (Tan & Le, 2019) brings a balanced way of scaling the network depth, width, and resolution for better performance with fewer resources. These architectures have shown exceptional performance for complex classification work, such as detecting cancer or recognizing neurological disorders. In many imaging modalities and across several anatomical regions, these architectures have shown high accuracy in classifying healthy and diseased tissue.

Moreover, specialized architectures have been engineered for precise anatomical localization and segmentation tasks to meet the unique requirements of medical imaging. In scenarios requiring accurate boundary demarcation, such as tumor segmentation or organ delineation, U-Net (Ronneberger, Fischer, & Brox 2015), characterized by its unique encoder-decoder architecture and skip connections, has established itself as the benchmark for medical image segmentation. Mask R-CNN enhances object detection by offering both bounding box localization and pixel-level segmentation masks (Kaiming He et al. 2017), rendering it especially useful for applications that require simultaneous detection and accurate segmentation of multiple pathological regions within a single diagnostic image.

The examination of volumetric medical imaging data, such as three-dimensional MRI acquisitions or CT scans, requires specialized methodologies that can holistically capture spatial correlations across many dimensions. Three-dimensional CNNs are utilized to exploit the extensive spatial context residing in volumetric data, facilitating a more thorough analysis of anatomical structures and pathological alterations that may be hidden in individual two-dimensional slices (Zhao, Jia, & Greenspan 2018). These three-dimensional renderings are especially significant in applications like brain imaging, where the connection patterns are essential for precise diagnosis.

In addition to the traditional supervised learning approaches for classification and segmentation, recent years have seen a gradual yet steady shift towards the use of deep generative models in medical imaging. This has been to address the long-standing challenge of limited availability of large, well-annotated datasets. Variational Autoencoders (VAEs) (Kingma and Welling 2013) and Generative Adversarial Networks (GANs) (Goodfellow et al. 2014) have been effectively employed for sophisticated data augmentation, enabling the creation of synthetic medical images that preserve both the statistical distribution and the clinical relevance of real patient data, thereby expanding the diversity of training samples. Such strategies significantly improve the robustness and generalization of models in clinical scenarios where annotated data is scarce or certain disease conditions are rare (Frid-Adar et al. 2018). Moreover, these generative approaches facilitate the generation of privacy-preserving synthetic datasets, encouraging collaborative research while also ensuring the privacy of patients' health information.

Transformer-Based Vision Models

Recently, self-attention architectures initially used in natural language processing have been repurposed for computer vision and medical imaging. Vision Transformers (ViT) learn global context without convolution, and when pre-trained on large natural-image corpora followed by domain-specific fine-tuning, they now match or exceed CNN performance on tasks such as pneumothorax detection and retinal-disease grading (Dosovitskiy et al. 2021; Philippi, Rothaus, & Castelli 2023). Hierarchical variants like Swin Transformer provide multi-scale features that integrate smoothly into encoder-decoder designs; the Swin-UNETR achieved competitive performance in multiple BraTS tumor-segmentation sub-tasks in 2023 (Hatamizadeh et al. 2022). These models are particularly attractive for institutions with limited labeled data because a single, foundation-level pre-training step can be reused across modalities.

Illustrative Use-Case: Chest-X-Ray Classification

One widely adopted public benchmark for thoracic-pathology detection is the CheXpert dataset, which contains 224 316 de-identified radiographs labelled for 14 findings, including pneumonia (Irvin et al. 2019). While early baselines that fine-tuned an ImageNet-pre-trained *ResNet-50* with focal loss attained an AUROC of 0.768 for the pneumonia task—already on par with a panel of board-certified radiologists—the leaderboard has since been surpassed by the *Large-scale Robust Deep AUC Maximization* (AUC-Margin) approach of Yang *et al.* (T. Yang et al. 2021). Their top-ranked submission (overall AUROC 0.930) averages the predictions of five independently trained backbones:

- DenseNet-121
- DenseNet-161
- DenseNet-169
- DenseNet-201
- Inception-ResNet-v2

Each model is fine-tuned with the novel *AUC-Margin loss*, and simple arithmetic ensembling of their logits delivers a robust boost on all 14 labels.

To illustrate how such research-grade models translate into clinical practice, a simple illustrative system in production can look like the following example. Such systems must handle the complex data formats and workflows that characterize real-world medical imaging environments, where standardization and interoperability are paramount.

Example DICOM header

Medical imaging data is typically stored and transmitted using DICOM (Digital Imaging and Communications in Medicine), a comprehensive standard that defines both the file format for medical images and the network protocol for their exchange between medical devices and systems.

```
   (0008,0060) Modality           = "DX"          # Digital
Radiography
   (0010,0020) PatientID          = "123456"
   (0008,103E) SeriesDescription  = "CHEXPERT STUDY"
   (0020,4000) ImageComments      = "Label: PNEUMONIA_POS"
   (7FE0,0010) PixelData          =  <binary>
```

Figure 2. Chest-X-ray classification pipeline with ResNet fine-tuning and real-time deployment

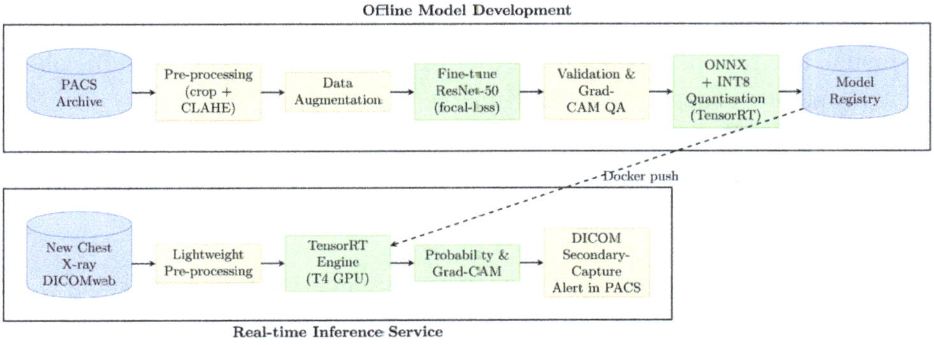

The pipeline consists of two main phases: offline model development where chest radiographs from the PACS archive undergo pre-processing (crop and contrast-limited adaptive histogram equalisation), data augmentation (rotation/flip, mix-up), and ResNet-50 fine-tuning with focal loss, followed by validation and Grad-CAM quality assurance before the model is exported to ONNX, quantized with TensorRT, and stored in the model registry. The real-time inference service processes new chest X-rays arriving via DICOMweb through lightweight pre-processing and scoring by the TensorRT engine on a T4 GPU. The pneumonia probability and Grad-CAM visualization are then packaged as a PAR (Secondary-Capture PAR object—a lightweight DICOM secondary-capture image whose sole purpose is Presentation And Review, where the AI service burns the model's probability map into a new single-frame image, tags it as "secondary capture," and sends it back to PACS as a non-destructive, PACS-friendly carrier for AI results) alert in the PACS for radiologist review.

To operationalize such a system in clinical practice, several key implementation considerations must be addressed. The following breakdown provides a detailed technical roadmap for deploying this pipeline in a production healthcare environment.

Possible implementation

1. **Data flow.** DICOM images are exported from the PACS, de-identified with the RSNA CTP toolkit, converted to 512 × 512 PNG, and augmented on-the-fly (small rotations, flips, contrast-limited adaptive histogram equalisation (CLAHE)).
2. **Training set-up.** ResNet-50 (ImageNet weights) is fine-tuned for 10 epochs on two NVIDIA A100 GPUs using mixed-precision AdamW; focal-loss and patient-level 80/10/10 split follow the CheXpert protocol.
3. **Model optimisation.** The converged network is exported to ONNX and INT8-quantised with TensorRT, achieving ~20 ms inference per image on a Tesla T4.
4. **Serving.** A FastAPI micro-service fetches DICOM via DICOMweb, runs the model, and returns probability + Grad-CAM overlay as a DICOM SR object; the container is deployed under Kubernetes.
5. **Monitoring.** Prometheus scrapes latency/drift metrics; a nightly Airflow task recalibrates output probabilities if AUROC on a running validation buffer falls by >5 %.

Taken together, these steps illustrate how a research-grade deep-learning model can be transformed into a regulated, real-time radiology decision aid—moving seamlessly from raw PACS images to clinically interpretable outputs at the point of care.

Personalized Medicine and Genomics

Personalized medicine represents a paradigmatic shift from the traditional "one-size-fits-all" approach to healthcare, embracing instead a precision-based methodology that tailors medical treatments to individual patient characteristics, genetic profiles, and environmental factors. Artificial Intelligence serves as the technological catalyst that enables this transformation, providing the computational power and analytical sophistication necessary to process and interpret the vast, complex datasets that underpin precision medicine approaches.

This field exemplifies the convergence of genomics, computational biology, and machine learning to create treatment strategies that are optimized for individual patients rather than population averages.

The foundation of AI-enabled precision medicine rests on sophisticated predictive modeling techniques applied to integrated genomic and clinical datasets, where the goal is to identify patterns and relationships that can inform personalized treatment decisions (Libbrecht & Noble 2015). This approach requires the analysis of multi-dimensional data structures that encompass not only traditional clinical variables but also high-throughput genomic data, proteomic profiles, metabolomic signatures, and environmental exposure information. The complexity and scale of these datasets necessitate advanced machine learning approaches capable of handling high-dimensional data while extracting clinically meaningful insights.

Traditional machine learning algorithms have been extensively adapted and optimized for genomic applications, with each approach offering distinct advantages for different aspects of precision medicine. Random forests demonstrate particular utility in genomic association studies due to their ability to handle high-dimensional data, capture non-linear relationships, and provide feature importance rankings that can identify the most relevant genetic variants for specific phenotypes. Support vector machines (SVMs) excel in high-dimensional genomic classification tasks, particularly when dealing with scenarios where the number of features (genetic variants) far exceeds the number of samples (patients), a common challenge in genomic medicine.

More recently, graph neural networks (GNNs) have emerged as powerful tools for analyzing the complex network structures inherent in biological systems. These sophisticated architectures are specifically applied to genomic interaction networks, protein-protein interaction networks, and metabolic pathways to identify candidate biomarkers for drug targeting and therapeutic intervention (Zitnik, Agrawal, & Leskovec 2018). GNNs excel at capturing the relational information between genes, proteins, and other biological entities, enabling the identification of previously unknown therapeutic targets and the prediction of drug efficacy based on network topology and node characteristics.

The rapidly advancing domain of treatment optimization has increasingly adopted reinforcement learning approaches in the design of personalized treatment procedures. In such methods, clinical interventions are evaluated and adjusted gradually over time with the aim of improving long-term patient outcomes. The focus is not only on immediate therapeutic gains but also on sustained benefits (Komorowski et al. 2018). This approach is especially relevant in complex procedures such as oncology. Clinicians find themselves taking treatment decisions in a sequential manner, guided by patient responses, disease progression patterns, and tolerance to medications, and such approaches can critically augment human care abilities by learning from patient-specific treatment responses along with population-level observations. Within reinforcement learning, policy gradient methods have emerged as an important class for modeling such sequential decision-making. These have been applied in high-stakes scenarios like chemotherapy dosing. In these cases, the challenge lies in striking a careful balance between therapeutic effectiveness and the risk of treatment-related toxicity over extended durations.

The combination of several multi-omics datasets constitutes a significant challenge and opportunity in personalized medicine, necessitating advanced computational methodologies capable of effectively synthesizing information from genomics, transcriptomics, proteomics, metabolomics, and other high-throughput biological data sources. Multi-view learning algorithms and attention-based fusion models have been developed specifically to address this challenge (Xu, Tao, & Xu 2013), where models learn joint representations of heterogeneous biological data sources that capture both individual data type characteristics and cross-modal relationships. These integrated approaches enable the creation of comprehensive patient profiles that incorporate multiple layers of biological information comprehensively, leading to more accurate predictions of drug response, disease progression, and treatment outcomes.

The clinical validation of these multi-omics approaches has demonstrated their superior predictive performance compared to single-modality predictors, particularly in complex tasks such as drug response prediction where multiple biological pathways and mechanisms contribute to therapeutic outcomes (Baltrušaitis, Ahuja, & Morency 2018). This enhanced predictive capability translates directly into improved clinical decision-making, enabling clinicians to select treatments that are most likely to be effective for individual patients while minimizing the risk of adverse reactions and treatment failures.

Case Study — Individualised Warfarin Dosing

A seminal example of algorithmic personalisation is warfarin therapy. Using data from more than 5 000 patients in the International Warfarin Pharmacogenetics Consortium, researchers trained a random-forest model that combines *CYP2C9* and *VKORC1* genotypes with demographic and clinical variables (Consortium 2013). Randomised and observational studies have suggested potential relative reductions in hospitalisations or major bleeding ranging from 33 % to 50 % when genotype-guided dosing is used, though results vary across studies and patient populations (Epstein et al. 2010; Franchini et al. 2014).

Implementation highlights:

1. **Data set:** Multi-centre dataset: genotype calls (SNP rs1057910, rs9923231), INR values, age, weight, race and co-medications.
2. **Pre-processing:** One-hot encode genotypes; winsorise extreme INRs; impute missing weight with sex-specific medians.
3. **Model:** Random Forest (500 trees, max-depth = 8) tuned by nested five-fold CV; feature importance exported for clinical interpretability.
4. **Integration:** REST micro-service listens for new genotype results; predicted dose is written to the e-prescribing module with a confidence band.
5. **Monitoring:** Rolling calibration curves (INR_{target} ±0.2) trigger re-training if deviation > 10 %.

Collectively, this pipeline shows how classical ensemble learning plus modest genomic features can deliver measurable bedside benefit while remaining transparent and maintainable.

Illustrative Use-Case: Graph-Based Prediction of Drug-Drug Interactions

Precision medicine increasingly requires reasoning over networks of biological entities. Graph Convolutional Networks (GCNs) applied to a heterogeneous drug-protein-disease graph (~5 million edges) predict 964 possible side-effects for any pair of drugs with an AUROC of 0.872 on the TWOSIDES benchmark (Zitnik, Agrawal, and Leskovec 2018). These predictions already inform clinical-pharmacy dashboards by flagging hazardous polypharmacy combinations during order entry.

Implementation highlights:

- **Graph construction:** Nodes: drugs, proteins, diseases; edges: Drug-Target (DrugBank), Drug-Disease (SIDER), Protein-Protein (STRING).
- **Model:** Two-layer GCN with 128-dim hidden units; message passing aggregates chemical fingerprints and protein embeddings.
- **Training:** Multi-label binary cross-entropy; negative sampling maintains 1:4 positive-negative ratio; trained on four V100 GPUs.
- **Serving:** Batch-inference jobs pre-compute scores for 28 000 approved drug pairs; top-risk quartile cached in a Postgres table queried by the CPOE (Computerized Physician Order Entry) interface.
- **Alert logic:** An interaction is surfaced to clinicians only if (i) predicted probability > 0.8 and (ii) no existing alert in FDA Adverse Event Reporting System—minimising alert fatigue. The risk table is refreshed nightly so that model updates propagate automatically without interrupting clinical workflow.

Figure 3. Graph-convolutional workflow for drug-drug interaction prediction.

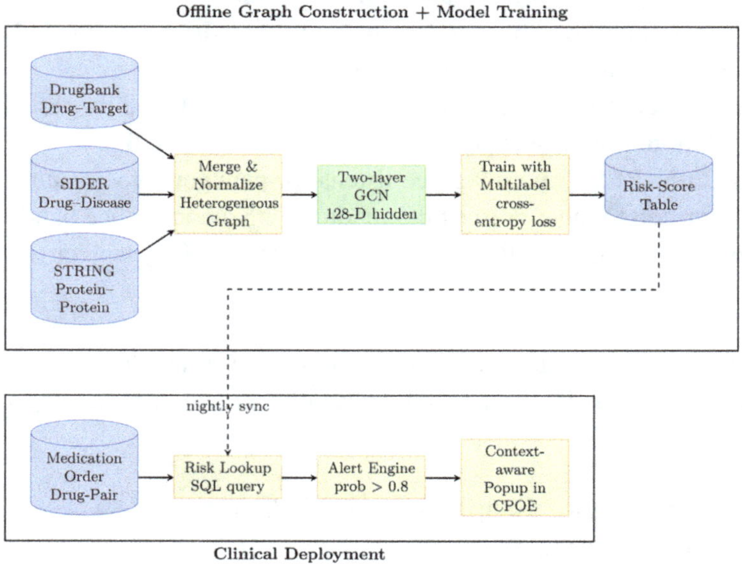

By embedding rich biomedical graphs into a production CPOE workflow, hospitals can move beyond static rule sets and deliver context-aware, data-driven medication safety checks.

Remote Monitoring and Telehealth

Remote monitoring and telehealth represent rapidly expanding domains in modern healthcare delivery, fundamentally reshaping the traditional paradigm of episodic, clinic-based care toward continuous, patient-centered monitoring and intervention. These technologies have become increasingly critical in addressing healthcare accessibility challenges, particularly for patients in rural or underserved areas, elderly populations with mobility limitations, and individuals with chronic conditions requiring frequent monitoring.

The integration of AI into remote monitoring systems enables the transformation of vast streams of continuously collected physiological data into actionable clinical insights, facilitating early intervention and preventive care strategies that can significantly improve patient outcomes while reducing healthcare costs.

Advanced time-series modeling approaches that can handle and make sense of the constant streams of physiological data that modern wearable devices and home monitoring equipment produce are a big part of the basic framework of intelligent remote patient monitoring systems. The data streams include a lot of physiological parameters, like heart rate variability, blood pressure patterns, activity levels, sleep quality metrics, and a lot of biomarkers. These are all collected over lengthy periods of time with great temporal resolution. The problem is not only acquiring data but also carefully looking at complex time patterns to find clinically important changes that require medical attention.

Long Short-Term Memory (LSTM) networks (Hochreiter & Schmidhuber 1997) and Gated Recurrent Units (GRUs) (Cho et al. 2014) are two types of recurrent neural networks (RNNs) that started as very powerful neural network layers for analyzing sequential physiological data streams. They are especially useful because they can capture long-term temporal dependencies without the problems that regular RNNs have with vanishing gradients. RNNs have been effective at learning intricate patterns and then detecting such patterns in the physiological data streams that even clinicians might not detect due to the sheer volume of data and non-obvious patterns. This makes it easier to spot early warning signs of clinical deterioration or the start of an acute medical crisis. These models work well because they can learn each patient's normal patterns and spot changes from those patterns that are specific to that patient based on their unique attributes and medical history.

In recent years, temporal modeling has seen notable progress, with temporal convolutional networks (TCNs) emerging as a preferred choice for remote monitoring tasks. Compared to conventional recurrent architectures, TCNs offer superior parallelization capabilities and achieve higher computational efficiency (Bai, Kolter, & Koltun 2018). Their design is significantly more effective for handling multiple streams of physiological signals in real time than RNNs, enabling monitoring systems to detect abnormal patterns with greater sensitivity and to issue alerts promptly when concerning trends are observed. Such architectures have shown extreme value in critical care environments, where timely recognition of a patient's deterioration can be pivotal in preventing adverse consequences.

The identification of anomalous patterns in physiological data streams represents one of the core components of remote monitoring systems. Thus, sophisticated anomaly detection techniques that can distinguish between normal physiological variations and clinically significant deviations are enablers for monitoring patient care. Traditional statistical approaches, including control charts and statistical process control methods (Montgomery 2007), provide baseline capabilities for detecting statistical outliers and trend changes in physiological parameters. However, more advanced approaches utilizing deep learning, particularly autoencoder-based anomaly detection systems (Sakurada & Yairi 2014), offer enhanced sensitivity and specificity by learning compressed representations of normal physiological patterns and flagging data points that cannot be accurately reconstructed from these learned representations.

These advanced methods for finding anomalies are meant to find early symptoms of a patient's decline so that doctors can step in before serious medical events happen. This skill has a lot more clinical importance than just alerting systems. It includes full risk classification and predictive modeling that can help with clinical decision-making and care coordination. These kinds of technologies are especially useful for keeping an eye on patients with long-term illnesses like heart failure, diabetes, or chronic obstructive pulmonary disease. Timely identification of developing conditions can facilitate early intervention and reduce the risk of extreme outcomes on a patient's health.

In the broader context of telehealth and virtual healthcare delivery, recent progress in transformer-based language models has shown significant potential in building intelligent conversational systems that actively engage patients and offer an initial perspective on their health status. Such advanced natural language processing frameworks now drive applications including smart symptom checkers, adaptive triage assistants, and AI-enabled chat interfaces capable of conducting preliminary consultations, gathering comprehensive symptom descriptions, and suggesting initial therapeutic measures (Xu et al. 2020). The integration of these conversational AI tools within telehealth platforms can lead to more efficient utilization of healthcare resources, as they are able to autonomously evaluate patients and guide them toward the appropriate level of care based on symptom severity and other relevant clinical risk indicators.

These AI-powered conversational systems are impressive at comprehending medical terms, symptom descriptions, and patient concerns stated in everyday language. This permits these systems to provide individualized health advice and support. Nevertheless, it is essential to assess these systems carefully for patient safety, and to design them such that, in situations where the case is complex or the model's confidence is low, they advise seeking further attention from qualified healthcare professionals.

Illustrative Use-Case: Predictive Wearables and Anomaly Detection

Continuous single-lead ECG streamed from consumer-grade patches can be analysed on-device by an LSTM auto-encoder trained on the MIT-BIH Arrhythmia corpus. In prospective testing the system demonstrated early detection capabilities, reportedly flagging ventricular tachycardia a median of 34 minutes before standard bedside monitors, with sensitivity and specificity values of 0.91 and 0.88 respectively (Faust et al. 2020). Complementing ECG, Google's open-source *CardioSignal* framework fuses accelerometer, gyroscope and photoplethysmography (PPG) data in a temporal convolutional network to output a five-level atrial-fibrillation risk score every 15 minutes, enabling proactive tele-cardiology outreach (Tison et al. 2018).

Implementation highlights:

1. **Data capture:** Raw ECG sampled at 250 Hz; motion and PPG at 50 Hz via Bluetooth Low-Energy.
2. **Pre-processing:** Baseline wander removed (0.5-40 Hz band-pass); signals segmented into 5-s windows and z-normalised.
3. **Model:** 1-D LSTM auto-encoder (128 hidden units) trained unsupervised on "normal" beats; anomaly score = reconstruction error.
4. **On-device inference:** Model converted to TensorFlow-Lite Micro; average latency 12 ms on an ARM-M4 MCU with 250 kB RAM.
5. **Cloud monitoring:** Reconstruction-error histograms streamed nightly; threshold re-tuned if false-alert rate drifts by >15 %.

Figure 4. Wearable ECG anomaly detection pipeline

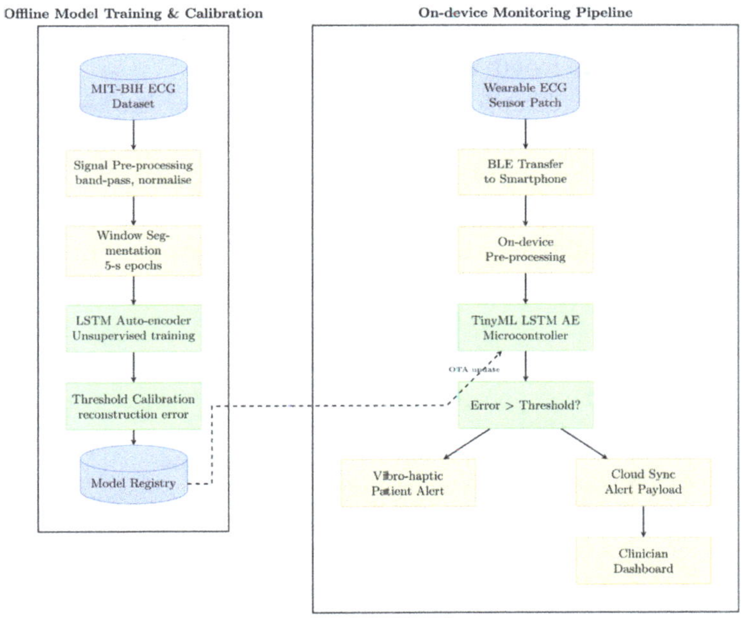

Offline: MIT-BIH signals are pre-processed, windowed, and used to train an LSTM auto-encoder; the quantised model and threshold are stored in a registry. *Online:* A patch streams ECG via BLE; the microcontroller runs the TinyML auto-encoder, triggering a patient vibration and cloud alert when reconstruction error exceeds the calibrated limit. Over-The-Air (OTA) updates keep the on-device model current.

Together, lightweight sequential models plus multi-sensor fusion illustrate how consumer wearables can evolve from passive trackers into early-warning medical devices.

Illustrative Use-Case: Conversational AI for Virtual Triage

On many telehealth portals a fine-tuned ClinicalBERT encoder followed by a BiGRU classifier now triages free-text symptom descriptions. In a corpus of 25,000 chat transcripts the model achieved a top-3 accuracy of 0.82 for routing patients to self-care, primary care or the emergency department, and showed potential for reducing nurse call volume by approximately 18 % while maintaining patient satisfaction levels (Chen et al., 2022).

Implementation highlights:

1. **Dataset:** De-identified chat logs manually labelled with one of three triage levels; class imbalance handled by focal loss.
2. **Tokenisation:** SentencePiece (32 k vocab) preserves medical abbreviations; maximum sequence length 256 tokens.
3. **Model:** ClinicalBERT encoder → BiGRU (hidden = 256) → softmax; fine-tuned for 4 epochs on two A100 GPUs.
4. **Serving:** gRPC micro-service (FastAPI) processes ≈40 requests s^{-1} with p95 latency <100 ms; returns top-3 triage suggestion + confidence scores.
5. **Governance:** Monthly audit compares model decisions with clinician overrides; a retraining pipeline is triggered if weighted F1 falls below 0.78.

This example shows how transformer-based language models can be embedded in telehealth front-end components to automate low-risk triage while keeping clinicians in the loop.

Table 1. Representative Applications of AI in Healthcare

Application Domain	AI Techniques	Potential Benefits
Clinical Decision Support	Logistic regression; random forests; XGBoost; Bayesian networks; NLP (BERT, BioBERT)	Evidence-based recommendations; reduced diagnostic errors; probabilistic risk scoring
Medical Imaging	CNNs (ResNet, DenseNet, EfficientNet); U-Net; Mask R-CNN; GAN/ VAE augmentation	Faster interpretation; high diagnostic accuracy; robust models with limited labels
Personalized Medicine	Random forests; SVMs; GNNs; reinforcement learning	Tailored drug dosing; novel target discovery; sequential treatment optimisation
Remote Monitoring	LSTMs; GRUs; TCNs; auto-encoder anomaly detection; transformer chatbots	Early intervention; reduced admissions; scalable virtual triage

Table 1 provides a comprehensive summary of representative applications, their associated AI techniques, and the substantial potential benefits they offer to modern healthcare delivery.

CHALLENGES AND ETHICAL CONSIDERATIONS

Integrating artificial intelligence into healthcare systems is a groundbreaking opportunity to enhance patient outcomes and the delivery of treatment. However, it also comes with major obstacles that need to be dealt with in a systematic way. These problems include a wide range of areas, including technological, ethical, legal, and social and economic issues. They go beyond only implementation issues to include basic questions about how medicine should be practiced in a world where AI is used (Mittelstadt et al. 2016).

Healthcare AI works in a special setting where choices have a direct effect on people's lives, trust and openness are very important, and mistakes in algorithms can have terrible effects. In this high-stakes situation, it's important to think carefully about how AI technologies fit into clinical workflows, serve a wide range of people, and uphold the moral standards that guide medical practice. To make sure that healthcare systems are trustworthy, fair, and clinically successful, technologists, doctors, ethicists, and policymakers must work together to solve these problems. This is necessary to unlock AI's full potential.

Data Quality and Availability

Healthcare AI systems rely on high-quality, representative training data to work effectively, and therefore quality issues with data are one of the biggest barriers to successful adoption (Beam & Kohane 2018) (Ghassemi et al. 2020). Healthcare data has its own unique set of challenges, like having different formats, being dependent on time, and having systems that are spread out across institutions. All of these factors have immediate impacts on model performance, generalizability, and clinical correctness.

1. **Data Heterogeneity**: Across institutions, EHRs differ incredibly in organization, coding schemes, and documentation workflows, making model generalization difficult (Shickel et al. 2017). This heterogeneity in data standards worsens portability: transferring a model from one site to another typically requires additional adaptation steps followed by cross-site validation and evaluation procedures.
2. **Incomplete and Noisy Data**: Healthcare datasets commonly contain missing fields, entry errors, and measurements that are not mutually consistent. Such data issues reduce model performance due to data quality issues, resulting in limited clinical utility (Hripcsak & Albers 2011). For predictable clinical quality guarantees, rigorous preprocessing and thorough validation are required to detect and mitigate these issues in model inputs.
3. **Limited Labeled Data:** Expert annotation of medical data requires significant amount of time and financial resources, which becomes a hurdle in the development process, especially for rare diseases and unique applications (Litjens et al. 2017). Self-supervised learning offers a practical pathway to avert these constraints. With continued progress, such approaches could reshape systems, in practice, intended for low-prevalence scenarios or areas with limited expert knowledge limiting dataset curation.

Addressing these data challenges demands continued work on standardization and close collaboration across institutions. Equally important are practical and clinically workable data-collection methods that meet required quality standards.

Bias and Fairness

Algorithmic bias is a substantial ethical problem in healthcare AI since biased models can compound health disparities in society, which hurts vulnerable groups the most (Obermeyer et al. 2019). The challenge originates from a number of reasons, such as past unfairness, not enough representation in training data, and algorithmic design choices that may render methodologies to operate sub-optimally for under-represented groups.

1. **Sampling Bias**: The underrepresentation of specific populations in training data results in diminished accuracy for those groups, potentially worsening health inequities. This occurs due to past study omissions and disparate healthcare access patterns.
2. **Label Bias**: Historical biases embedded in medical records can lead to bias behaviors in AI systems when trained using such data (Chen et al. 2021). Systematic differences in treatment patterns across different population groups due to inherent societal differences can lead to the creation of biased training data.
3. **Algorithmic Bias**: Model architectures and optimization goals can unintentionally make differences worse by using feature selection, training methods, or assessment metrics that benefit the majority of people (Barocas, Hardt, & Narayanan 2016).

To deal with bias in AI systems, one needs to do a full fairness assessment utilizing measures like demographic parity, equalized odds, and predictive parity (Hardt, Price, and Srebro 2016). Hence, in order to develop fair system monitoring, diverse development teams, and institutional commitments to uniform health access are required.

Interpretability and Explainability

Before employing AI systems' suggestions in clinical workflows, healthcare professionals need to understand how these systems function (Gunning & Aha 2019). This requirement comes from the need for physicians to keep an eye on medical decision-making, as well as professional and legal obligations. However, the majority of high-performing AI models, especially deep learning systems, work like *black boxes*, which makes it hard to balance how well they forecast and how explainable their outputs are.

1. **Post-hoc Explainability**: Techniques such as LIME (Ribeiro, Singh, & Guestrin 2016) and SHAP (Lundberg & Lee 2017) approximate local feature contributions by analyzing how input changes affect model predictions. While valuable for understanding complex models, these methods may not always accurately reflect true decision-making processes.
2. **Interpretable-by-Design Models**: Decision trees, generalized additive models (GAMs), and rule-based systems provide inherent interpretability (Caruana et al. 2015). These approaches offer transparent reasoning but may sacrifice some predictive accuracy for complex medical patterns.

Achieving effective explainability calls for close cooperation between AI researchers, medical practitioners, and interface designers so that explanation methods can be tailored to the specific needs of the clinical domain while remaining technically sound and clinically meaningful. This involves designing user interfaces that are easy to understand for end-users and establishing clear benchmarks for evaluating the quality of explanations within real clinical settings.

Privacy and Security

Patient data privacy and security remain of paramount importance in healthcare AI, as breaches may result in serious consequences such as identity theft, discrimination, and erosion of trust in healthcare institutions (Greenwood & Stopczynski 2019). The central challenge lies in achieving a balance between enabling broad data accessibility and enforcing stringent privacy.

1. **Data Privacy Laws**: Regulatory frameworks, such as HIPAA in the United States and GDPR in the European Union, mandate strict guidelines for the collection, storage, and usage of patient information (Rosenbaum 2017). Achieving such compliance by AI systems demands well-structured consent mechanisms, role-based access controls, and comprehensive audit trails.
2. **Anonymization Challenges**: Even after data undergoes de-identification, the risk of re-identification persists due to the possibility of linking datasets from different sources (Narayanan & Shmatikov 2008). To solve for this, methods such as differential privacy and synthetic data generation offer promising mitigation strategies. However, they involve inherent trade-offs between maintaining privacy and preserving the utility of data for AI model performance.
3. **Security Threats**: Advanced adversarial attacks targeting medical imaging AI models can potentially lead to diagnostic errors by introducing imperceptible perturbations (Finlayson et al. 2019). Such manipulations, often invisible to the human eye, can mislead even high-performing models into producing clinically dangerous outcomes. Addressing these threats requires the deployment of resilient defense mechanisms and the establishment of continuous model and system monitoring processes.

Privacy-preserving learning approaches, including federated learning (Yang et al. 2019) and differential privacy (Dwork & Roth 2014), have emerged as practical solutions, enabling collaborative AI model development without compromising the confidentiality and security of sensitive patient data. For healthcare AI to be adopted at a much larger scale, the integration of such methods is becoming essential.

Regulatory and Legal Barriers

The clearance process for AI-based medical devices and clinical decision-support systems presents substantial challenges, as regulatory structures are in a phase of ramping up to address the demands of evolving technologies like artificial intelligence (Benjamens, Dhunnoo, and Mesko 2020). Traditional approval mechanisms, originally designed for fixed-function devices, must now be redesigned to address the unique demands of AI, including its ability for continuous learning, dependency on varied data sources, and complex reasoning pathways.

1. **Validation Standards**: Regulatory agencies mandate thorough clinical validation prior to deployment (U.S. Food and Drug Administration 2019). However, conventional validation methodologies may not fully capture the performance variability of AI systems that are perceptive to data distribution shifts and demonstrate heterogeneous outcomes across different patient groups. This situation calls for the development of fresh evaluation strategies tailored for such dynamic systems.
2. **Dynamic Learning Systems:** AI solutions that continuously update their models introduce difficulties for rigid regulatory systems (Weber et al., 2020). There is a need for adaptive approval routes that can accommodate model evolution to support innovation while ensuring patient safety.

Addressing these regulatory concerns demands coordinated action among regulators, clinical information specialists, and AI innovators, with the shared objective of formulating policies that preserve safety while enabling timely access to beneficial technological advancements.

Socio-technical Integration

AI solutions must integrate effectively within healthcare workflows, ensuring that they do not add unnecessary complexity for physicians or compromise patient safety (Sendak et al., 2020). Achieving this requires maintaining a careful balance between technological capabilities and human factors, organizational culture, and established clinical processes.

1. **Usability**: Interfaces or alert systems that are poorly designed can aggravate clinician workload and even affect patient safety. For AI systems to be effective, they should employ intuitive designs that convey information in formats that are clinically relevant, reduce the incidence of false alerts, and integrate smoothly into existing workflows without causing disruption.

2. **Trust**: The level of trust clinicians place in AI systems depends significantly on the transparency of the models, the strength of validation evidence, and most importantly, the degree to which outputs align with clinical reasoning. Building trust requires demonstrating tangible value in practice, delivering adequate training for users, and defining clear mechanisms for feedback and observability.

To make AI systems that are both technically advanced and usable in the real world, technologists, physicians, and implementation scientists need to work together across disciplines. This is only possible if organizations are committed to change management and continuous development.

FUTURE DIRECTIONS AND EMERGING TRENDS

The area of AI in healthcare is at an exciting point at present because of new algorithms, more efficient and powerful compute, and increasingly high-quality healthcare information becoming available. These factors coming together are making it possible to create healthcare solutions that are more advanced, integrated, and tailored than anything seen before.

Emerging trends show a shift toward holistic, patient-centered approaches that promise to greatly improve clinical practice over the next ten years (Topol, 2019). These changes don't only make small gains; they tackle major problems in healthcare by using novel methods for individualized therapy, data integration, and making advanced medical knowledge available to everyone.

As these trends evolve, one of the key areas of focus is the integration of diverse data sources to create more comprehensive healthcare solutions.

Multimodal Learning

The next promising direction is to enhance pattern extraction from patients' data to combine multiple data modalities to assess the scenario more holistically. In one of the studies, multimodal embeddings combining vision, text, time-series, and tabular data enhanced the quality of chest pathology diagnosis prediction (Ye et al., 2024) (Soenksen et al., 2022).

Multimodal learning aims at the integration of patient's healthcare data ranging from structured and unstructured FHIR resources, including imaging, time series, tabular, and unstructured note data. Along with formal sources, augmenting personal care devices' sensor data and genomics would further enhance physiological understanding of the patient's health. In principle this is achieved by integrating signals from different modalities into a shared embedding space, followed by model training through a machine learning task (Baltrušaitis, Ahuja, and Morency 2018). Recent advancements in transformer architecture to extract meaningful features in several modalities have enabled the feasibility of such applied techniques in practice (Tsai et al., 2019).

Self-Supervised Learning

Self-supervised learning (SSL) is an emerging branch of training strategy that reduces the dependency on labels required in a supervised training regime. This is achieved by carefully crafting the training setup such that the model predicts information that is already available in the raw data itself, obliterating the need for supervised labels. This strategy enables training models that have been conventionally dependent on the labels due to limited resources and furthermore allows AI to leverage extreme volumes of data to learn clinically relevant patterns without human intervention. Self-supervised learning has allowed for engineering stronger domain adaptation and better performance on enhanced patient demographics since models get access to a higher magnitude of data (Jing & Tian, 2020).

Large Language Models like BioGPT (Luo et al., 2022) and Med-PaLM (Singhal et al., 2023) have successfully utilized SSL-based training strategies to learn from medical literature, including cutting-edge research and established knowledge repositories, in a semi-supervised manner. Once these models learn patterns autonomously from the large set of data, they are later exposed to downstream tasks with fine-tuning. This is only feasible as the models get to learn core concepts in the first phase, known as pre-training, and then can be nudged on specific downstream tasks with minimal fine-tuning.

Contrastive learning, a specific style of self-supervised learning, utilizes the concept of similarity among data points to successfully discover and learn patterns from datasets. SimCLR (Chen et al., 2020) and MoCo (He et al., 2020) train to achieve robust visual feature representations by augmenting views of the same sample to learn that the different views mean the same concept. This idea has helped models to become clinically robust to natural variations and noise.

Privacy-Preserving AI

In today's world, a substantial amount of data gets generated in medical organizations that cater to patients on a day-to-day basis. This far exceeds the volume of all the research-oriented datasets and thus has representations across various demographics that usually get omitted in the curation of research-specific datasets. However, compliance mechanisms like HIPAA and GDPR prevent the data owners from sharing with research organizations, creating bottlenecks in building models that lack generalization and may be biased due to limited training data. Federated learning specifically solves these challenges by allowing AI models to be trained over the data from different owners without the patient data ever leaving the owner organization's infrastructure. This empowers the AI community across organizations to innovate by collaborating in the data domain while maintaining privacy compliance. Each organization trains a local copy of the model and shares only the learned parameters of the machine learning model with each other to aggregate a model with much superior performance (Yang et al., 2019).

Homomorphic encryption when integrated with federated learning, takes the success even further to provide even superior privacy guarantees by ensuring individual data sample updates cannot be reconstructed. This class of encryption makes it possible to perform operations directly on the encrypted items, as if the operations were performed on the original items, hence allowing the execution of operations without ever observing the raw data (Bonawitz et al., 2017).

Such strategies hold critical importance in multinational research that will result in generalized models trained on much more diverse population data. This also marks the beginning of a broader democratization of the clinical AI innovation.

Explainable and Causally-Aware AI

Several state-of-the-art models are powered by deep learning that primarily rely on post-hoc methods of explainability. This limitation is fundamentally due to the black box nature of such models. This is now being overcome by introducing transparent and causally aware modules within learnable models (Pearl & Mackenzie, 2018). With the advent of such strategies, clinical systems are becoming more transparent and enhancing adoption since clinicians get a clear reasoning trail, increasing trust in AI systems. To build such components that learn and depict cause-effect relationships, causal inference frameworks utilize feature extractors along with causal discovery algorithms by leveraging directed acyclic graphs.

Additionally, counterfactual reasoning, which predicts patient outcomes under hypothetical interventions (Schwab et al., 2022), enables the creation of personalized treatment plans by illustrating potential scenarios based on different clinical interventions. These advanced models hold the potential to significantly improve the personalization of treatment strategies by offering clear, well-grounded explanations of causal pathways. This promotes enhanced collaboration between human experts and AI systems, which has long been the goal of such systems.

CONCLUSION

The advancements in clinical decision-making, assisted medical imaging, and personalized healthcare devices have transformed healthcare with a wave of technology and artificial intelligence. This change has enhanced every stage of healthcare, ranging from patient care to pre-care discoveries or post-care population-level analysis. The development of advanced multimodal learning systems that combine data from different modalities and sources to discover new drugs is a great example of how AI advances have expedited the pre-care stage, a process that required far more time and resources.

This chapter covered applications ranging from CNN-powered diagnostic imaging to transformer-based clinical assistants, demonstrating how AI can greatly augment human knowledge in healthcare workflows. These systems have benefited healthcare processes from learning patterns in significantly large populations and consequently applying personalized strategies to individual care. This accessibility of high-quality specialized medical knowledge has been celebrated in the clinical community, thus, AI has seen widespread adoption in recent years.

However, realizing AI's full potential requires solving challenges like data quality, algorithmic bias, and interpretability. The path forward requires methods that ensure fairness across diverse populations in the dataset creation step, provide explainability, and allow regulatory adaptations in the methods that can balance innovation with patient safety. Privacy-preserving and explainable AI systems will become key enablers for adoption in healthcare, as they help build trust among clinicians during usage.

The trajectory for the future is set on a positive slope, as artificial intelligence is set to become an essential companion in healthcare delivery as a result of upcoming policy frameworks that prioritize population equity and responsible innovation. Guidelines from the U.S. Food and Drug Administration (FDA) on clinical decision support systems are also setting an example for other regulatory bodies (Food and Drug Administration, 2022).

Ultimately success depends on technologists, physicians, ethicists, and policymakers working together to develop an AI that is based on clinical needs, ethical values, and fairness. As the healthcare technology space moves forward, the use of AI in healthcare must be based on strict testing, constant monitoring, and a strong commitment to patient safety and making healthcare available to everyone across the world.

REFERENCES

Bai, S., Kolter, J. Z., & Koltun, V. (2018). "An Empirical Evaluation of Generic Convolutional and Recurrent Networks for Sequence Modeling." *arXiv Preprint arXiv:1803.01271*.

Baltrušaitis, T., Ahuja, C., & Morency, L.-P. (2018). Multimodal Machine Learning: A Survey and Taxonomy. *IEEE Transactions on Pattern Analysis and Machine Intelligence*, *41*(2), 423–443. DOI: 10.1109/TPAMI.2018.2798607

Barocas, S., M. Hardt, and A. Narayanan. (2016). *Fairness and Machine Learning*. fairmlbook.org.

Beam, A. L., & Kohane, I. S. (2018). Big Data and Machine Learning in Health Care. *Journal of the American Medical Association*, *319*(13), 1317–1318. DOI: 10.1001/jama.2017.18391

Benjamens, S., Dhunnoo, P., & Mesko, B. (2020). The State of Artificial Intelligence-Based FDA-Approved Medical Devices and Algorithms: An Online Database. *NPJ Digital Medicine*, *3*(1), 118. DOI: 10.1038/s41746-020-00324-0

Bonawitz, K., Ivanov, V., Ben Kreuter, A. M., McMahan, H. B., Patel, S., Ramage, D., Segal, A., & Seth, K. (2017). "Practical Secure Aggregation for Privacy-Preserving Machine Learning." In *Proceedings of the 2017 ACM SIGSAC Conference on Computer and Communications Security*, 1175–91. CCS '17. New York, NY, USA: Association for Computing Machinery. DOI: 10.1145/3133956.3133982

Caruana, R., Lou, Y., & Gehrke, J.. (2015). "Intelligible Models for Healthcare: Predicting Pneumonia Risk and Hospital 30-Day Readmission." In *Proceedings of the 21th ACM SIGKDD International Conference on Knowledge Discovery and Data Mining*, 1721–30. DOI: 10.1145/2783258.2788613

Chen, I. Y., Pierson, E., Rose, S., Joshi, S., Ferryman, K., & Ghassemi, M. (2021). Ethical Machine Learning in Healthcare. *Annual Review of Biomedical Data Science*, *4*(May), 123–144. DOI: 10.1146/annurev-biodatasci-092820-114757

Chen, Q., Raghu, M., Yedidia, B., Luo, Y., & Tan, W. (2022). "Transformer-Based Patient Triage from Conversational Text." In *Proceedings of the Conference on Empirical Methods in Natural Language Processing*.

Chen, T., Kornblith, S., Norouzi, M., & Hinton, G. (2020). "A Simple Framework for Contrastive Learning of Visual Representations." In *International Conference on Machine Learning*, 1597–607.

Cho, K., van Merriënboer, B., Gulcehre, C., Bahdanau, D., Bougares, F., Schwenk, H., & Bengio, Y. (2014). "Learning Phrase Representations Using RNN Encoder-Decoder for Statistical Machine Translation." In *Proceedings of EMNLP*, 1724–34. DOI: 10.3115/v1/D14-1179

Consortium, International Warfarin Pharmacogenetics. (2013). "Estimation of the Warfarin Dose with Clinical and Pharmacogenetic Data." *New England Journal of Medicine* 360 (8): 753–64. .DOI: 10.1056/NEJMoa0809329

de Dombal, F. T., Leaper, D. J., Staniland, J. R., McCann, A. P., & Horrocks, J. C. (1972). Computer-Aided Diagnosis of Acute Abdominal Pain. *BMJ (Clinical Research Ed.)*, *2*(5804), 9–13. DOI: 10.1136/bmj.2.5804.9

Dosovitskiy, A., Beyer, L., Kolesnikov, A., Weissenborn, D., Zhai, X., Unterthiner, T., & Dehghani, M.. (2021). "An Image Is Worth 16x16 Words: Transformers for Image Recognition at Scale." In *International Conference on Learning Representations (ICLR)*. https://arxiv.org/abs/2010.11929

Dwork, C., & Roth, A. (2014). The Algorithmic Foundations of Differential Privacy. *Foundations and Trends in Theoretical Computer Science*, *9*(3–4), 211–407. DOI: 10.1561/0400000042

Eapen, B. R., Sartipi, K., & Archer, N. (2020). Serverless on FHIR: Deploying Machine-Learning Models for Healthcare on the Cloud. *Journal of Digital Health*, *6*(1), 1–11. https://arxiv.org/abs/2006.04748

Epstein, R. S., Moyer, T. P., Aubert, R. E., Teagarden, J. R., Chen, Y., Symm, B., Zhou, Z., Larrabee, P., Razouki, Z., & Sack, J. S. (2010). Warfarin Genotyping Reduces Hospitalization Rates: Results from the Medco–Mayo Warfarin Effectiveness Study. *Journal of the American College of Cardiology*, *55*(25), 2804–2812. DOI: 10.1016/j.jacc.2010.03.009

Esteva, A., Kuprel, B., Novoa, R. A., Ko, J., Swetter, S. M., Blau, H. M., & Thrun, S. (2017). Dermatologist-Level Classification of Skin Cancer with Deep Neural Networks. *Nature*, *542*(7639), 115–118. DOI: 10.1038/nature21056

Faust, O., Shenfield, A., Kareem, M., Brechtl, J., & Ciaccio, E. J. (2020). Anomaly Detection for Short Single-Lead ECG Using Autoencoders. *Computers in Biology and Medicine*, *124*, 103922. DOI: 10.1016/j.compbiomed.2020.103922

Finlayson, S. G., Bowers, J. D., Ito, J., Zittrain, J. L., Beam, A. L., & Kohane, I. S. (2019). Adversarial Attacks on Medical Machine Learning. *Science*, *363*(6433), 1287–1289. DOI: 10.1126/science.aaw4399

Food and Drug Administration. (2022, September 28). Clinical Decision Support Software: Guidance for Industry and Food and Drug Administration Staff. U.S. Department of Health and Human Services.

Franchini, M., Mengoli, C., Cruciani, M., Bonfanti, C., Pizzolo, G., Veneri, D., Lippi, G., & Favaloro, E. J. (2014). Effects on Bleeding Complications of Pharmacogenetic Testing for Initial Dosing of Vitamin k Antagonists: A Systematic Review and Meta-analysis. *Journal of Thrombosis and Haemostasis*, *12*(9), 1480–1490. DOI: 10.1111/jth.12647

Frid-Adar, M., Klang, E., Amitai, M., Goldberger, J., & Greenspan, H. (2018). "Synthetic Data Augmentation Using GAN for Improved Liver Lesion Classification." In *2018 IEEE 15th International Symposium on Biomedical Imaging (ISBI 2018)*, 289–93. DOI: 10.1109/ISBI.2018.8363576

Ghassemi, M., Naumann, T., Schulam, P., Beam, A. L., Chen, I. Y., & Ranganath, R. (2020). A Review of Challenges and Opportunities in Machine Learning for Health. [American Medical Informatics Association.]. *Proceedings of the AMIA Joint Summits on Translational Science*, *2020*, 191–200.

Goodfellow, I., Pouget-Abadie, J., Mirza, M., Xu, B., Warde-Farley, D., Ozair, S., Courville, A., & Bengio, Y. (2014). "Generative Adversarial Networks." In *Advances in Neural Information Processing Systems (NIPS)*, 27:2672–80.

Greenwood, D., & Stopczynski, A. (2019). Privacy and Human Behavior in the Age of Information. *Science*, *347*(6221), 509–514.

Gunning, D., & Aha, D. W. (2019). DARPA's Explainable Artificial Intelligence (XAI) Program. *AI Magazine*, *40*(2), 44–58. DOI: 10.1609/aimag.v40i2.2850

Hardt, M., Price, E., & Srebro, N. (2016). Advances in Neural Information Processing Systems: Vol. 3315–23. *Equality of Opportunity in Supervised Learning*.

Hatamizadeh, A., Nath, V., Tang, Y., Yang, D., Roth, H., & Xu, D. (2022). "Swin UNETR: Swin Transformers for Semantic Segmentation of Brain Tumors in MRI Images." *arXiv Preprint arXiv:2201.01266*. https://arxiv.org/abs/2201.01266 DOI: 10.1007/978-3-031-08999-2_22

He, K., Fan, H., Wu, Y., Xie, S., & Girshick, R. (2020). "Momentum Contrast for Unsupervised Visual Representation Learning." In *Proceedings of the IEEE/CVF Conference on Computer Vision and Pattern Recognition*, 9729–38. DOI: 10.1109/CVPR42600.2020.00975

He, K., Gkioxari, G., Dollar, P., & Girshick, R. (2017). "Mask r-CNN." In *Proceedings of the IEEE International Conference on Computer Vision (ICCV)*, 2961–69. DOI: 10.1109/ICCV.2017.322

He, K., Zhang, X., Ren, S., & Sun, J. (2016). "Deep Residual Learning for Image Recognition." In *Proceedings of the IEEE Conference on Computer Vision and Pattern Recognition (CVPR)*, 770–78. DOI: 10.1109/CVPR.2016.90

Hochreiter, S., & Schmidhuber, J. (1997). Long Short-Term Memory. *Neural Computation*, *9*(8), 1735–1780. DOI: 10.1162/neco.1997.9.8.1735

Hripcsak, G., & Albers, D. J. (2011). Correlating Electronic Health Record Concepts with Healthcare Process Events. *Journal of the American Medical Informatics Association : JAMIA*, *18*(4), 509–513. DOI: 10.1136/amiajnl-2011-000153

Huang, G., Liu, Z., van der Maaten, L., & Weinberger, K. Q. (2017). "Densely Connected Convolutional Networks." In *Proceedings of the IEEE Conference on Computer Vision and Pattern Recognition (CVPR)*, 4700–4708. DOI: 10.1109/CVPR.2017.243

Irvin, J., Rajpurkar, P., Ko, M., Yu, Y., Ciurea-Ilcus, S., Chute, C., Marklund, H., Haghgoo, B., Ball, R., Shpanskaya, K., Seekins, J., Mong, D. A., Halabi, S. S., Sandberg, J. K., Jones, R., Larson, D. B., Langlotz, C. P., Patel, B. N., Lungren, M. P., & Ng, A. Y. (2019). CheXpert: A Large Chest Radiograph Dataset with Uncertainty Labels and Expert Comparison. *Proceedings of the AAAI Conference on Artificial Intelligence*, *33*(1), 590–597. DOI: 10.1609/aaai.v33i01.3301590

Jing, L., & Tian, Y. (2020). Self-Supervised Visual Feature Learning with Deep Neural Networks: A Survey. *IEEE Transactions on Pattern Analysis and Machine Intelligence*, *43*(11), 4037–4058. DOI: 10.1109/TPAMI.2020.2992393

Johnson, A. E. W., Bulgarelli, L., Pollard, T. J., Horng, S., Celi, L. A., & Mark, R. G. (2023). MIMIC-IV, a Freely Accessible Electronic Health Record Dataset. *Scientific Data*, *10*(1), 23. DOI: 10.1038/s41597-023-02136-9

Kingma, D. P., & Welling, M. 2013. "Auto-Encoding Variational Bayes." In *International Conference on Learning Representations (ICLR), arXiv Preprint.*

Komorowski, M., Celi, L. A., Badawi, O., Gordon, A. C., & Faisal, A. A. (2018). The Artificial Intelligence Clinician Learns Optimal Treatment Strategies for Sepsis in Intensive Care. *Nature Medicine*, *24*(11), 1716–1720. DOI: 10.1038/s41591-018-0213-5

Ledley, R. S., & Lusted, L. B. (1959). Reasoning Foundations of Medical Diagnosis. *Science*, *130*(3366), 9–21. DOI: 10.1126/science.130.3366.9

Lee, J., Yoon, W., Kim, S., Kim, D., Kim, S., So, C. H., & Kang, J. (2020). BioBERT: A Pre-Trained Biomedical Language Representation Model for Biomedical Text Mining. *Bioinformatics (Oxford, England)*, *36*(4), 1234–1240. DOI: 10.1093/bioinformatics/btz682

Libbrecht, M. W., & Noble, W. S. (2015). Machine Learning Applications in Genetics and Genomics. *Nature Reviews. Genetics*, *16*(6), 321–332. DOI: 10.1038/nrg3920

Litjens, G., Kooi, T., Bejnordi, B. E., Setio, A. A. A., Ciompi, F., Ghafoorian, M., van der Laak, J. A. W. M., van Ginneken, B., & Sánchez, C. I. (2017). A Survey on Deep Learning in Medical Image Analysis. *Medical Image Analysis*, *42*, 60–88. DOI: 10.1016/j.media.2017.07.005

Lucas, P. J. F., van der Gaag, L. C., & Abu-Hanna, A. (2004). Bayesian Networks in Biomedicine and Health-Care. *Artificial Intelligence in Medicine*, *30*(3), 201–214. https://doi.org/https://doi.org/10.1016/j.artmed.2003.11.001. DOI: 10.1016/j.artmed.2003.11.001

Lundberg, S. M., & Lee, S.-I. (2017). "A Unified Approach to Interpreting Model Predictions." In *Proceedings of the 31st International Conference on Neural Information Processing Systems*, 4768–77. NIPS'17. Red Hook, NY, USA: Curran Associates Inc.

Luo, R., Sun, L., Xia, Y., Qin, T., Zhang, S., Poon, H., & Liu, T.-Y. (2022). BioGPT: Generative Pre-Trained Transformer for Biomedical Text Generation and Mining. *Briefings in Bioinformatics*, *23*(6), bbac409. DOI: 10.1093/bib/bbac409

Miller, R. A.Jr, Pople, H. E.Jr, & Myers, J. D. (1982). INTERNIST-1, an Experimental Computer-Based Diagnostic Consultant for General Internal Medicine. *The New England Journal of Medicine*, *307*(8), 468–476. DOI: 10.1056/NEJM198208193070803

Mittelstadt, B. D., Allo, P., Taddeo, M., Wachter, S., & Floridi, L. (2016). The Ethics of Algorithms: Mapping the Debate. *Big Data & Society*, *3*(2), 2053951716679679. DOI: 10.1177/2053951716679679

Montgomery, D. C. (2007). *Introduction to Statistical Quality Control* (5th ed.). John Wiley & Sons.

Narayanan, A., & Shmatikov, V. (2008). "Robust de-Anonymization of Large Sparse Datasets." In *2008 IEEE Symposium on Security and Privacy*, 111–25. DOI: 10.1109/SP.2008.33

Obermeyer, Z., Powers, B., Vogeli, C., & Mullainathan, S. (2019). Dissecting Racial Bias in an Algorithm Used to Manage the Health of Populations. *Science*, *366*(6464), 447–453. DOI: 10.1126/science.aax2342

Pearl, J., & Mackenzie, D. (2018). *The Book of Why: The New Science of Cause and Effect*. Basic Books.

Philippi, D., Rothaus, K., & Castelli, M. (2023). A Vision Transformer Architecture for the Automated Segmentation of Retinal Lesions in Spectral-Domain Optical Coherence Tomography Images. *Scientific Reports*, *13*(1), 517. DOI: 10.1038/s41598-023-27616-1

Ponce-Bobadilla, A. V., Schmitt, V., & Valderrama, D. (2024). Practical Guide to SHAP Analysis: Explaining Supervised Machine-Learning Model Predictions in Drug Development. *Clinical and Translational Science*, *17*(11), e70056. DOI: 10.1111/cts.70056

Rajkomar, A., Dean, J., & Kohane, I. (2019). Machine Learning in Medicine. *The New England Journal of Medicine*, *380*(14), 1347–1358. DOI: 10.1056/NEJMra1814259

Ribeiro, M., Singh, S., & Guestrin, C. (2016). ""Why Should i Trust You?": Explaining the Predictions of Any Classifier." In *Proceedings of the 22nd ACM SIGKDD International Conference on Knowledge Discovery and Data Mining*, 1135–44. DOI: 10.1145/2939672.2939778

Ronneberger, O., Fischer, P., & Brox, T. (2015). "U-Net: Convolutional Networks for Biomedical Image Segmentation." In *Medical Image Computing and Computer-Assisted Intervention – MICCAI 2015*, 9351:234–41. DOI: 10.1007/978-3-319-24574-4_28

Rosenbaum, L. (2017). Data Sharing and the Ethics of Public Trust. *The New England Journal of Medicine*, *377*(5), 401–403. DOI: 10.1056/NEJMp1005897

Sakurada, M., & Yairi, T. (2014). "Anomaly Detection Using Autoencoders with Nonlinear Dimensionality Reduction." In *Workshop on Machine Learning for Sensory Data Analysis (MLSDA)*, 4–11. DOI: 10.1145/2689746.2689747

Schwab, P.. (2022). Counterfactuals and Causal Inference for Clinical Prediction Models. *Nature Machine Intelligence*, *4*, 813–824.

Sendak, M. P., D'Arcy, J., & Kashyap, S.. (2020). A Path for Translation of Machine Learning Products into Healthcare Delivery. *EMJ Innovations*, *4*(1), 56–64.

Shickel, B., Tighe, P. J., Bihorac, A., & Rashidi, P. (2017). Deep EHR: A Survey of Recent Advances in Deep Learning Techniques for Electronic Health Record (EHR) Analysis. *IEEE Journal of Biomedical and Health Informatics*, *22*(5), 1589–1604. DOI: 10.1109/JBHI.2017.2767063

Shortliffe, E. H. (1976). *Computer-Based Medical Consultations: MYCIN*. Elsevier.

Singhal, K., Tu, T., & Gottweis, J.. (2023). Large Language Models Encode Clinical Knowledge. *Nature, 620*(7972), 172–180. DOI: 10.1038/s41586-023-06291-2

Soenksen, L. R., Ma, Y., Zeng, C., Boussioux, L., Carballo, K. V., Na, L., Wiberg, H. M., Li, M. L., Fuentes, I., & Bertsimas, D. (2022). Integrated Multimodal Artificial Intelligence Framework for Healthcare Applications. *NPJ Digital Medicine*, *5*(149), 149. Advance online publication. DOI: 10.1038/s41746-022-00689-4

Tan, M., & Le, Q. (2019). "EfficientNet: Rethinking Model Scaling for Convolutional Neural Networks." In *Proceedings of the 36th International Conference on Machine Learning*, edited by Kamalika Chaudhuri and Ruslan Salakhutdinov, 97:6105–14. Proceedings of Machine Learning Research. PMLR. https://proceedings.mlr.press/v97/tan19a.html.

Tison, G. H., Sanchez, J. M., Ben Ballinger, M. S., Olgin, J. E., Pletcher, M. J., & Vittinghoff, E.. (2018). Passive Detection of Atrial Fibrillation Using a Commercially Available Smartwatch. *JAMA Cardiology*, *3*(5), 409–416. DOI: 10.1001/jamacardio.2018.0136

Topol, E. J. (2019). *Deep Medicine: How Artificial Intelligence Can Make Healthcare Human Again*. Basic Books.

Tsai, Y. H. H., Bai, S., & Yamada, M.. (2019). "Multimodal Transformer for Unaligned Multimodal Language Sequences." In *Proceedings of the Annual Meeting of the Association for Computational Linguistics*. DOI: 10.18653/v1/P19-1656

U.S. Food and Drug Administration. (2019). "Proposed Regulatory Framework for Modifications to Artificial Intelligence/Machine Learning-Based Software as a Medical Device." FDA.gov.

Weber, G. M.. (2020). Regulating AI in Health Care: An Overview of Proposed U.S. and EU Approaches. *Health Affairs*, *39*(11), 2024–2031.

Xu, C., Tao, D., & Xu, C. (2013). "A Survey on Multi-View Learning." *arXiv Preprint arXiv:1304.5634*.

Xu, J., Hu, Z., Miao, J., Li, K., & Liu, L. (2024). Machine Learning for Predicting Hemodynamic Deterioration of Patients with Intermediate-Risk Pulmonary Embolism in the Intensive Care Unit. *Shock (Augusta, Ga.)*, *61*(1), 51–58. DOI: 10.1097/SHK.0000000000002261

Xu, R., Brown, L., Ahuja, V., & Li, N. (2020). Towards Natural Language Processing for Symptom Checking: A Digital Dispatcher Study. *Translational International Medical Applications Science*, *12*(1), 109–114.

Yang, Q., Liu, Y., Chen, T., & Tong, Y. (2019). Federated Machine Learning: Concept and Applications. *ACM Transactions on Intelligent Systems and Technology*, *10*(2), 12. DOI: 10.1145/3298981

Yang, T., Zhang, W., & Wang, W.. (2021). "Large-scale Robust Deep AUC Maximization: A New Surrogate Loss and Empirical Studies on Medical Image Classification." In *Proceedings of the IEEE/CVF Conference on Computer Vision and Pattern Recognition.*

Ye, J., Woods, D., Jordan, N., & Starren, J. (2024). The Role of Artificial Intelligence for the Application of Integrating Electronic Health Records and Patient-Generated Data in Clinical Decision Support. *AMIA Joint Summits on Translational Science Proceedings AMIA Summit on Translational Science*, ***•••***, 459–467.

Zhao, L., Jia, L., & Greenspan, H. (2018). Multi-Scale Coarse-Fine Fusion for Volumetric Segmentation of Medical Images. *Medical Image Analysis*, *47*, 20–30. DOI: 10.1016/j.media.2018.08.002

Zhou, N., Zhang, C. T., Lv, H. Y., Hao, C. X., Li, T. J., Zhu, J. J., ... & Zhang, X. C. (2019). Concordance study between IBM Watson for oncology and clinical practice for patients with cancer in China. The oncologist, 24(6), 812-819.

Zitnik, M., Agrawal, M., & Leskovec, J. (2018). Modeling Polypharmacy Side Effects with Graph Convolutional Networks. *Bioinformatics (Oxford, England)*, *34*(13), i457–i466. DOI: 10.1093/bioinformatics/bty294

Chapter 3
A Bibliometric Analysis of IoT Based Applications in Smart Healthcare Systems

Aashka H. Thakkar
https://orcid.org/0000-0003-3629-7603
Parul University, India

Axita R. Thakkar
https://orcid.org/0000-0001-7592-3537
Parul University, India

ABSTRACT

The Internet of Things highlights the forthcoming detection, computing, and connections trends. The Internet of Things has transformed the field of health care. Growing numbers of chronic patients as well as an ageing population provide significant challenges with respect to current healthcare and medical services, demanding the incorporation of medical care alongside modern information and communication technologies (ICT) in order to enhance treatments. As a result, the breakthroughs studied in the present research will capitalize on healthcare systems powered by the Internet of things.A systematic bibliometric review is conducted by using the data of 4950 articles derived from the scopus database. The present study has used VOS viewer and R Biblioshiny package for the bibliometric analysis. The results pointed out that the Internet of Things can help policymakers improve societal health and economy.

DOI: 10.4018/979-8-3373-5727-0.ch003

INTRODUCTION

The transmission, administration, and understanding of health services have undergone a paradigm shift in the past few decades due to a convergence of technological innovation and healthcare. The Internet of Things is one of the most cutting-edge innovations propelling this development. IoT builds the groundwork to establish what are today known as Smart Healthcare Systems by facilitating smooth exchange of information throughout the ecosystem of healthcare through interrelated networks of equipment, devices, sensors, and software systems. To boost the treatment of patients, cut down on inefficiencies in operations, and raise the standard of care generally, these systems make use of continuous surveillance, automation, predictive analytics, and personalized healthcare. A significant evolution from conventional responsive therapy paradigms is represented by intelligent healthcare. IoT-enabled solutions allow for instantaneous choices, remote diagnostics, and ongoing health surveillance in place of periodic medical attention based on diagnostic presentation (Yuehong et al., 2016; Li et al., 2024). IoT is the process of connecting various physical objects, devices, and sensors over the Internet so they may exchange information as well as communicate with one another. Technological advancements have changed practically every aspect of existence, transforming our way of life in the process. (Chopade et al., 2023). As the number of people grows rapidly, older adults' and children's attention to sanitation and health is becoming increasingly important. As a result, the requirements for healthcare for wellbeing have increased, and patients and consumers are also making it harder for medical professionals to deliver high-quality services. Considering this component, the healthcare industry requires that technicians help clients live smarter lives in order to improve the efficacy and precision of medical diagnosis and treatment. (Baker, Xiang, & Atkinson, 2017)

IoT technologies have been successfully implemented in the healthcare sector to enhance patient care and results, providing new avenues for efficient healthcare delivery, individualized regimens for treatment, and distant monitoring. The Internet of Things is the result of sophisticated technology for communications and information technology. It could enhance the standard of living for urban residents. The necessity for developing affordable healthcare organizations that can effectively oversee and deliver a broad spectrum of medical treatments while lowering the overall cost continues to rise as a result of the alarming rate at which global population is expanding and the rising incidence of chronic medical conditions. Through simplified procedures, routine automation of tasks, substantially a reduction in the need for costly measures, the Internet of Things has resulted in the ability to significantly and significantly decrease healthcare expenses while simultaneously enhancing the quality of care for patients. By making real-time information collecting and monitoring possible, it improves patient outcomes and reaction times, which in turn improves patient care. Secondly, IoT technology simplify healthcare operations by streamlining administrative responsibilities (Yuehong et al., 2016; Li et al., 2024). This minimizes the necessity for a large team of administrators and saves time, which lowers expenses. Although studies in the same domains have demonstrated the feasibility of remote health surveillance, the advantages it may offer in some situations may be more significant. By keeping an eye on non-critical patients at home instead of in a hospital setting, monitoring their health remotely might ease the burden on hospital accommodations and medical staff. It might be employed to help senior citizens stay independent in their homes more frequently or for offering individuals in remote locations better access to healthcare. In essence, it can help people always have a greater influence over their individual health and increase access to medical services while lessening the burden on healthcare providers. Remote monitoring of health has a handful of drawbacks. The main drawbacks are the security hazards associated with storing a lot of private information in one database, the possible need for frequent sensor measurements to guarantee accurate monitoring, and the potential for interruptions from medical services in instances in which an individual was out of mobile data can vary. Furthermore, machine learning may be helpful make choices and therapeutic management (Yuehong et al., 2016; Li et al., 2024). Machine learning algorithms may produce therapy suggestions based on the unique features of each patient by assessing vast databases of medical guidelines, results of treatment, and patient records. This can help medical professionals maximize the treatment of patients and select the best course of action. Additionally, biological information could consist of novel characteristics and associations that machine learning software algorithms may find, which might result in breakthrough discoveries and advances in medical research. Machine learning Yuehong et al., 2016; Li et al., 2024) can assist in the identification DNA indicators, biological markers, and possible targets for therapy through the analysis of extensive genetic code, proteomics, and other types of -omics evidence. This can speed up the drug clinical trial process, resulting in developing of more specialized and individualized therapies. It is crucial to remember that there are difficulties in applying algorithmic learning algorithms in the healthcare sector. Additionally, robots with detectors and machine learning algorithms can be used for online healthcare and remote surveillance. These robots might move through the patient rooms and medical facility areas taking measurements, communicating with medical professionals, and facilitating virtual consultations. This makes it possible for medical personnel to keep an eye on patients in real time and respond quickly, particularly in isolated or underprivileged locations (Baker, Xiang, & Atkinson, 2017). For jobs like medicine distribution, inventory control, and hospital environment sterilisation, robotic assistants can be used. Healthcare professionals can concentrate on providing greater patient care thanks to this technology, which also increases accuracy and lessens staff effort (Chopade et al., 2023). Utilising natural language processing may assist to improve collaboration and patient involvement. NLP-powered chatbots and virtual assistants might interact with patients, responding to their enquiries, offering informational tools, and helping to make consultations. Such virtual assistants are capable of understanding and responding to natural language examinations thanks to NLP algorithms, which improves the efficiency and usability of the interactions. NLP can also help with learning discovery and clinical investigation algorithms may extract relevant data, spot trends, and help academics investigate new theories and findings by assessing vast amounts of scientific research and study papers. The result can improve awareness of medicine and speed up the research effort. (Ali et al., 2025; Amaraweera & Halgamuge, 2019)

ROLE OF IOT IN HEALTHCARE

In over the past two decades, the worldwide IoT sector has witnessed substantial development. (Bhuiyan et al., 2021; Dhanvijay & Patil, 2019)According to Markets and Markets, for example, the business sector IoT sector is expected to increase at an average yearly growth rate of 7.4% between 2020 and 2025, exceeding 110.6 billion USD. Additionally, it is anticipated that by the year 2030, there will likely be up to 50 billion IoT devices as well as commodities in utilize globally, forming an extensive network of connected gadgets that includes everything from cellphones to home appliances. This development, in conjunction with the continually decreasing costs of networks and gadgets, have contributed to the expansion of IoT applications in the field of healthcare. The growing use of technologically advanced healthcare solutions in hospitals and medical center's is a major factor in India's explosive growth. In order to enhancement overall quality of healthcare, accelerate medical treatments, and increase real-time tracking of patients, these organizations are setting up IoT-enabled devices. (Baker, Xiang, & Atkinson, 2017) Conventional healthcare procedures have been transformed into more efficient and adaptable systems by means of the implementation of connected technologies. Remote surveillance for medical purposes has been made possible by internet of things gadgets, which led to less time in the hospital and fewer returning patients. The growing number of chronic illnesses, which need for ongoing surveillance and medical treatment, represents one of the main factors propelling the expansion of healthcare IoT. Frequent tracking of symptoms along with additional health data is made possible by IoT-connected medical equipment and remote surveillance systems as a whole which can assist to improve outcomes for patients. Governments and privately owned companies are investing in research because the increasing number of elderly people in affluent nations is placing more strain on their medical systems. Surprisingly IoT-derived solutions are transforming healthcare from a conventional hub-centered model to more personalized healthcare systems because of the rapid adoption of cellphones and wearable devices. Individualized medical treatments are now being offered through e-health to satisfy people's requirements for medical care. In the big data age, IoT represents a significant advancement that supports a number of timely technological software programs to optimize services. These days, medical systems use IoT data analytics as a consumer source of data to gather more information, identify illnesses early, and establish serious circumstances for improved quality of life. (Chopade et al., 2023)

The growing emphasis on treating patients first is another factor propelling the expansion of the medical IoT. Internet of Things devices can help patients take an active role in their own personal wellness management by giving them access to real-time physiological data and allowing them to monitor their development achieving health objectives. (Yuehong et al., 2016; Li et al., 2024). In healthcare systems, health information which include blood pressure, temperature, heart rate, ECG, and EEG can be collected from the individual's body using monitors that are potentially covered or implanted in the human body. Medical professionals can finally monitor vitals of patients, give real-time measures, and enhance managing chronic illnesses remotely because to a rise of RPM enabled by IoT technology. (Chopade et al., 2023) Integrating IoT-powered surveillance devices has become essential due to the growing number of chronic illnesses, increased patient loads, and a worldwide scarcity of healthcare professionals. It has been proven that the use of RPM has improved the treatment of chronic illnesses by 25% and decreased readmissions to hospital by 50% (Bhuiyan et al., 2021; Dhanvijay & Patil, 2019). It is also possible for recording environmental data like humidity, temperature, as well as the date. The information gathered aid in developing accurate and significant conclusions about individual patients' medical situations. Considering a lot of data is gathered and documented from a variety of systems sensors, cellphones, email, computer software, and apps, its storage and retrieval are also fundamental in IoT systems. (Dhanvijay & Patil, 2019)

It is projected that IoT-based medical facilities will enhance customer experience, save expenses and increase their quality of life. IoT might assist healthcare professionals schedule regular and ongoing activities more effectively by reducing device downtime. IoT improves the treatment of patients by making the most of extremely constrained healthcare resources. It is projected that IoT-based medical facilities will enhance customer experience, save expenses and increase their quality of life. IoT might assist healthcare professionals schedule regular and ongoing activities more effectively by reducing device downtime. IoT improves the treatment of patients by making the most of extremely constrained resources in healthcare.

IOT IN HEALTH CARE: THE GLOBAL INSIGHTS

The majority of the technology and electronic devices in the healthcare sector are connected via IoT, which also enables internet-based information preservation and transfer. The industry is anticipated to increase as a result of the rising demand for automated remote patient treatment and linked medical devices. Many developed nations' medical systems have been moving away from physical clinical documentation and embracing digital health records for the past couple of years (Dhanvijay & Patil, 2019). Government is urging businesses to invest in IoT by the speed at which technology is developing in order to make healthcare services accessible from a distance. Therefore, it is anticipated that government programs promoting technological advancement and adoption would propel the significant expansion of IoT in the healthcare sector. Having revenues anticipated to be slightly over 2.42 billion by 2025, the healthcare and Internet of things sector in India is expected to grow significantly. The increasing adoption as well as integration of IoT technology in the Indian healthcare industry is reflected throughout this projection. Throughout 2025 to 2029, this sector is anticipated to develop at a CAGR of 12.54%, maintaining a robust and consistent development trend. The worldwide market is predicted to surpass an approximate value of about 3.89 billion by 2029 if this tendency persists, demonstrating a steady expanding path. On the other hand, with an expected $33.04 bn in revenue generating in 2025 by itself, the US is well-positioned to lead the worldwide digital health market. As a result of marketplace size, technical advancement, and dedication to intelligent healthcare infrastructure, these factors put the United States in the very forefront globally. (Baker, Xiang, & Atkinson, 2017)

Throughout the COVID-19 pandemic, whenever online healthcare and home care were required to reduce medical and clinic visits, the implementation of IoT technology has proven particularly advantageous. Intelligent cardiac pacemakers that monitor rhythms in the heart while offering real-time information to medical professionals are another prominent application case. It enables prompt medication modifications as well as early identification of anomalous cardiac activit (Yuehong et al., 2016; Li et al., 2024) y. Implantable delivering medications devices, which provide medications like chemotherapy or analgesics at varying rates, are another creative use. This type of technology allows for accurate dosage depending on the patient's the body's functions real-time input. Finally, digital implants are advancing significantly by improving convenience and practicality by autonomously modifying their motions in response to human behaviour. Wearable technology health and fitness monitors with Internet of Things capabilities supply continuous surveillance and customised treatment. Comprehensive health data to assist with early chronic illness detection and exercise improvement is provided by gadgets like fingerprint scanners, Wireless connection, and guided by artificial intelligence analytics. The global marketplace for wearable fitness devices is expected to grow at a CAGR of 18.6% from 2024 to 2030, reaching USD 32.6 billion (Yuehong et al., 2016; Li et al., 2024). For instance, health indicators are tracked by smartwatches and fitness bracelets, which give medical experts valuable information about illnesses like arrhythmias and sleep patterns. These devices are cutting-edge preventative healthcare technology that can identify diseases like diabetes and heart disease early. Additionally, electronic gadgets and artificial intelligence are getting integrated progressively to offer individualised health guidance and exercise instruction.

In 2023, the highest market penetration for the Internet of Things related to healthcare was found in Asia Pacific (Chopade et al., 2023). In years to come, it is projected that this specific area will keep on growing at an increasing CAGR. The government concentrate on digital health solutions that cater to the diverse demographics of nations like China, India, South Korea, and other countries. The Chinese government has begun using telemedicine and online medical services to offer assistance in rural and outlying regions. During the course of the estimated time frame, India is anticipated to have an increased rate of growth. India's rapidly increasing population and government-funded massive expenditures in technological advancements are creating enormous healthcare future possibilities. To assist the healthcare sector, the leading companies like Amazon, Inc, Cisco Systems, Inc, General Electric Company, IBM Corporation, Intel Corporation, Microsoft Corporation, SAP SE, and Wipro in the market is providing a range of IoT platforms and services. In order to offer a connected IoT solution that provides appropriate handling of devices, patient care, enhanced workflow in hospitals, ongoing patient monitoring, and additionally, these firms are spending money on development and research efforts. To enhance linked technologies and strategies, the major companies are proactively forming partnerships and working together (Yuehong et al., 2016; Li et al., 2024). Numerous intangible advantages, such as improved experience for patients, employee retention, and creating a more pleasant and healthy atmosphere for everyone, are made possible by IoT systems.

Serious disorders may be remotely managed and monitored in live time using smart medical gadgets. Pacemakers, glucose delivery systems, and neurostimulators are examples of equipment that provide automated functioning and information collecting. Furthermore, the international market for smart healthcare implants is expected to grow at a CAGR of 15.99% to achieve USD 21.9 billion by 2033 (Dang et al., 2023). Healthcare facilities dominate the marketplace with approximately 52.2% of the sophisticated implants the marketplace share in 2024, especially due to their crucial role in medical treatment. Additionally, businesses provide a comprehensive collection of medical databases that patients, doctors, and healthcare providers may utilize to gain insightful information. In the year 2023, A healthcare technology framework for companies with an emphasizes on healthcare as well as medical sciences was developed by Samsung's subsidiary HARMAN. This startup believes that this will enable companies to provide better consumer experiences. Amazon Health Lake, a HIPAA-eligible answer for medical care and biotechnology entities, was introduced by Amazon Inc in the year 2020 (Greco et al., 2020), their subsidiary Health Lake facilitates enterprises to convert, safeguard, while evaluating all of their personal information in the cloud. Improved IoT-related including hospitals and other sectors were introduced by Intel Corporation with the 11th Gen Core. With its cutting-edge healthcare imaging equipment, it provides high-quality displays and automated diagnostics. Microsoft Inc. and ZEISS Group set up an alliance in 2020, to improve treatment of patients by using driven by information healthcare services .(Amaraweera & Halgamuge, 2019). Through the use of a new digitally linked remedy, the firms want for enhancing clinical efficiency, administration, and the management of devices. There are numerous challenges related to embedding digital medical instruments into medical facilities. Compatibility is one of the most significant challenges. Many clinics and hospitals continue to employ conventional IT infrastructures that aren't built to handle interpreting information from IoT devices in real time. Other great problem is cybersecurity. Each linked gadget represents an emerging threat as medical care hacks increase. Ransomware attacks on medical connected devices are projected to cost the sector more than $20 billion a year, according to a 2024 estimate (Ali et al., 2025). Health care professionals must make assured that smart gadgets are constructed with multiple layers authentication, safe accessibility restrictions, and secured data transfer in order to combat insecurity. It's not unexpected that the healthcare industry is going to reinforce its cyber protections by investing a total of $125 billion between 2020 and 2025. (Baker, Xiang, & Atkinson, 2017)

BIBLIOMETRIC ANALYSIS

A systematic quantitative technique to identify trends in academic journals is a bibliometric evaluation. It is essential for assessing the effectiveness of empirical papers, identifying research themes, and interpreting the conceptual framework of an area of study. Researchers may do thorough bibliometric analysis using the R programming environment and robust tools like Bibliometrix. A thorough examination of bibliographical information has been made possible by this software, which imports data from prominent databases such as Scopus. The ability of R's bibliometric examination to produce Visualization. Outputs which transform unstructured publishing data to comprehensible visual forms, is just one of its major benefits. These consist of geographic collaboration illustrations, co-authorship connections, citations visualizations, keyword occurrence graphs, and theme progression mapping. Data from 4950 publications in the database of Scopus is used to perform a rigorous bibliometric review. For the bibliometric analysis in this work, the R Biblioshiny package and the VOS viewer were implemented. The findings demonstrated how the Internet of Things might assist regulators in enhancing the economy and overall health of society.

Annual Scientific Production

During the course of a three-year span from 2023 to 2025, this chart shows the publishing pattern for studies concerning IoT-based applications in smart medical care. Given an estimated number of about 2,900 documents in 2024, it shows a notable increase in academic curiosity. Given the aftermath of the pandemic technological shift and increased expenditures in health-tech facilities such steep increase most likely reflects the growing need for smarter medical solutions. Approximately 850 papers were published in 2023, signifying a modest start to scholarly investigation on the incorporation of Internet of Things technology into healthcare systems. The sharp rise in 2024 is a result of increased international research activities aimed at using IoT for predictive analytics, virtual diagnostics, real-time monitoring, and sustainable hospital ecological systems.

Figure 1. Annual Scientific Production

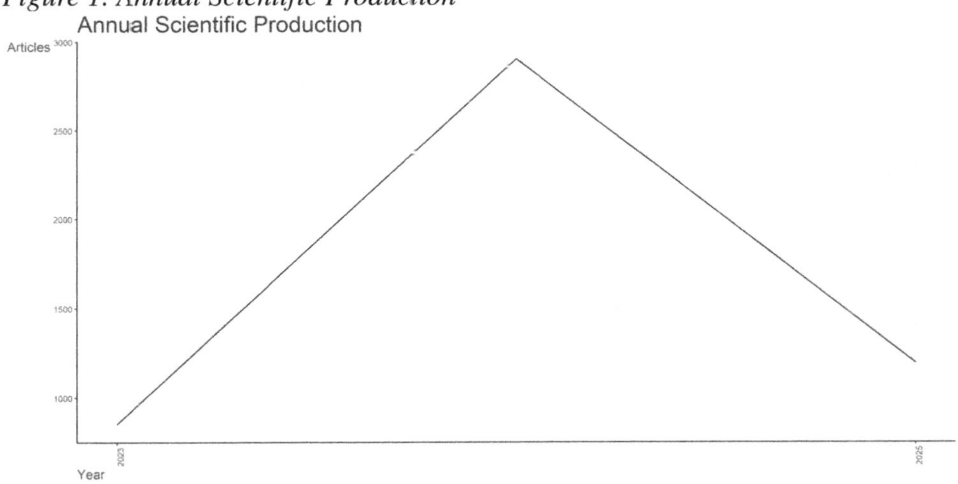

Interestingly, considered as an entire phenomenon, the graph shows that IoT in smart healthcare continues to remain a vibrant and significant field of study that demands constant technological breakthroughs and multidisciplinary cooperation to tackle challenging issues related to healthcare.

Global Scientific Production

The globally distributed map displays the geographical breakdown of research efforts and shows the worldwide scientific output relating to IoT-based applications for smart healthcare. The total number of publications is represented by the degree of the blue pigment; greater volume is indicated by darker colours. The deepest shade of blue indicates a significant national commitment on leveraging IoT technology across healthcare systems, indicating that India has emerged as the world leader through this field. China, United States, and parts of Europe, such as Germany and the United Kingdom, are also making key contributions following India. They also offer robust legislative assistance for digital health advances and well-developed technology facilities, which stimulate a high volume of research output.

Figure 2. Publication as per Country

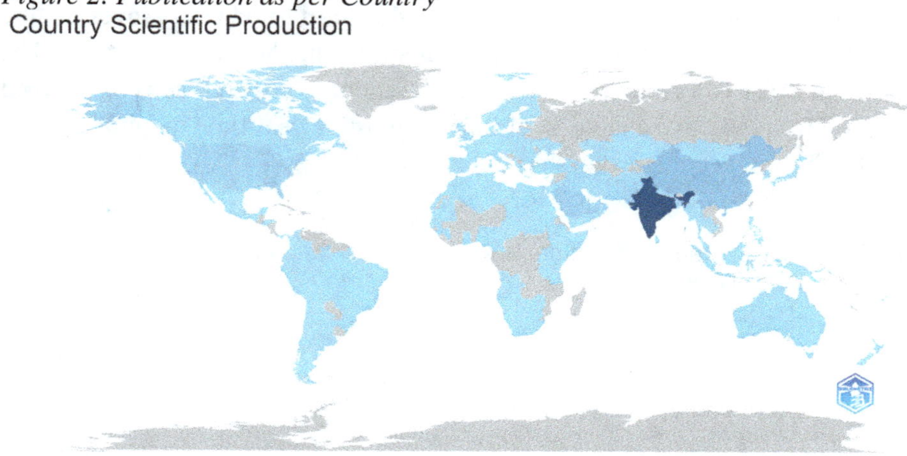

Countries like Brazil, Australia, South Africa, and the Southeast Asia region indicates the research has grown more proactive, probably as a result of the demand for scalable, accessible, as well as, affordable electronic therapies as well as combating the healthcare challenges. Grey-shaded areas, like an extensive portion of Central Africa and the Middle East, indicate barely any noticeable research output, perhaps because of a lack of financial resources or willingness to accept cutting-edge technologies. In order to promote inclusive and equitable growth of smart healthcare, global collaboration and financing are crucial, because this map highlights the worldwide dynamism and discrepancies in IoT-driven advancements in healthcare.

Bradford's Law

Bradford's Law is employed to figure out the premier journals and conference papers that make a substantial contribution to the research in the field of IoT-based applications in smart health care. As stated by Bradford's Law, a majority of papers on a given subject are published in a comparatively limited number of publications. The outlined section designated "Core Sources," which contains highly effective publications like Lecture Notes in Networks, IEEE Internet of Things Journal, and Sensors, exemplifies this idea. These key journals provide focused, highly influential empirical findings on smart healthcare technologies, and they operate as the cornerstones of scholarly discussion.

Figure 3. Publication Sources by Bradfords Law

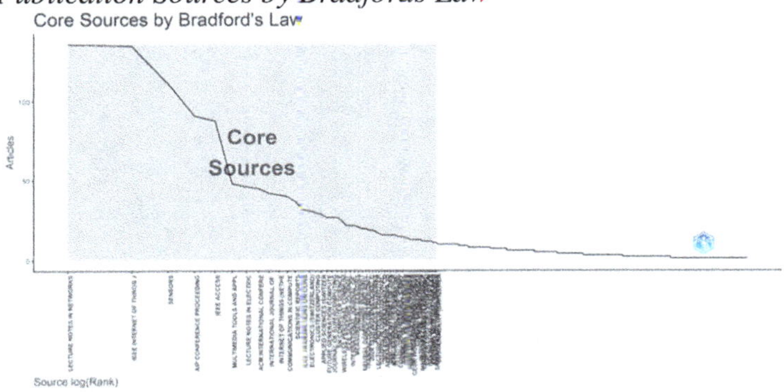

These cover multidisciplinary fields, including data science, artificial intelligence, healthcare, with Internet of Things, highlighting their subject's diversity. Although numerous journals touch on the subject, only a small number of them are significant production centres, as seen by the noticeable decrease in publication contributions seen in journals outside of the shaded zone. For researchers intending to gain recognition or remain up to date in the subject, this visualisation is essential. In addition to increasing exposure and scholarly influence, concentrating on these key domains gives access to the most recent developments in IoT-driven smart healthcare solutions

Clusters by Document Coupling:

The data visualisation shows the intellectual framework of investigations on IoT-based applications for smart health care by representing clustering by documentation coupling. The graphic illustrates important conceptual clusters and their effect within the subject area. The internet of medical things (IOMT) and IoT clusters are located in the upper-right quadrant, which denotes significant effects and prominence. Since these problems constitute the intellectual core of the field and stimulate both theoretical and pragmatic breakthroughs, their placement underlines the vital role that they play in shaping the discipline.

Figure 4. Cluster Visualization

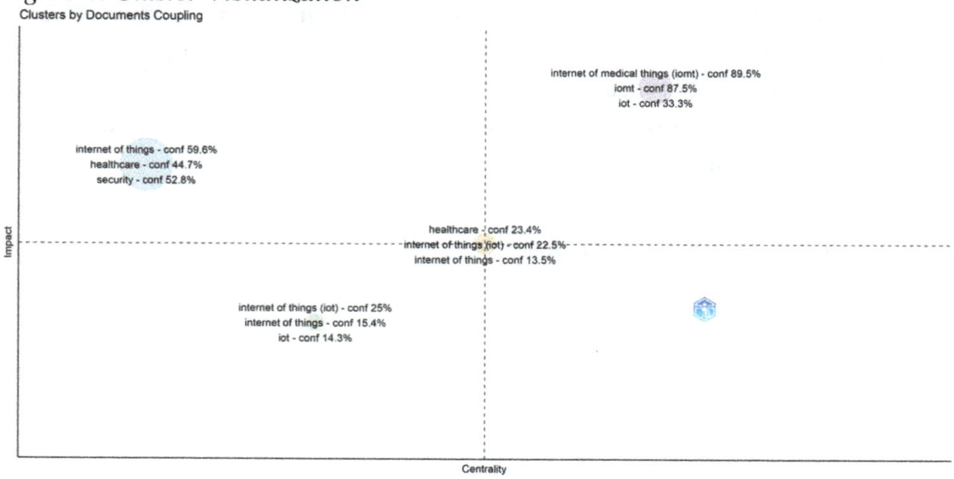

Scattered concepts like “IoT” and broad allusions to “Internet of Things” are reflected in the lower-left the first quadrant, indicating that they may be fundamental but undeveloped in the field of healthcare. The following chart, which illustrates how linked medical technology and intelligent infrastructure are continuously changing the way in which healthcare is offered, highlights how important and influential IoMT-specific investigation.

Lotka's Law

The inverse correlation between the number of authors with the total number of documents they generate has been recognised as Lotka's Law. This breakdown illustrates the changing pattern of academic contributions in this multifaceted discipline in relation to connected to the internet of applications for smart medical care. The Y-axis shows the proportion of authors, while the X-axis shows the total number of papers written. There is a sharp slope at the very beginning, with nearly eighty percent of writers having written an article. This supports Lotka's theory that while most writers generate very little, and a handful of scholars provide significantly. A small percentage of writers are accountable for a proportionally significant number of publications, as seen by the graph's long tail as well as diminishing tendency.

Figure 5. Corelation of Authors as per Lotkas Law:

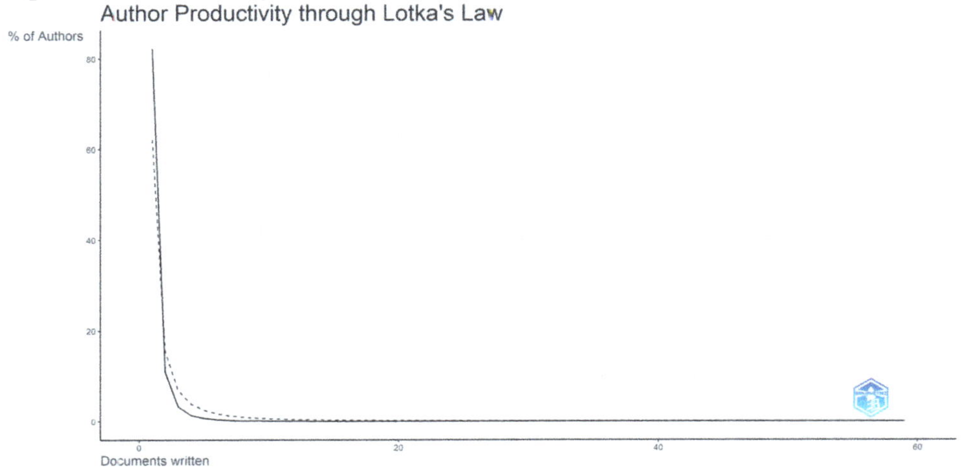

This key group of researchers may be responsible for a great deal of breakthrough discoveries and innovative thinking in smart healthcare IoT, influencing the discipline's theoretical and applied directions. Finding significant contributions and possible partners requires a comprehending of this tendency. Additionally, it highlights the necessity of increasing involvement and supporting up-and-coming researchers, particularly in a field that is developing quickly like intelligent healthcare, where multidisciplinary contributions from the Internet of Things, biomedical engineering, data sciences, and healthcare are crucial for adaptable, pragmatic outcome.

Three-Field Plot Visualization

Three-Field Plot integrates three essential sections: Citated References Authors, as well as Keywords to demonstrate the scientific framework of studies based on IoT-based technologies for smart health care. Each of these fields is represented by a vertical block, and its depth is determined by the total number of connections or occurrence. The lines of text that connect these segments indicate how often writers use specific sources and correlate their articles to specific keywords.

Figure 6. Three-Field Plot

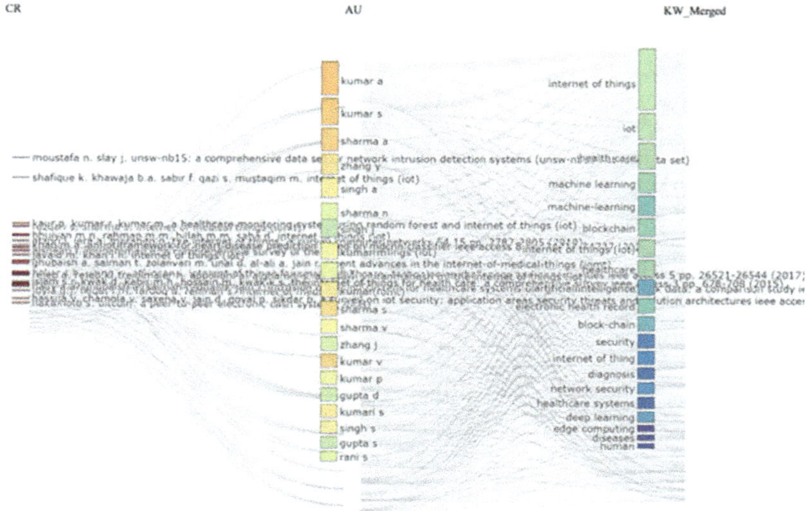

Fundamental research is provided by commonly mentioned authors like Moustafa et al. and Shafique et al. on the left, especially in fields like detection of intrusion and IoT safety. Prominent writers that have made significant contributions in this field, like Zhang J., Sharma R., and Kumar A., are highlighted in the central the block. As an indication of the cross-disciplinary combination of information, technology, and physical well-being, buzzwords such as Internet of Things, IoT, Machine Learning, Information Security, Blockchain, and Healthcare Systems dominated on the left side. The graphical representation is essential for charting important works of literary work, locating important scholars, and seeing emerging trends. In addition to providing insight into the changing research environment, it highlights the confluence of IoT, AI, and robust healthcare facilities, facilitating the creation of extensible, data-driven, and patient-oriented intelligent health care solutions.

Tree Map

The treemap conveys a useful summary of major academic issues by depicting the term distribution of frequencies in the field of IoT-based solutions for smart healthcare. The dimensions and colour for each circle, which stands for a term, indicate how frequently that keyword appears in works of literature. Internet of Things (2462 references, 14%) is the most popular keyword, highlighting the basic function that IoT is playing in reshaping healthcare. Related words like Internet of Things (IoT) (5%), IoT, and Smart Healthcare highlight its multifarious significance. Terms associated with the healthcare industry, such as Healthcare, Health Care, Electronic Health Record, Patient Treatment, and Remote Patient Monitoring, suggest a significant emphasis on clinical procedure optimisation and patient treatment digitalisation. The increasing integration of intelligent and secure systems for data-driven diagnosis and making choices is reflected in technical themes such as Cloud Computing, Deep Learning, Artificial Intelligence, Machine Learning, Blockchain, and Edge Computing.

Figure 7. Tree Map

Furthermore, the use of phrases like Data Privacy, Cybersecurity, and Authentication draws attention to important issues with the protection of personal healthcare data. In order to create individualised, preventative, and pre-emptive healthcare systems, this treemap makes it abundantly evident that academic interest in IoT-based smart health care is not only evolving rapidly additionally growing more multidisciplinary. The concept of "internet of things" makes up the largest part of the visualization. This shows its important role in today's tech developments and academic discussions. The growing use of IoT in healthcare services, especially in digital evolution and patient care, is seen in terms like IoT, health care, healthcare, and electronic medical records. The rising use of terms like deep neural networks, blockchain, computational intelligence, and machine learning highlights other innovations that work with IoT to improve automated processes, data handling, and decision-making in healthcare systems.

Specific healthcare-related terms like diagnosis, remote patient tracking, mhealth, and patient care show how IoT supports real-time patient monitoring and remote diagnostics. Additionally, the use of phrases like confidentiality of information, the field of cybersecurity, and authentication suggests that people are becoming increasingly concerned about the ethical and security ramifications of using IoT in healthcare settings. Considering the fragile nature of health care information as well as the potential hazards of information theft or exploitation, these issues are crucial. All things considered; the visualisation shows that IoT is an innovative force in health care automation as well as a technical advancement.

Thematic Relevance of Keywords

The word cloud demonstrates how often and thematically relevant terms are used in the field of Internet of Things-connected intelligent health care technologies. The phrase "Internet of Things" dominates the visualisation, emphasising its pivotal significance in modern healthcare innovation.

Figure 8. Thematic Relevance of Keywords:

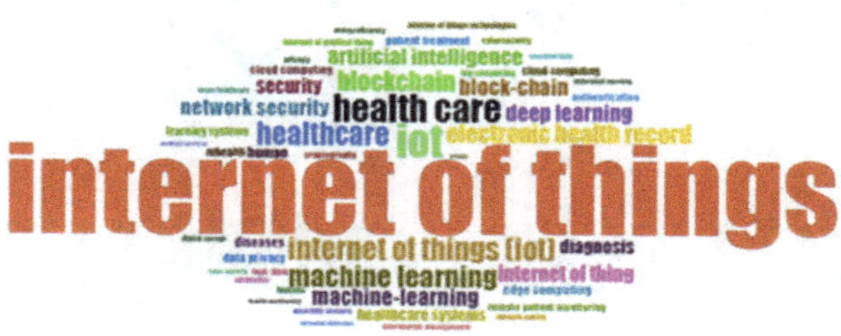

The terms Healthcare, IoT, Health Care, Electronic Health Record, Diagnosis, and Remote Patient Monitoring emerge together in rapid succession, a number of which indicate the increasing use of linked technology in healthcare workflows and the administration of patients. Phrases like Machine Learning, Artificial Intelligence, Deep Learning, and Edge Computing illustrate how IoT and sophisticated analysis of information are coming together to provide personalised therapeutic strategies, immediate choice-making, and predictive diagnoses. Terminologies like "Cloud Computing," "Wearable Sensors," as well as "Smart Healthcare" refer to continuous advancements in the provision of healthcare services and all-encompassing surveillance. An interconnected study trend that combines computer science, biomedical engineering, cybersecurity, and medical care is shown by the word cloud's popularity and broad range of terms. This visualisation demonstrates that IoT-enabled smart healthcare is a dynamic, rapidly emerging sector with a broad range of programs, deep scientific understanding, and significant socioeconomic significance.

Conceptual Structure Map

A landscape of published work regarding IoT-based applications in smart health care systems is provided by the Conceptual Structure Map above, which was produced by applying Multiple Correspondence Analysis. Every phrase stands for a keyword which was derived from research papers and arranged with respect to conceptual closeness in the two dimensions, which taken together account for a sizable amount of variance.

Figure 9. Conceptual Map

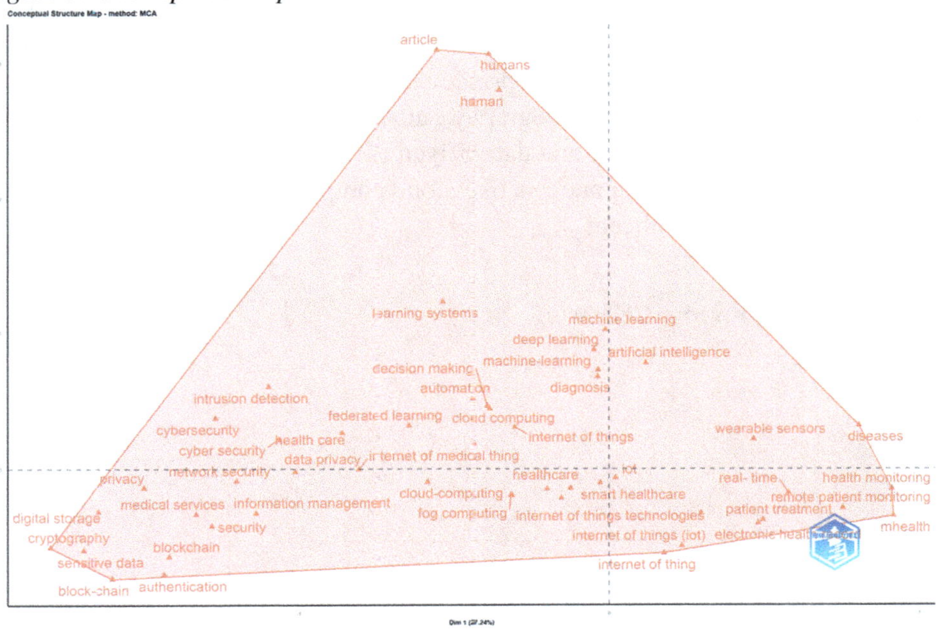

Incorporating intelligence into health care systems becomes essential, as demonstrated by the core cluster, which shows integration around key innovations and practices such as cloud computing, decision-making, centralised learning, and the Internet of Things (IoT). With a strong emphasis on patient-focused and real-time applications, the right-hand cluster illustrates crucial topics such as wearable sensors, remote patient surveillance, health tracking, and mHealth. The concurrent issue with protecting confidential medical information in interconnected structures is highlighted by a separate theme part on the left that emphasises on cybersecurity, blockchain, privacy, and data protection. In contrast, broad themes like "humans," "articles," and "learning systems" are reflected in the main region, suggesting their wider intellectual inclusiveness. This visualisation highlights an extensive, multi-disciplinary research area where medical value, data security, and technological breakthroughs converge to create the basis of IoT-powered innovative health care.

Dendrogram Cluster Analysis

A hierarchical cluster study concerning significant perspectives in the field of Internet of Things-based applications for intelligent healthcare systems is displayed in the dendrogram image. Using co-occurrence derived from bibliometric information this type of clustering technique illustrates the degree of relationship between different phrases. While the y-axis shows the degree of difference across clusters, each of the nodes on the x-axis symbolises a unique phrase. Three main clusters may be seen in the structure. The primary cluster, which is located in the lower left corner, focusses on security and privacy infrastructures involving fundamental technologies like blockchain, cryptography, authentication, and information management. With the computational and data-driven core of IoT healthcare systems, the central cluster incorporates concepts like cloud computing, data privacy, federated learning, and smart healthcare.

Figure 10. Dendrogram Clustering

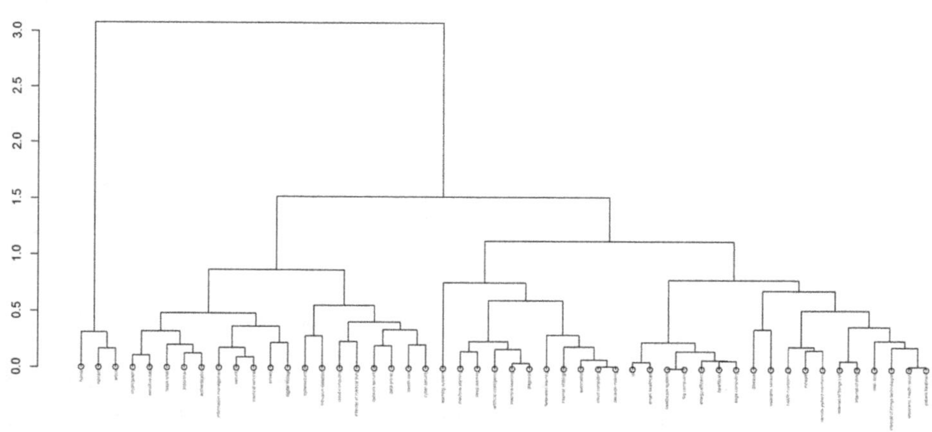

Average Citations Per Year

A line chart showing the gradual increase in scientific influence for publications about IoT-based applications in smart healthcare systems from 2023 to 2025 is presented in the data visualisation Average Citations per Year. The horizontal axis shows the year of publication as well as whereas the vertical axis shows an average amount of citations per work. The graph clearly shows an overall decline in citations statistics.

Figure 11. Average Citations YoY

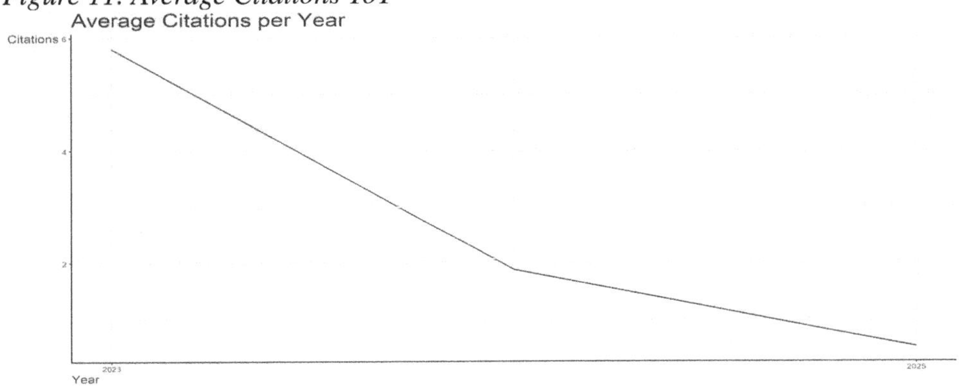

With an average of over six citations per publication in 2023, academic attention peaked, maybe as a result of greater interest in research and following the pandemic energy propelling innovation regarding digital healthcare. The average number of citations fell to about two by 2024, and then to less than one in 2025. There are a number of reasons for this reduction. First, especially in 2025, the recency bias always results in lower citations for recently released papers. Second, early groundbreaking articles get more citations as the area develops, but later studies could be more progressive. Finally, citation concentration across multiple articles may be diluted due to the fast growth of investigations in the IoT-healthcare topic. The results highlight that in order to maintain high effect, future research must concentrate on innovative, multidisciplinary, and clinically transferable technologies.

Coordination and deliberate distribution may additionally boost long-term academic exposure and references success. A thorough co-occurrence chain of terms associated with Internet of Things-based programs for intelligent healthcare systems is displayed in the visualisation above, which was created with VOS viewer. The structure of the network is divided into many thematic zones, every one of which is signified by a distinct colour. The size of the nodes in the network reflects how frequently a term appears in the literature. Strong phrases like "internet of things," "IoT," "artificial intelligence," "machine learning," and "healthcare" appear at the exact centre of the diagram, demonstrating their importance and significance in the field. The green clusters mostly relate to smart healthcare applications, such as digital medical records, healthcare for patients, remote patient monitoring, and wellness monitoring.

Keyword's Co- Occurrence

The Internet of Things is the most significant and most integrated node at the focal point of the networks, making it the obvious choice for a major study topic. Key words like artificial intelligence, machine learning, healthcare, remote patient monitoring, and security are closely related to it. This demonstrates how important IoT is to the development of intelligent, networked health systems. To illustrate how Internet of Things gadgets like wearables and automated sensors are transforming diagnostics, telemedicine, and real-time patient care, focusses on patient monitoring, biomedical engineering, and the delivery of healthcare. The significance of IoT in treating chronic diseases like diabetes and cardiovascular problems is demonstrated by the red cluster, which links IoT with disease forecasting, heart rate, blood sugar tracking, and investigations.

Figure 12. Keyword Cooccurrence

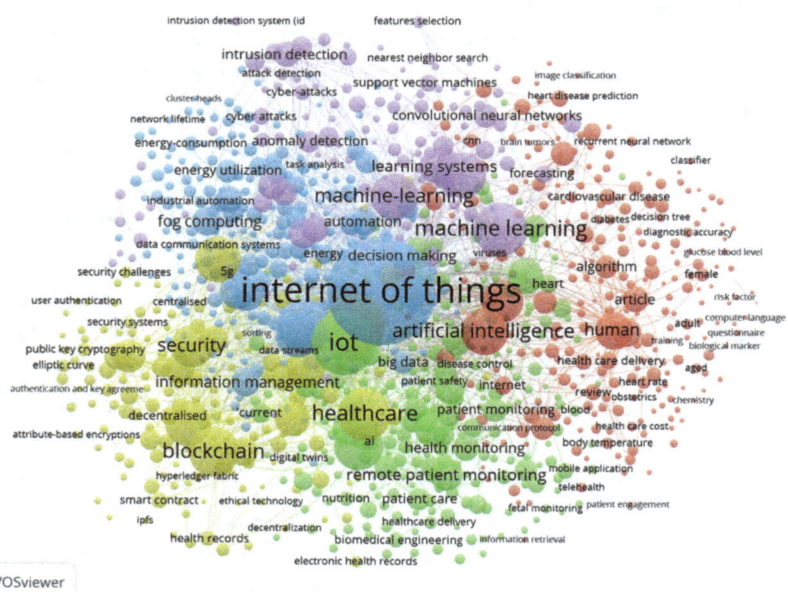

Contrary to that, the yellow cluster handles significant challenges in IoT-based health care environments by concentrating on blockchain technology, privacy, and data security. The overall visualisation demonstrates how IoT is not merely a breakthrough in technology but also a catalyst for change in the field of health care, allowing for forward-thinking, information-driven, and customised medical treatment but also demanding innovation in security of data and ethical artificial intelligence application.

Thematic Map

In order to demonstrate the level of sophistication and strategic significance of subjects in IoT-based digital health education, the thematic diagram arranges research groups along the two axes of relevance -"centrality" and advancement density". The discipline's stronghold cluster is made up of the Internet of Things, IoT, and Internet of Things (IoT). They have the greatest concentration and density scores, indicating that technology ecosystems and connection infrastructures are the main sources for innovation in healthcare technology. Thus, the any high-impact contribution needs to concentrate on IoT design advancements, seamless integration, and widespread implementation concerns.

Figure 13. Thematic Visualization

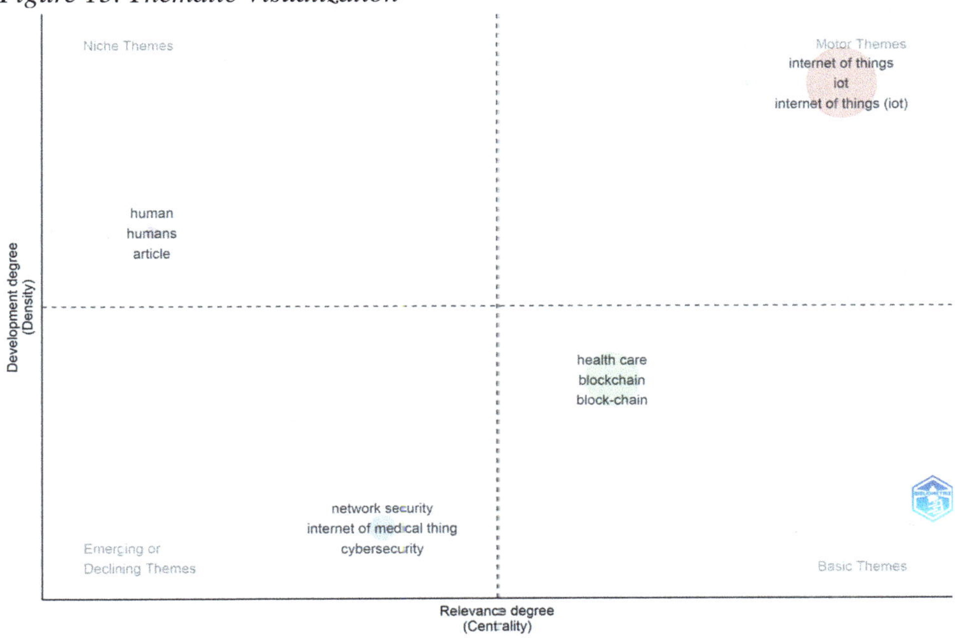

Fundamental themes cover Blockchain and health care show small density but strong centrality. They are laying the groundwork but are still coming together. This cluster can be quickly moved higher from "basic" to "motor" by research that combines IoT with blockchain-enabled provenance, consent management, or tamper-proof electronic health records. The emerging themes includes, Internet of Medical Things (IoMT), the safety of networks, and online safety all receive poor ratings on both axes, indicating either a lack of research or a disconnect between the need and academic interest. A straightforward route for significant, foreseeable research is provided by fortifying cyber-physical safety models, zero confidence frameworks, and IoMT-specific regulations. Niche Themes, although are methodologically mature, highly dense but only partially linked nodes indicate specialised, self-contained investigations, typically clinical or demographic that have had little impact on the larger conversation. The complete social effect might be improved by incorporating user-centric results arising from these niches into system-level IoT solutions. All things considered, the landscape shows that improving security and blockchain research while keeping IoT at the forefront is crucial to developing robust, patient-centred intelligent healthcare networks.

CONCLUSION

The Internet of Things is shaping the future of healthcare and not simply a trend. IoT offers substantial advantages whether or not you're monitoring medical equipment, treating chronic diseases, or protecting newborns. It enhances human lives, simplifies processes, and keeps you ahead in a field that is always evolving. Considering significant ramifications, the Internet of Things will transform healthcare the way we perceive today. We have moved to an entirely novel frontier with respect to regarding how people, technologies, and applications interact to provide healthcare solutions. We now have emerging instruments and an entirely novel perspective for a cohesive healthcare ecosystem since the advent of the Internet of Things, which has significantly improved the overall quality of healthcare. Healthcare processes that initially took a long time and had human error probability are now feasible thanks to the Internet of Things. The new age of smart healthcare is finally here largely a result of the Internet of Things' (IoT's) implementation with healthcare facilities, which has sparked a revolution across healthcare institutions throughout the world. In order to outline the cognitive, conceptual, and geographical landscape of IoT-based solutions for healthcare, the present study carried out a thorough bibliometric analysis of 4,950 academic publications from the Scopus database. According to the report, curiosity regarding research increased exponentially, especially between 2023 and 2024. This surge was a result of both the post-COVID-19 technological acceleration and the growing problems in healthcare brought on by people getting older, chronic diseases, and the demand for remote patient care. According to the data, India leads the world in scientific production, followed by China, the US, and major European nations. This indicates that technologically sophisticated and densely populated countries place significant emphasis on investigation. Bradford's Law highlighted a core group of publications that shape the conversation and advancement in this field, including IEEE Internet of Things Journal and Sensors. High-impact studies focusses on the Internet of Medical Things (IoMT), smart monitoring, and predictive analytics, while also highlighting cross-cutting issues like cybersecurity, data privacy, and interoperability, according to the theme and coupling analyses. Only a handful of productive authors contribute significantly to the area, indicating outstanding intellectual dominance that stimulates innovation, according to the study's implementation of Lotka's Law.

But this additionally suggests that in order to increase the variety and influence of the discipline, there has to be more cooperation and the involvement of up-and-coming academics. Important technologies including blockchain technology, cutting-edge computing, learning federation, and artificial intelligence have come together with healthcare applications like telemedicine, which e-health records, and remote monitoring of patients, according to analysis of keywords using term clouds and tree maps. The conceptual and thematic maps also showed that although IoT continues to be the theoretical and practical foundation of smart healthcare, important but little-studied topics like cyber-physical protection, lack of confidence systems, and patient-centric digital ethics offer promising avenues for further research. In conclusion, the study indicates that IoT-based smart healthcare is a rapidly developing interdisciplinary field that additionally serves as a vital force behind patient-centred, accessible, and affordable systems for providing healthcare. Its greatest potential, however, depends on resolving issues with data security, legal compliance, technology standardisation, and accessibility equity globally. This bibliometric study provides a road map for academics, decision-makers, and technologists to traverse the current academic landscape, pinpoint high-impact areas, and participate in the collaborative development of smart and powerful, and morally sound IoT-powered healthcare infrastructures. Intelligent technology and design that prioritises humans are going to intersect to determine healthcare in the decades to come, because the Internet of Things is at the very heart of this shift.

REFERENCES

Ali, T. E., Ali, F. I., Dakić, P., & Zoltan, A. D. (2025). Trends, prospects, challenges, and security in the healthcare internet of things. *Computing*, *107*(1), 28. DOI: 10.1007/s00607-024-01352-4

Amaraweera, S. P., & Halgamuge, M. N. (2019). Internet of things in the healthcare sector: overview of security and privacy issues. Security, privacy and trust in the IoT environment, 153-179.

Baker, S. B., Xiang, W., & Atkinson, I. (2017). Internet of Things for Smart Healthcare: Technologies, Challenges, and Opportunities. *IEEE Access : Practical Innovations, Open Solutions*, *5*, 26521–26544. DOI: 10.1109/ACCESS.2017.2775180

Bhuiyan, M. N., Rahman, M. M., Billah, M. M., & Saha, D. (2021). Internet of things (IoT): A review of its enabling technologies in healthcare applications, standards protocols, security, and market opportunities. *IEEE Internet of Things Journal*, *8*(13), 10474–10498. DOI: 10.1109/JIOT.2021.3062630

Chopade, S. S., Gupta, H. P., & Dutta, T. (2023). Survey on sensors and smart devices for IoT enabled intelligent healthcare system. *Wireless Personal Communications*, *131*(3), 1957–1995. DOI: 10.1007/s11277-023-10528-8 PMID: 37360143

Dang, V. A., Vu Khanh, Q., Nguyen, V. H., Nguyen, T., & Nguyen, D. C. (2023). Intelligent healthcare: Integration of emerging technologies and Internet of Things for humanity. *Sensors (Basel)*, *23*(9), 4200. DOI: 10.3390/s23094200 PMID: 37177402

Dhanvijay, M. M., & Patil, S. C. (2019). Internet of Things: A survey of enabling technologies in healthcare and its applications. *Computer Networks*, *153*, 113–131. DOI: 10.1016/j.comnet.2019.03.006

Greco, L., Percannella, G., Ritrovato, P., Tortorella, F., & Vento, M. (2020). Trends in IoT based solutions for health care: Moving AI to the edge. *Pattern Recognition Letters*, *135*, 346–353. DOI: 10.1016/j.patrec.2020.05.016 PMID: 32406416

Yuehong, Y. I. N., Zeng, Y., Chen, X., & Fan, Y. (2016). The internet of things in healthcare: An overview. *Journal of Industrial Information Integration*, *1*, 3–13.

ADDITIONAL READING

Jayachitra, S., Prasanth, A., Hariprasath, S., Benazir Begam, R., & Madiajagan, M. (2023). AI enabled internet of medical things in smart healthcare. In *AI models for blockchain-based intelligent networks in IoT systems: Concepts, methodologies, tools, and applications* (pp. 141–161). Springer International Publishing. DOI: 10.1007/978-3-031-31952-5_7

Krishnamoorthy, S., Dua, A., & Gupta, S. (2023). Role of emerging technologies in future IoT-driven Healthcare 4.0 technologies: A survey, current challenges and future directions. *Journal of Ambient Intelligence and Humanized Computing*, *14*(1), 361–407. DOI: 10.1007/s12652-021-03302-w

Li, C., Wang, J., Wang, S., & Zhang, Y. (2024). A review of IoT applications in healthcare. *Neurocomputing*, *565*, 127017. DOI: 10.1016/j.neucom.2023.127017

Pradhan, B., Bhattacharyya, S., & Pal, K. (2021). IoT-based applications in healthcare devices. *Journal of Healthcare Engineering*, *2021*(1), 6632599. PMID: 33791084

Taloba, A. I., Elhadad, A., Rayan, A., Abd El-Aziz, R. M., Salem, M., Alzahrani, A. A., Alharithi, F. S., & Park, C. (2023). A blockchain-based hybrid platform for multimedia data processing in IoT-Healthcare. *Alexandria Engineering Journal*, *65*, 263–274. DOI: 10.1016/j.aej.2022.09.031

Tiwari, S., Nahak, K., & Mishra, A. (2023). Revolutionizing healthcare: The power of IoT in health monitoring. *Journal of Data Acquisition and Processing*, *38*(2), 2416.

Tyagi, A. K., Kumari, S., & Tiwari, S. (2024). Internet of things for smart healthcare: A survey. In *Future of AI in Medical Imaging* (pp. 19–41). IGI Global Scientific Publishing. DOI: 10.4018/979-8-3693-2359-5.ch002

Chapter 4
An Explainable AI-Based Transfer Learning Ensemble Model for Brain Tumor Classification

Savir Ali
https://orcid.org/0000-0001-6830-1027
Aligarh Muslim University, India

Mohammad Ubaidullah Bokhari
Aligarh Muslim University, India

ABSTRACT

The early discovery and accurate classification of brain tumors are crucial for improved recovery and prognosis due to their high lethality. The study introduces an ensemble model utilizing transfer learning, incorporating EfficientNetV2B3 and ResNetRS100 to identify four categories of brain tumors: meningioma, glioma, pituitary and No tumor. Grad-CAM is used to localize important MRI regions while feature importance analysis explains each model's contribution to the final decision, enhancing trust and interpretability. The proposed ensemble model had an accuracy of 98.95%, precision of 98.60%, recall of 98.80% and F1-score of 98.70% demonstrating superior performance compared to alternative approaches on a publicly accessible dataset. This innovative approach to prediction generation utilizes transfer learning, explainable artificial intelligence and ensemble modeling. It produces intelligible predictions that are efficacious for healthcare decision-making.

DOI: 10.4018/979-8-3373-5727-0.ch004

1) INTRODUCTION

Brain tumors are severe and challenging medical disorders that necessitate a thorough diagnosis prior to the implementation of appropriate treatment. The predominant method for detecting a brain tumor is through an MRI (Wahlang et al., 2022). It surpasses other modalities for internal visualization, such as CT, PET, and X-ray. Radiologists often engage in extensive manual analysis of MRI data, leading to variability in interpretations among various practitioners. The greater the number of images you examine manually, the higher the probability of misinterpreting them. It is essential to develop automated and accurate methods for classifying brain tumors to enable physicians to determine the specific type of brain tumor a patient possesses (Akter et al., 2024). The utilization of machine learning, particularly deep learning, for automation has become increasingly prevalent due to its capability to differentiate among various types of brain tumors. Numerous methods exist for utilizing deep learning models, particularly Convolutional Neural Networks (CNNs) to classify brain tumors. Convolutional Neural Networks (CNNs) have proven to be highly effective in analyzing medical images. Researchers have using AI models in several approaches to identify brain tumors. Some individuals employed a singular model utilizing transfer learning, others implemented specialized convolutional neural networks, while some adopted a combination of both approaches. There are a great deal of difficulties associated with these methods (Wang et al., 2024). An ensemble of algorithms, or a hybrid algorithm, can combine the predictions or inputs of many algorithms. For the purpose of transfer learning and other independent approaches, the characteristics of a single model are utilized entirely here. When constructing your own CNN models, it is vital to train them for prolonged durations from the beginning and provide a bigger number of examples for learning. This is because CNN models acquire knowledge through experience. Hybrid or ensemble models are advantageous as they do not rely solely on the characteristics or forecasts of a single model (Arumugam et al., 2024). One final consideration is that these models must ensure their outcomes are unequivocal, particularly in the medical sector, where inaccurate diagnoses could have detrimental consequences and prolong treatment. Brain tumors are perilous and hard to treat. A comprehensive diagnosis must be established prior to the commencement of treatment. An MRI is the predominant method for detecting a brain tumor. The research sought to refine algorithms for the detection of diverse malignant brain tumors by integrating deep learning, ensemble techniques (Sakib & Siddiqui, 2023) (Sakib & Mustajab 2025), and explainable artificial intelligence to augment their clarity and applicability (Bruton et al., 2020). The proposed approach employs ensemble and transfer learning to identify glioma, meningioma, and pituitary tumors, which are among the most prevalent forms of brain tumor. We trained and evaluated the model using an accessible dataset. The proposed method employed pre-trained models ResNetRS100 and EfficientNet-V2B3, removing their classifiers layers in favor of a freshly designed classifier layers specifically for this task. We employed gradient-weighted class activation maps (Grad-CAM) (Sánchez-Moreno et al., 2025) to identify the regions of the image that most influenced the projected class. The new strategy outperformed the previous ones, and incorporating explainable techniques into the ensemble model enhanced its comprehensibility and reliability.

The principal conclusions of this study are as follows:

i. An explainable AI mechanism utilizing Grad-CAM was incorporated into the ensemble model to visually identify the regions of MRI images impacted by the tumor for prediction verification.
ii. We developed an ensemble model integrating EfficientNetV2B3 and ResNetRS100, employing transfer learning to identify and categorize four types of brain tumors: no tumor, meningioma, glioma, and pituitary. The proposed model surpassed current techniques and attained the highest accuracy on a widely used public MRI dataset.
iii. The suggested model was tested against well-known architectures as MobileNetV3, GoogLeNet and VGG19. It regularly did better than all the others on important evaluation parameters.

The structure of this research article is as follows: Section 2 outlines the dataset description and preprocessing techniques; Section 3 presents a detailed explanation of the proposed methodology; Section 4 highlights the experimental results and Section 5 concludes the study with conclusion and future scope.

2) RELATED WORK

Amin et al., (Amin et al., 2022) developed a method that has proven to be effective in accurately diagnosing brain tumors. Additionally the MobileNet architecture is utilized by this strategy. An area under the curve (AUC) value of 0.98 refers to glioma, 0.97 refers to meningioma, and 1.00 refers to pituitary tumors from the framework. It indicates that it can distinguish them proficiently. The research outcomes show that may function as an efficient strategy for medical decision making and for the recognition of emerging problems. Saeedi et al., (Saeedi et al,. 2023) propose that identifying brain tumors in MRI data necessitates the utilization of two deep learning models: a 2D CNN and a convolutional auto-encoder, in conjunction with traditional machine learning algorithms. Their techniques allow the identification among gliomas, meningiomas, pituitary tumors, and normal the brain tissue. The area under the curve (AUC) of the model is quite near to 1.00, and we may estimate that it is approximately 96.47% accurate. Due to its ease of use and real-world performance, the 2D CNN is superior than earlier techniques. Agarwal et al., (Agarwal et al., 2024) have devised a bifurcated hybrid approach for the diagnosis and categorization of brain tumors. The ODTWCHE method enhances low-contrast MRI images, while the Inception V3 algorithm is employed for classification purposes. The model surpasses all previously employed methods in accuracy, achieving a rate of 98.89%. This is a reliable, automated CAD method for the early identification of brain tumors appropriate for therapeutic use. Kollem et al., (Kollem et al., 2023) discuss AlexNet-NDTL a modified iteration of the 32-layer AlexNet model that employs IWGAN for deep transfer learning and Lipschitz-based data augmentation. The model demonstrates an accuracy of 99.77%, sensitivity of 99.92% and precision of 97.45% when evaluated on the BRaTS 2020 and Figshare datasets. It outperforms previous models indicating its potential utility in clinical environments for the swift and accurate detection of brain cancers. Abd El-Wahab et al., (Abd El-Wahab et al., 2023) developed BTC-fCNN a compact convolutional neural network with 13 layers and 583,613 parameters capable of distinguishing between glioma, meningioma, and pituitary brain tumors in MRI data. The model achieves an accuracy of 98.86% with 1×1 convolution and internal transfer learning and its execution time is minimal. It outperforms the majority of deep models. It is highly precise, comprehensive, and beneficial for identifying issues in real time. Ramdas Vankdothu and colleagues assert that MRI scans could be utilized to develop a computer algorithm for categorizing brain cancers. It incorporates a recurrent convolutional neural network (RCNN), adaptive filtering, enhanced K-means clustering and GLCM-based feature extraction. The model surpasses BP and U-Net, with a 95.17% accuracy rate on a Kaggle dataset. It may accurately identify gliomas, meningiomas, pituitary tumors, and non-malignant conditions for clinical diagnosis. Batool et al., (Batool and Byun 2024) investigate both conventional and novel approaches for the detection of brain tumors utilizing CT, PET, and MRI imaging techniques. The discussion encompasses the advantages and disadvantages of deep learning, feature fusion and hybrid methodologies, alongside challenges that frequently arise such as dataset imbalance, generalizability, and real-time application. The research demonstrates methods to enhance the accuracy, comprehensibility and use of brain tumor diagnostic tests in the future. Sharma et al., (Sharma et al., 2023) propose an enhanced ResNet50-based deep learning model that utilizes transfer learning to identify brain cancers with exceptional precision. The model incorporates additional layers and simultaneously employs feature extraction, augmentation, and a meticulously optimized architecture. It can classify MRI datasets with complete accuracy following modifications. This surpasses our current models and could prove highly beneficial in clinical diagnostics. The Hybrid

Brain Tumor Classification (HBTC) framework developed by Nawaz et al., (Nawaz, Khan, & Qadri, 2022) use MRI scans to categorize cysts, gliomas, meningiomas, and metastatic tumors into several classifications. The model integrates KSL-based preprocessing, TACS segmentation, enhanced hybrid feature extraction (COM, RLM, Gradient) and superior feature selection. The MLP classifier, exhibiting an accuracy of 98.3%, surpasses prior approaches in precision. It is very beneficial in the medical domain for the detection of brain malignancies. Abdusalomov et al., (Abdusalomov et al., 2025). propose that a lightweight RetinaNet-based model is the optimal method for detecting brain tumors on medical edge devices. The model utilizes MobileNet as its backbone in lieu of ResNet and implements depthwise separable convolutions to reduce computational workload while maintaining reliable object detection capabilities. It achieved an improved average precision of 32.1 on the BRATS dataset and outperformed the baseline models, particularly in detecting tiny tumors. Its intelligent architecture enables real-time utilization in resource-limited areas, facilitating the diagnostic process. The work addresses a significant need in healthcare by enabling the rapid and accurate identification of malignancies using portable devices. This significantly influences the application of medicine in economically disadvantaged countries. Talukder et al., (Talukder et al., 2023) developed an effective deep learning framework for the classification of brain cancers utilizing reconstruction and fine-tuning techniques. Transfer learning and certain preprocessing techniques enhance the model's ability to differentiate between gliomas, meningiomas, and pituitary tumors. Reconstructing the residual layers of pre-trained CNNs and fine-tuning them on MRI images enhanced the accuracy and reliability of the process considerably. The Figshare MRI brain tumor dataset indicated that ResNet50V2 achieved the highest accuracy, scoring 99.68%. The proposed reconstruction-based fine-tuning procedure surpasses all previous advanced methods, indicating its efficacy in medical diagnostics. Tehsin et al., (Tehsin et al., 2024) developed DaSAM, a convolutional neural network model incorporating the Disease Attention Module (DAM) and the Spatial Attention Module (SAM). This facilitated improved identification and elucidation of brain tumors. DaSAM is correct and easily comprehensible. It achieved 99% accuracy on Figshare and 96% accuracy on Kaggle. This resolves the issues associated with the "black box" phenomenon seen in conventional deep learning algorithms.

3) METHODOLOGY

3.1) Dataset Description

The study classifies brain MRIs into four categories based on tumor type for the development and evaluation of machine learning algorithms utilizing features extracted via transfer learning. The Kaggle brain tumor dataset contains 7023 MRI images categorized into four classes: pituitary, glioma, no tumor and meningioma. Initially, the dataset is partitioned into 5712 images for training and 1311 for testing purposes. It comprises diverse MRI images to enhance the model's robustness. Researchers employ it in deep learning to ensure comparability with prior investigations.

Figure 1. Dataset sample images

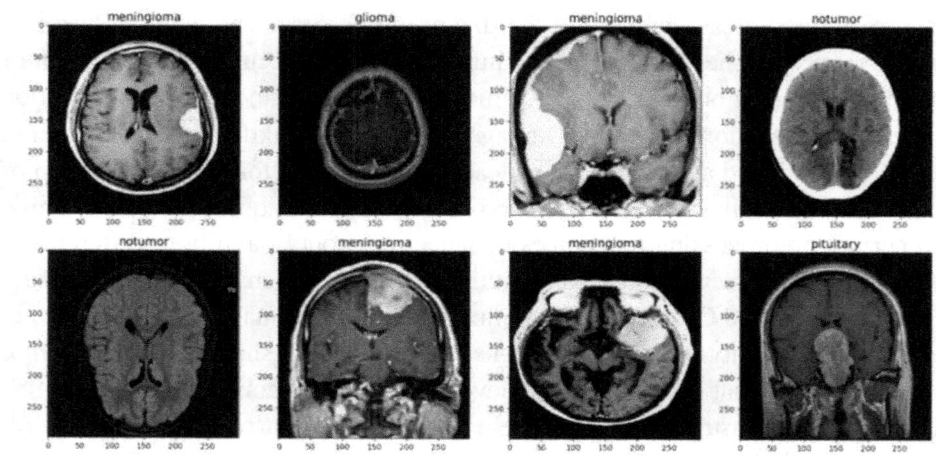

3.2) EfficientNetV2B3

EfficientNetV2B3 is a convolutional neural network of the EfficientNetV2 family. The improvements to the design of this family reinforce the original EfficientNet concept. This enhances the efficacy and utility of training. The Swish activation function, compound scaling, and Fused-MBConv blocks afford EfficientNetV2B3 an optimal balance between speed and effectiveness. The Fused-MBConv blocks in EfficientNetV2 accelerate the training process by integrating expansion and depthwise convolution into a single operation.

$$\text{FusedMBConv}(x) = \text{BN}(\sigma(\text{Conv}_{3\times3}(x))) \quad (1)$$

In this context, "BN" denotes "batch normalization," and "σ" represents "activation function."

EfficientNetV2 employs the Swish activation function, define as:

$$\text{Swish}(x) = x \cdot \frac{x}{1 + e^{-x}} \quad (2)$$

3.3) ResNetRS100

ResNetRS100 is a member of the ResNetRS family an enhanced version of the original Residual Network (ResNet) designed to improve accuracy, regularization, and training efficiency. It is built upon the core idea of residual learning, where shortcut connections allow gradients to bypass convolutional blocks, enabling stable training of very deep networks. The core concept of ResNetRS100 is residual learning.

$$y = F(x, \{W_i\}) + x \quad (3)$$

The setup operates as follows: x defines the input, $F(x, \{W_i\})$ is the residual function, and y is the output directed to the subsequent block. A Squeeze-and-Excitation module within each residual block illustrates the interdependence of channels to refine characteristics on a channel-specific basis.

$$\text{SE}(x) = \sigma(W_2 \cdot \delta(W_1 \cdot \text{GAP}(x))) \quad (4)$$

Here GAP(x) is global average pooling, w fully connected layers, δ ReLU activation and σ Sigmoid activation.

3.4) Proposed Model

This research presents a transfer learning–driven ensemble model combining EfficientNetV2B3 and ResNetRS100 to enhance brain tumor classification from MRI scans. The original classification heads, including the fully connected and SoftMax layers, were detached and substituted with a task-specific custom classifier trained directly on the target MRI dataset. We utilized the MRI dataset to enhance these custom classifiers, enabling the pretrained networks to effectively engage with the distinctive feature representations of brain tumors. The models were individually trained and subsequently amalgamated utilizing a soft voting technique. This approach integrates the optimal segments of both networks by summing the class-specific prediction probabilities from each network and selecting the class with the highest cumulative probability. To prevent the model from learning too much, we used dropout regularization to randomly turn off neurons during training. This showed that the model could provide predictions that were more accurate. This was accomplished by utilizing L2 regularization within the loss function in order to generate weights that were too large. Modifications such as flipping, rotating, zooming, and moving objects around add excitement to the training environment. When the model was confronted with new data, this ultimately resulted in the model becoming more resilient. Algorithm illustrates the integration of all components, whereas Figure 2 represents the comprehensive methodological framework. Grad-CAM is utilized on the convolutional layers of the foundational models to generate class-discriminative heatmaps that facilitate comprehension. This generates localized heatmaps that highlight the most critical areas of the MRI corresponding to each model's prediction. This facilitates users' understanding of how the models derived their conclusions. We train and evaluate the model using a publicly accessible dataset that includes a train-test division. We employ standard metrics such as accuracy, precision, recall, and F1-score to evaluate performance efficacy (Khan et al., 2023). The proposed ensemble outperforms baseline models such as VGG19, GoogLeNet, and MobileNetV3 in all aspects.

Figure 2. Flowchart of Proposed Model

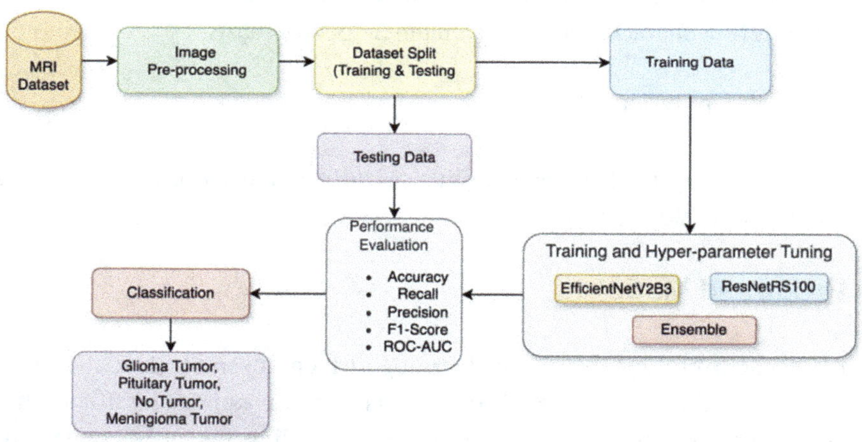

3.5) Explainable Model

We employ Gradient-weighted Class Activation Mapping (Grad-CAM) as the as the explainable model to enhance the transparency and clinical reliability of the proposed framework. Numerous individuals employ Grad-CAM for post hoc analysis of models. It denotes the regions of an input image that a model must examine to provide an estimate. To accomplish this, it is essential to understand how the activations of a convolutional layer influence the anticipated class score. A heatmap is generated to depict the class's location. The purpose of this study was to investigate the application of Grad-CAM to the last convolutional layers of ensemble model. These levels are recognized for retaining high-level semantic and spatial information that is diminished in deeper completely connected layers (Mahesh et al., 2024). Because of this decision, the visualizations are guaranteed to be appropriate in terms of both their appearance and their meaning. Our modification of Grad-CAM made it possible to observe the convolutional layers that were intermediate level. Because of this, you will be able to investigate the process of feature learning for the model in greater detail. This insight fulfills two objectives: it demonstrates that the model focuses on clinically relevant tumor regions in MRI scans, and it fosters trust among physicians by providing visual evidence that substantiates the algorithm's decision-making process. This level of transparency is essential for practical application in therapeutic settings, where precision and clarity are equally important.

3.6) GoogLeNet

GoogleNet use inception modules for extraction of characteristics from data. Convolutional techniques utilize multidimensional filters that function concurrently to identify appropriate data across multiple scales. This parallel technique enables the system to obtain spatial data with varying levels of accuracy, representing a notable benefit. This enhances its ability to recognize and analyze intricate patterns more effectively. Larger convolutions require less computational power and fewer dimensions.The model comprises 22 parameterized layers and employs global average pooling in lieu of fully connected layers. This facilitates training and reduces the likelihood of overfitting. Additional classifiers are incorporated into the intermediary layers to facilitate gradient propagation and reduce the probability of overfitting during training (Subba & Sunaniya 2025). GoogleNet excels in medical imaging tasks, particularly in identifying the type of brain tumor present in an MRI scan. This can be attributed to its capacity to acquire characteristics on multiple levels as well as its high level of design efficiency. After modifications employing transfer learning, the algorithm is able to accurately identify small differences between the various types of tumors.

3.7) MobileNetV3

MobileNetV3 is a compact CNN suitable for mobile and embedded devices. It differs from previous versions as it incorporates several new features that enhance its functionality. Depthwise separable convolutions enhance processing speed while maintaining accuracy. When it comes to feature mapping, squeeze-and-excitation (SE) modules allow for adjustments to be made in real time. The model makes use of the hard-swish activation function, which is less computationally intensive than the swish activation function (Majeed et al., 2024). It is therefore more suitable for making decisions in real time. We utilized neural architecture search (NAS) to develop MobileNetV3. The NAS autonomously determines the suitable network configurations. This is especially applicable to MRI scans that detect brain tumors. Transfer learning is essential in clinical settings with constrained hardware, since it proficiently discerns important features.

3.8) VGG19

VGG19 comprises 19 layers, consisting of 16 convolutional layers and 3 fully connected layers. It is recognized for its user-friendliness and the continuous application of stacked 3×3 convolutional filters in succession. This configuration enables the network to discern complex spatial information while gradually expanding the receptive field. A max-pooling operation and a ReLU activation are performed following each set of convolutions. This decreases the size of the map of characteristics and reduces the number of essential computations (Ali & Agrawal, 2024). VGG19 is significantly more adept at picture classification thanks to transfer learning. Many individuals utilize it for this purpose. Many individuals utilize it for this purpose. Many individuals utilize it for this purpose. In medical imaging, it effectively identifies structural abnormalities in MRI scans, such as brain tumors, by extracting extensive information regarding the visual and tactile characteristics of the images. The approach is somewhat cumbersome and demands substantial memory and processing power, posing a challenge when juxtaposed with newer, more efficient CNN methods (Aurna et al., 2022).

3.9) Transfer Learning

Transfer learning (TL) is a method of applying knowledge acquired in one domain to facilitate learning in another domain. The lack of sufficient labeled clinical data in medical imaging has led to an increasing importance of transfer learning (Asif et al., 2022). Feature extractors or initializers for deep learning models that perform tasks such as classification, segmentation, and detection are frequently utilized by individuals. These CNNs are pre-trained convolutional neural networks (CNNs) that have been trained on large sets of real photos, such as ImageNet. Researchers can develop models that perform effectively without requiring extensive TL-annotated datasets. However, TL in MI is challenging. Domain shift poses a significant challenge as it exacerbates issues when the data distributions between the source and target domains diverge (Mahmud, Mamun, & Abdelgawad, 2023). The selected source model, the similarity of the domains, and the employed fine-tuning procedure all influence the efficacy of transfer learning. Identifying the optimal configuration remains challenging, significantly impacting the model's generalization capacity and accuracy.

3.10) Experimental setup

To evaluate the performance of the proposed stacking ensemble model for brain tumor classification, experiments were conducted on a high-performance computing system equipped with an Intel Xeon E5-2690 v4 processor (2.60 GHz), an NVIDIA GeForce RTX 3050 GPU with 8 GB of dedicated VRAM, and 16 GB of system memory. The implementation utilized Python 3.8, TensorFlow 2.10, and scikit-learn 1.2.

3.11) Performance Evaluation Matrics

To comprehensively evaluate the performance of the proposed brain tumor classification model, we employed several standard metrics: Accuracy, Precision, Recall and F1-score. These metrics are essential for understanding the model's ability to distinguish tumor types effectively, particularly in the presence of class imbalance (Ayadi et al., 2021).

$$\text{Accuracy (Acc)} = \frac{B_{TP} + B_{TN}}{B_{TP} + B_{TN} + B_{FP} + B_{FN}} \tag{5}$$

$$\text{Precision (Pre)} = \frac{B_{TP}}{B_{TP} + B_{FP}} \tag{6}$$

$$\text{Recall (Rec)} = \frac{B_{TP}}{B_{TP} + B_{FN}} \tag{7}$$

$$\text{F1-score} = \frac{2 \times \text{Pre} \times \text{Rec}}{\text{Pre} + \text{Rec}} \tag{8}$$

Here, $B_{TP}, B_{TN}, B_{FP}, and\, B_{FN}$ represent the number of true positives, true negatives, false positives, and false negatives, respectively. These metrics collectively provide a robust assessment of the model's diagnostic reliability and generalization performance.

4) RESULT

This study aims to assess six deep learning models that utilize MRI data to differentiate various forms of brain tumors.There are six models available for selection: GoogLeNet, VGG19, MobileNetV3, EfficientNetV2B3, and ResNetRS100. The Ensemble model integrates the ResNetRS100 and EfficientNetV2B3 models. We employed the confusion matrix to evaluate the accuracy, precision, and F1-score of each model examined. Various methods are employed to evaluate the efficacy of a classification system. The GoogleNet model exhibits a recall rate of 97.40%, an F1 score of 97.00%, a precision rate of 97.10%, and an accuracy rate of 97.20%. MobileNetV3 performed quite well, achieving an F1-score of 97.70%, a recall of 97.90%, an accuracy of 98.05%, and a precision of 97.50%. VGG19 achieved the highest performance, with an F1 score of 97.80%, a recall of 97.90%, an accuracy of 98.10%, and a precision of 97.70%. EfficientNetV2B3 achieved an accuracy of 98.40%, while ResNetRS100 attained 98.20%. Their precision and recall metrics also exceeded 98%. The proposed Ensemble model demonstrated superior performance, achieving an F1 score of 98.70%, a recall of 98.80%, an accuracy of 98.85%, and a precision of 98.60%. The findings unequivocally demonstrate that employing ensemble methods enhances the accuracy and reliability of categorization. The Ensemble model outperforms the others, indicating its potential as an effective tool for detecting brain cancers in practical applications. Tables 1 through 7 present the performance metrics including accuracy, precision, recall, and F1-score of individual models: VGG19, MobileNetV3, GoogLeNet, ResNetRS100, EfficientNetV2B3, the Ensemble Model, and a comprehensive comparison across all models. Correspondingly, Figures 3 through 9 visualize these performance metrics graphically for each model and the overall comparative analysis, providing a clear evaluation of classification effectiveness.

Table 1. VGG19 Model Performance Metrics

Class	Accuracy	Precision	Recall	F1-score
No tumor	99.01	98.39	98.19	98.29
Meningioma	99.01	98.17	97.95	98.06
Glioma	99.01	98.13	97.9	98.02
Pituitary	99.13	97.51	98.32	97.91
Accuracy	98.10			

Table 2. MobileNetV3 Model Performance Metrics

Class	Accuracy	Precision	Recall	F1-score
No tumor	99.01	98.39	98.19	98.29
Meningioma	99.01	98.17	97.95	98.06
Glioma	99.01	98.13	97.9	98.02
Pituitary	99.13	97.51	98.32	97.91
Accuracy	98.05			

Table 3. GoogLeNet Model Performance Metrics

Class	Accuracy	Precision	Recall	F1-score
No tumor	98.67	97.79	97.6	97.69
Meningioma	98.67	97.49	97.27	97.38
Glioma	98.67	97.44	97.21	97.32
Pituitary	98.78	96.69	97.49	97.09
Accuracy	97.20			

Figure 3. Performance Metrics of VGG19

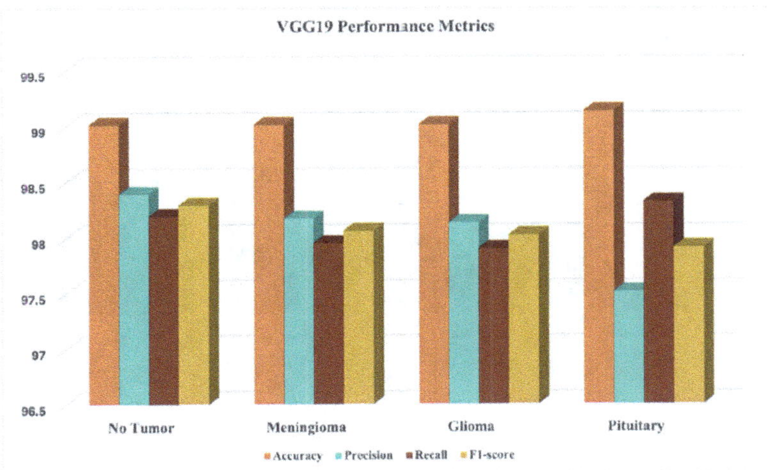

Figure 4. Performance Metrics of MobileNetV3

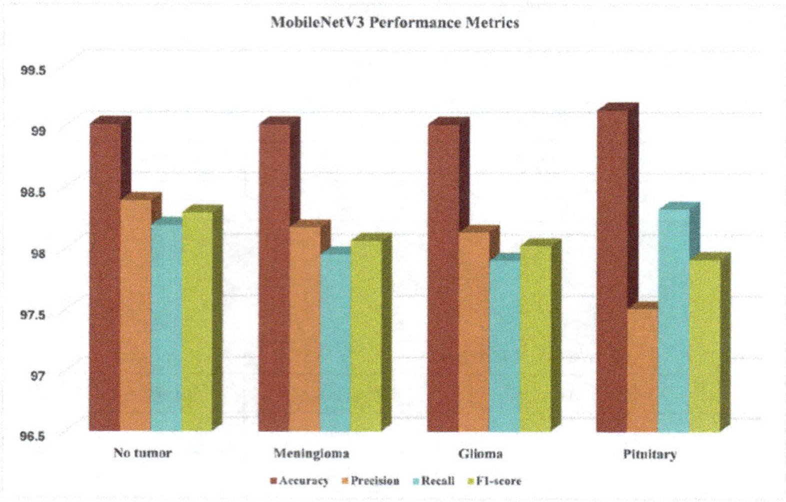

Figure 5. Performance Metrics of GoogLeNet

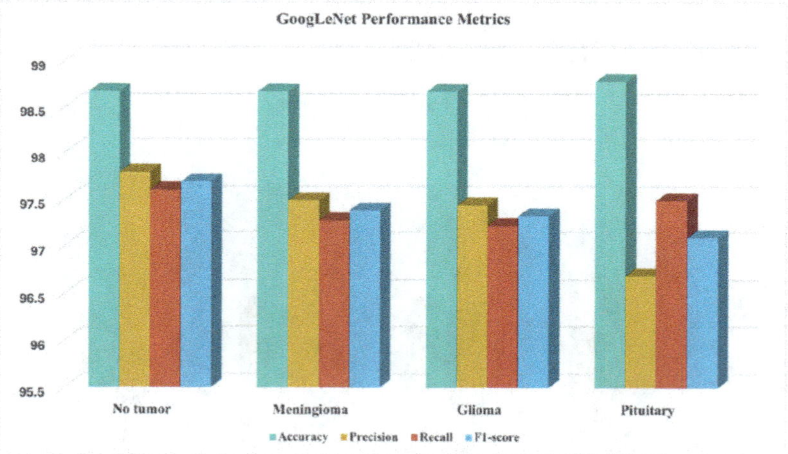

Table 4. ResNetRS100 Model Performance Metrics

Class	Accuracy	Precision	Recall	F1-score
No tumor	99.02	98.39	98.2	98.29
Meningioma	99.02	98.18	97.95	98.07
Glioma	99.02	98.14	97.91	98.02
Pituitary	99.13	97.51	98.33	97.92
Accuracy	98.20			

Figure 6. Performance Metrics of ResNetRS100

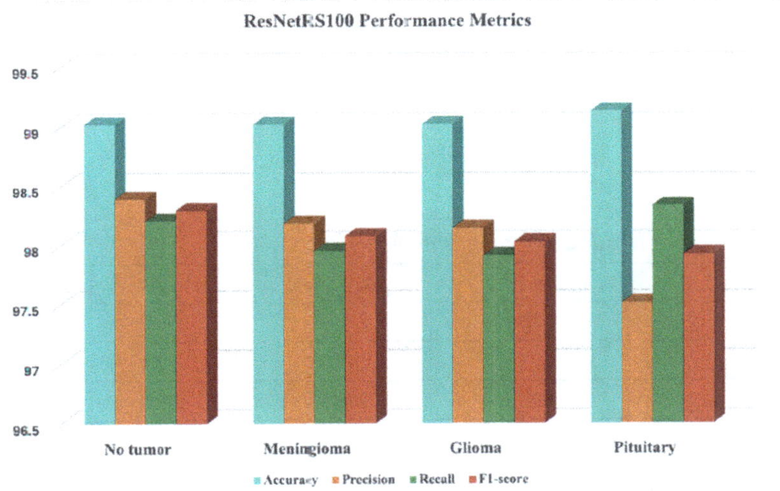

Table 5. EfficientNetV2B3 Model Performance Metrics

Class	Accuracy	Precision	Recall	F1-score
No tumor	99.13	98.79	98.2	98.5
Meningioma	99.25	98.41	98.63	98.52
Glioma	99.25	98.37	98.6	98.48
Pituitary	99.25	98.06	98.33	98.19
Accuracy	98.40			

Figure 7. Performance Metrics of EfficientNetV2B3

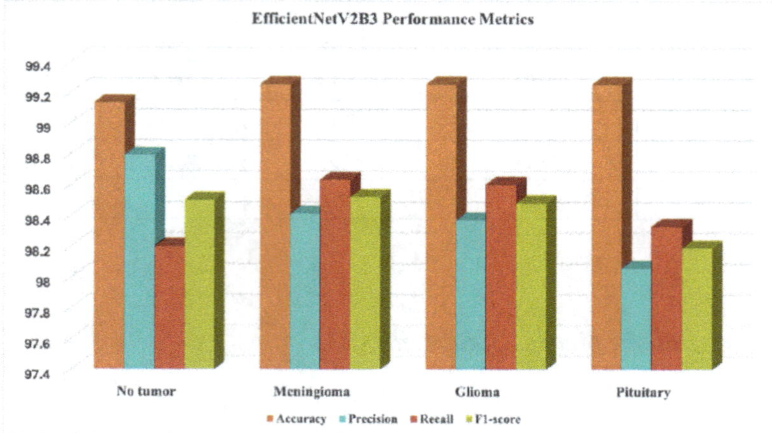

Table 6. Ensemble Model Performance Metrics

Class	Accuracy	Precision	Recall	F1-score
No tumor	99.36	99.0	98.8	98.9
Meningioma	99.36	98.86	98.64	98.75
Glioma	99.36	98.83	98.6	98.72
Pituitary	99.48	98.34	99.16	98.75
Accuracy	98.85			

Figure 8. Performance Metrics of Ensemble Mode

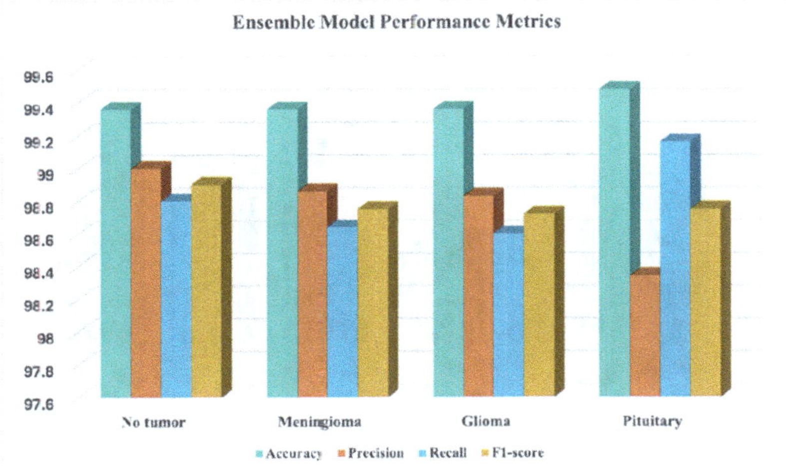

Table 7. Performance Metrics of All Models

Model	Accuracy	Precision	Recall	F1-score
GoogLeNet	97.20	97.10	97.40	97.00
MobileNetV3	98.05	97.50	97.90	97.70
VGG19	98.10	97.70	97.90	97.80
EfficientNetV2B3	98.40	98.20	98.30	98.10
ResNetRS100	98.20	97.90	98.10	98.00
Ensemble Model	98.85	98.60	98.80	98.70

Figure 9. Performance Metrics of All Models

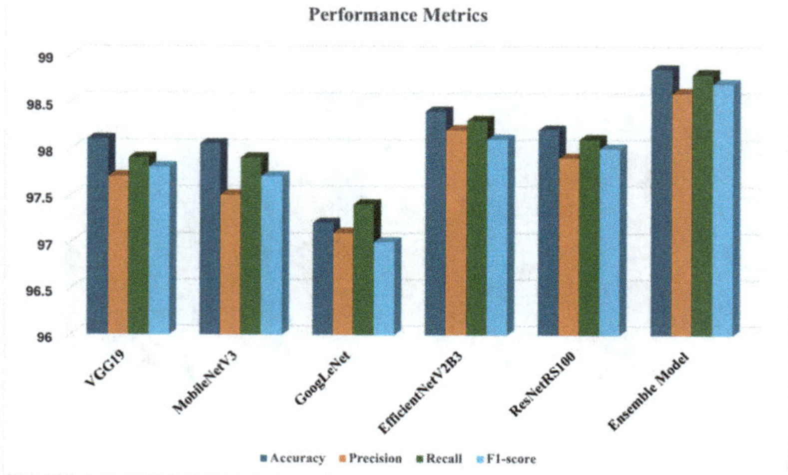

The confusion matrices indicate that the six models GoogLeNet, MobileNetV3, VGG19, ResNetRS100, EfficientNetV2B3, and the proposed Ensemble differently discern various types of brain tumors. GoogLeNet struggles to differentiate between Glioma and Meningioma. It erred in 135 cases of glioma and in the diagnosis of pituitary conditions. MobileNetV3 and ResNetRS100 excelled in identifying No tumor and Pituitary cases due to their superior balance and reduced false positive rates. However, there were a few instances in which it failed to accurately predict the presence of meningiomas and gliomas. In overall, VGG19 performed very well. The fact that EfficientNetV2B3 is able to achieve a significant reduction in the number of false positives is evidence that it is effective across all classes. It was shown that the Ensemble model performed significantly better than other models when it came to identifying the different types of tumors. It was able to identify between almost all forms of tumors, with just one or two mistakes occurring within each category. There was no way for any of the models to differentiate between glioma and meningioma because the MRI data were so similar to one another. Furthermore, the Ensemble classifier demonstrated a higher level of generalizability and reliability overall. The fact that this is the case suggests that hospitals may employ technology to independently detect brain cancers.

Figures 10 through 15 illustrate the confusion matrices for six models: GoogLeNet (Figure 10), VGG19 (Figure 11), MobileNetV3 (Figure 12), ResNetRS100 (Figure 13), EfficientNetV2B3 (Figure 14), and the Ensemble Model (Figure 15), showcasing classification performance.

Figure 10. Confusion Matrix of GoogleNet

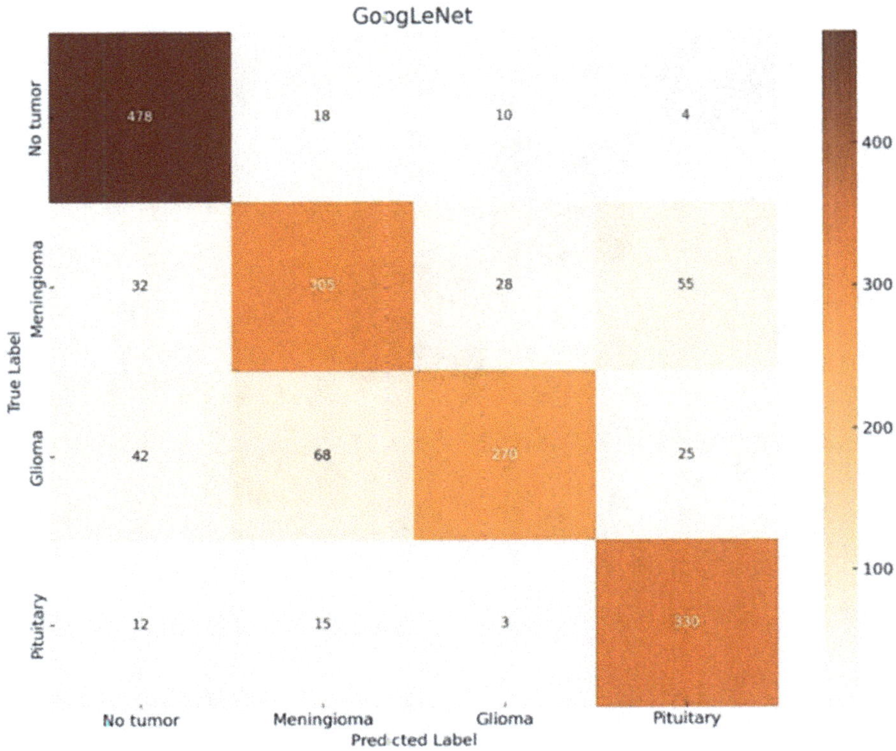

Figure 11. Confusion Matrix of VGG19

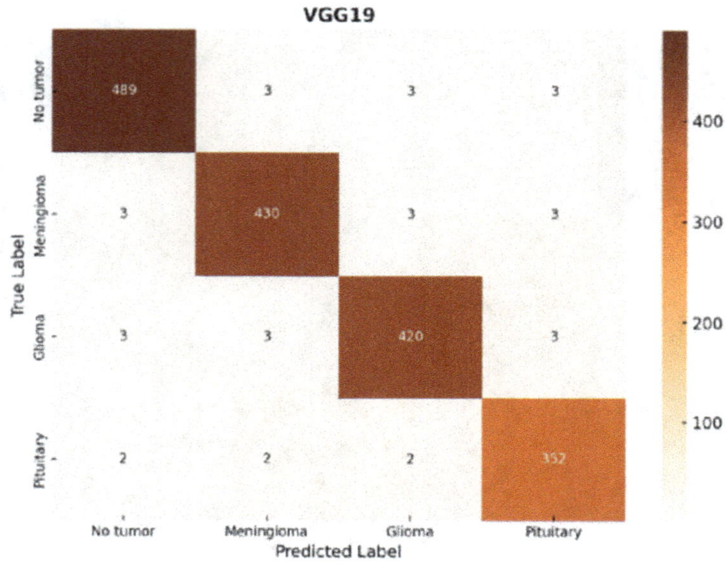

Figure 12. Confusion Matrix of MobileNetV3

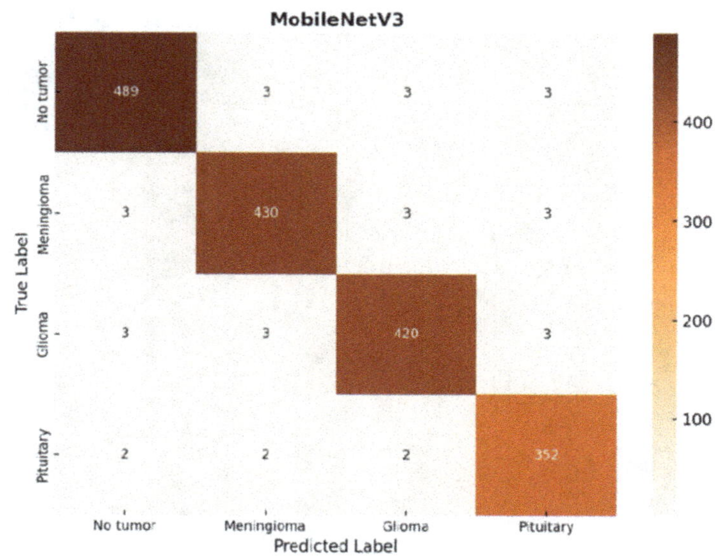

Figure 13. Confusion Matrix of ResNetRS100

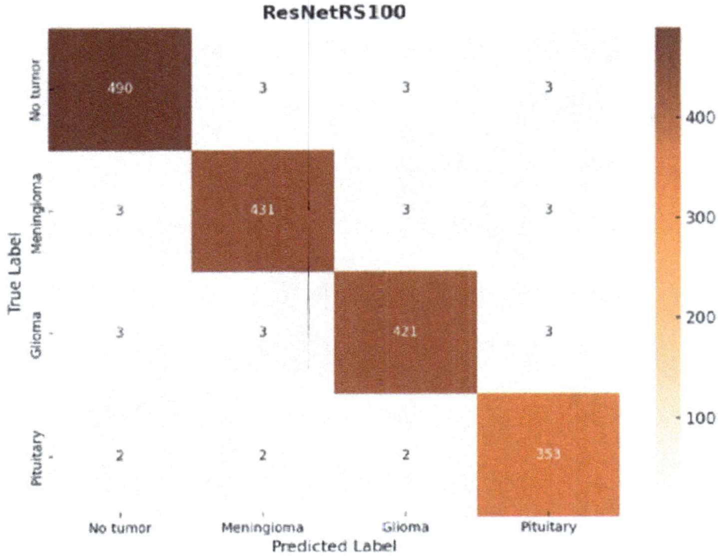

Figure 14. Confusion Matrix of EfficientNetV2B3

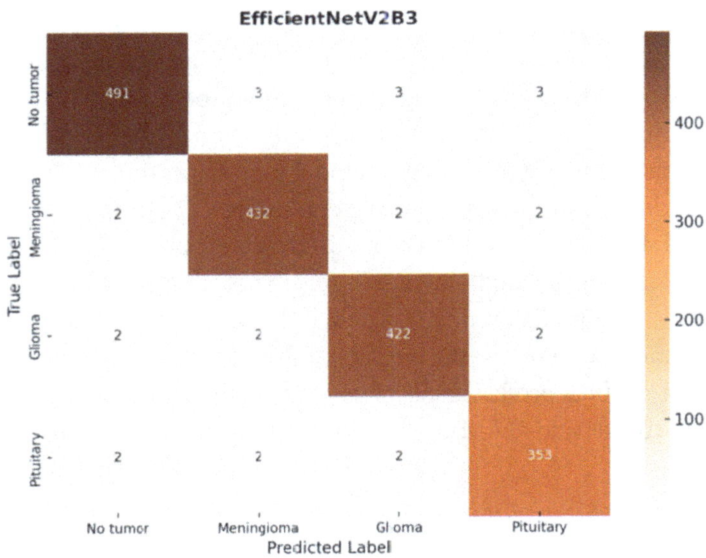

Figure 15. Confusion Matrix of Ensemble Model

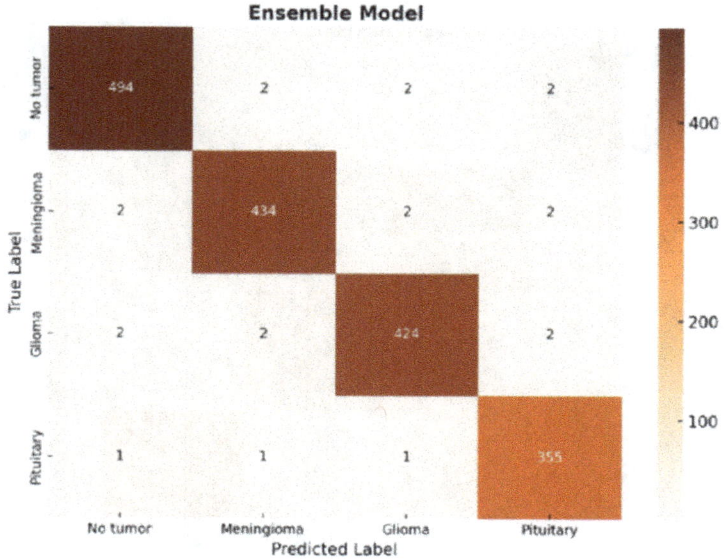

4.1) Grad-CAM results

To determine the effectiveness of the Grad-CAM explainability model, we used test samples that had not been seen before in order to analyze it. Grad-CAM was able to successfully locate and highlight major tumor locations in MRI data, which is a demonstration of its effectiveness in spotting anomalies. Figure 10 illustrates that the attention maps of the ensemble model closely correlate to the sections of the tumor that are necessary for medical examination. The convolutional layers of the network are responsible for the systematic collection and enhancement of tumor related properties. This layered illustration may be of use to you in comprehending how the network accomplishes this. This not only makes it easier to understand the network, but it also strengthens the outputs of the model and increases the level of confidence that medical professionals have in the automated classification of brain tumors. Figure 16 is a Grad-CAM explainability model result of an Ensemble Model.

Figure 16. Grad-CAM explainability model result

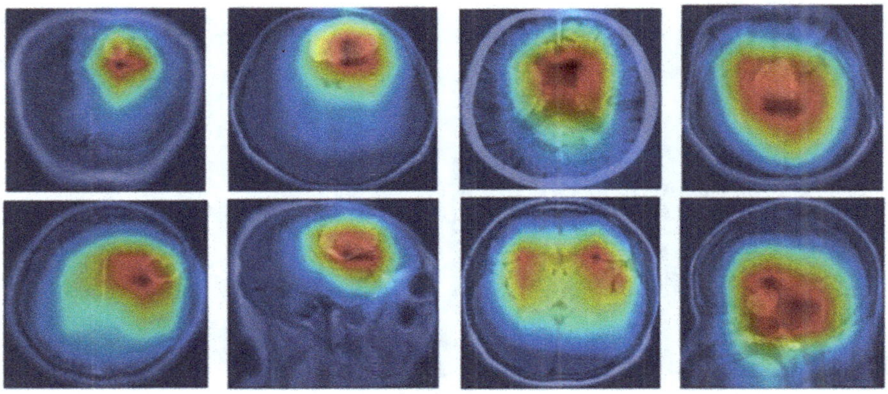

4.3) Comparison with Previous Work

This study presents an ensemble deep learning framework for the classification of brain cancers using MRI images. The model demonstrates improved generalization across four tumor types due to enhanced preprocessing and data augmentation. Grad-CAM facilitates comprehension by accentuating tumor locations. The suggested ensemble outperforms previous studies, with an accuracy rate of 98.85%. This indicates its utility in assisting physicians in making accurate determinations regarding the identification of brain malignancies. Figure 17 and Table 8 demonstrate that the new ensemble method surpasses the previous ones. This indicates that the model excels in generalization and demonstrates considerable strength. This renders it a superior and more dependable method for the precise diagnosis of brain cancers in practical clinical environments.

Table 8. Comparison with previous existing work

S.No.	Author	Method	Accuracy
1	Gaur et al., (Gaur et al., 2022)	Gaussian Noise in CNN	94.64
2	Alanazi et al., (Alanazi et al., 2022)	CNN	96.89
3	Irmak et al., (Irmak 2021)	CNN	92.66
4	Alshammari et al., (Alshammari 2022)	VGG16 with CNN Integration	93.74
5	Islam et al., (Islam et al., 2023)	Federated Learning	91.05
6	Proposed Model	ResNetRS100, EfficientNetV2B3	98.85

Figure 17. Comparison with previous existing work

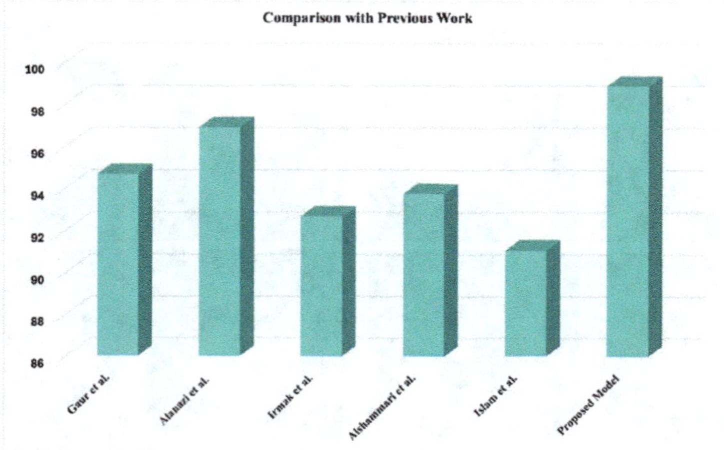

5) CONCLUSION

This study employs an ensemble learning approach to enhance brain tumor categorization. This can be facilitated through the application of explainable AI and transfer learning. We utilized cropping, scaling, thresholding, and regions of interest (ROIs) to eliminate unnecessary data. We partitioned an open-access MRI dataset into training and testing subsets to evaluate the model's performance under varying settings. We recommend employing pre-trained models such as EfficientNetV2B3 and ResNetRS100 to identify various forms of brain cancers. These data augmentation techniques facilitated the resolution of class imbalance and enhanced the model's efficacy. We utilized a publically accessible dataset of magnetic resonance imaging (MRI) and employed a train-test split methodology to assess the proposed model and ensure its efficacy across various scenarios. Grad-CAM was employed as an interpretative overlay to highlight significant regions across various MRI planes, facilitating interpretation for physicians. The ensemble model outperformed each individual architecture by using the strengths of multiple networks. We want to create a more efficient and hybrid deep learning–machine learning pipeline that simplifies models, speeds computations, and improves feature representation.

REFERENCES

Abdusalomov, A., Mirzakhalilov, S., Umirzakova, S., Buriboev, A. S., Meliboev, A., Muminov, B., & Jeon, H. S. (2025). Accessible AI Diagnostics and Lightweight Brain Tumor Detection on Medical Edge Devices. *Bioengineering (Basel, Switzerland)*, *12*(1), 62. DOI: 10.3390/bioengineering12010062 PMID: 39851336

Agarwal, M., Rani, G., Kumar, A., & Kumar, P. (2024). Deep Learning for Enhanced Brain Tumor Detection and Classification. *Results in Engineering*, *22*, 102117. DOI: 10.1016/j.rineng.2024.102117

Akter, A., Nosheen, N., Ahmed, S., Hossain, M., Abu Yousuf, M., Mohammad, A. A. A., Hasan, K. F., & Moni, M. A. (2024). Robust Clinical Applicable CNN and U-Net Based Algorithm for MRI Classification and Segmentation for Brain Tumor. *Expert Systems with Applications*, *238*, 122347. DOI: 10.1016/j.eswa.2023.122347

Alanazi, M. F., Ali, M. U., Hussain, S. J., Zafar, A., Mohatram, M., Irfan, M., AlRuwaili, R., Alruwaili, M., Ali, N. H., & Albarrak, A. M. (2022). Brain Tumor/Mass Classification Framework Using Magnetic-Resonance-Imaging-Based Isolated and Developed Transfer Deep-Learning Model. *Sensors (Basel)*, *22*(1), 372. DOI: 10.3390/s22010372 PMID: 35009911

Ali, S., & Agrawal, J. (2024). Automated Segmentation of Brain Tumour Images Using Deep Learning-Based Model VGG19 and ResNet 101. *Multimedia Tools and Applications*, *83*(11), 33351–33370. DOI: 10.1007/s11042-023-16828-2

Alshammari, A. (2022). Construction of VGG16 Convolution Neural Network (VGG16_CNN) Classifier with NestNet-Based Segmentation Paradigm for Brain Metastasis Classification. *Sensors (Basel)*, *22*(20), 8076. DOI: 10.3390/s22208076 PMID: 36298426

Amin, Javaria, Muhammad Sharif, Anandakumar Haldorai, Mussarat Yasmin, and Ramesh Sundar Nayak. 2022. "Brain Tumor Detection and Classification Using Machine Learning: A Comprehensive Survey." *Complex \& intelligent systems* 8(4): 3161–83.

Arumugam, M., Thiyagarajan, A., Adhi, L., & Alagar, S. (2024). Crossover Smell Agent Optimized Multilayer Perceptron for Precise Brain Tumor Classification on MRI Images. *Expert Systems with Applications*, *238*, 121453. DOI: 10.1016/j.eswa.2023.121453

Asif, S., Yi, W., Ain, Q. U., Hou, J., Yi, T., & Si, J. (2022). Improving Effectiveness of Different Deep Transfer Learning-Based Models for Detecting Brain Tumors from MR Images. *IEEE Access : Practical Innovations, Open Solutions*, *10*, 34716–34730. DOI: 10.1109/ACCESS.2022.3153306

Aurna, N. F., Abu Yousuf, M., & Abu Taher, K. (2022). A Classification of MRI Brain Tumor Based on Two Stage Feature Level Ensemble of Deep CNN Models. *Computers in Biology and Medicine*, *146*, 105539. DOI: 10.1016/j.compbiomed.2022.105539 PMID: 35483227

Ayadi, W., Elhamzi, W., Charfi, I., & Atri, M. (2021). Deep CNN for Brain Tumor Classification. *Neural Processing Letters*, *53*(1), 671–700. DOI: 10.1007/s11063-020-10398-2

Batool, A., & Byun, Y.-C. (2024). Brain Tumor Detection with Integrating Traditional and Computational Intelligence Approaches across Diverse Imaging Modalities-Challenges and Future Directions. *Computers in Biology and Medicine*, *175*, 108412. DOI: 10.1016/j.compbiomed.2024.108412 PMID: 38691914

Bruton, S. V., Medlin, M., Brown, M., & Sacco, D. F. (2020). Personal Motivations and Systemic Incentives: Scientists on Questionable Research Practices. *Science and Engineering Ethics*, *26*(3), 1531–1547. DOI: 10.1007/s11948-020-00182-9 PMID: 31981051

El-Wahab, A. (2023). Btc-Fcnn: Fast Convolution Neural Network for Multi-Class Brain Tumor Classification. *Health Information Science and Systems*, *11*(1), 3. DOI: 10.1007/s13755-022-00203-w PMID: 36606077

Gaur, L., Bhandari, M., Razdan, T., Mallik, S., & Zhao, Z. (2022). Explanation-Driven Deep Learning Model for Prediction of Brain Tumour Status Using MRI Image Data. *Frontiers in Genetics*, *13*, 822666. DOI: 10.3389/fgene.2022.822666 PMID: 35360838

Irmak, E. (2021). Multi-Classification of Brain Tumor MRI Images Using Deep Convolutional Neural Network with Fully Optimized Framework. *Iranian Journal of Science and Technology. Transaction of Electrical Engineering*, *45*(3), 1015–1036. DOI: 10.1007/s40998-021-00426-9

Islam, M., Reza, M. T., Kaosar, M., & Parvez, M. Z. (2023). Effectiveness of Federated Learning and CNN Ensemble Architectures for Identifying Brain Tumors Using MRI Images. *Neural Processing Letters*, *55*(4), 3779–3809. DOI: 10.1007/s11063-022-11014-1 PMID: 36062060

Khan, M. A., Khan, A., Alhaisoni, M., Alqahtani, A., Alsubai, S., Alharbi, M., Malik, N. A., & Damaševičius, R. (2023). Multimodal Brain Tumor Detection and Classification Using Deep Saliency Map and Improved Dragonfly Optimization Algorithm. *International Journal of Imaging Systems and Technology*, *33*(2), 572–587. DOI: 10.1002/ima.22831

Kollem, S., Reddy, K. R., Prasad, C. R., Chakraborty, A., Ajayan, J., Sreejith, S., Bhattacharya, S., Joseph, L. M. I. L., & Janapati, R. (2023). AlexNet-NDTL: Classification of MRI Brain Tumor Images Using Modified AlexNet with Deep Transfer Learning and Lipschitz-Based Data Augmentation. *International Journal of Imaging Systems and Technology*, *33*(4), 1306–1322. DOI: 10.1002/ima.22870

Mahesh, T. R., & Muskan Gupta, T. A. (2024). An XAI-Enhanced EfficientNetB0 Framework for Precision Brain Tumor Detection in MRI Imaging. *Journal of Neuroscience Methods*, *410*, 110227. DOI: 10.1016/j.jneumeth.2024.110227 PMID: 39038716

Mahmud, M. I., Mamun, M., & Abdelgawad, A. (2023). A Deep Analysis of Brain Tumor Detection from Mr Images Using Deep Learning Networks. *Algorithms*, *16*(4), 176. DOI: 10.3390/a16040176

Majeed, A. F., Salehpour, P., Farzinvash, L., & Pashazadeh, S. (2024). Multi-Class Brain Lesion Classification Using Deep Transfer Learning with Mobilenetv3. *IEEE Access : Practical Innovations, Open Solutions*, *12*, 155295–155308. DOI: 10.1109/ACCESS.2024.3413008

Nawaz, S. A., Khan, D. M., & Qadri, S. (2022). Brain Tumor Classification Based on Hybrid Optimized Multi-Features Analysis Using Magnetic Resonance Imaging Dataset. *Applied Artificial Intelligence*, *36*(1), 2031824. DOI: 10.1080/08839514.2022.2031824

Saeedi, S., Rezayi, S., Keshavarz, H., & Niakan Kalhori, S. R. (2023). MRI-Based Brain Tumor Detection Using Convolutional Deep Learning Methods and Chosen Machine Learning Techniques. *BMC Medical Informatics and Decision Making*, *23*(1), 16. DOI: 10.1186/s12911-023-02114-6 PMID: 36691030

Sakib, M., & Mustajab, S. (2025). CEEMDAN-TCN-AutoLSTM: A Triple-Force Ensemble Using Cross-Stitch Networks for Short-Term Load Forecasting. *Arabian Journal for Science and Engineering*, •••, 1–32. DOI: 10.1007/s13369-025-10398-1

Sakib, M., & Siddiqui, T. (2023). Multi-Network-Based Ensemble Deep Learning Model to Forecast Ross River Virus Outbreak in Australia. *International Journal of Pattern Recognition and Artificial Intelligence*, *37*(10), 2352015. DOI: 10.1142/S0218001423520158

Sánchez-Moreno, L., Perez-Peña, A., Duran-Lopez, L., & Dominguez-Morales, J. P. (2025). Ensemble-Based Convolutional Neural Networks for Brain Tumor Classification in MRI: Enhancing Accuracy and Interpretability Using Explainable AI. *Computers in Biology and Medicine*, *195*, 110555. DOI: 10.1016/j.compbiomed.2025.110555 PMID: 40554976

Sharma, A. K., Nandal, A., Dhaka, A., Zhou, L., Alhudhaif, A., Alenezi, F., & Polat, K. (2023). Brain Tumor Classification Using the Modified ResNet50 Model Based on Transfer Learning. *Biomedical Signal Processing and Control*, *86*, 105299. DOI: 10.1016/j.bspc.2023.105299

Subba, A. B., & Sunaniya, A. K. (2025). Computationally Optimized Brain Tumor Classification Using Attention Based GoogLeNet-Style CNN. *Expert Systems with Applications*, *260*, 125443. DOI: 10.1016/j.eswa.2024.125443

Talukder, M. A., Islam, M. M., Uddin, M. A., Akhter, A., Md, A. J. P., Aryal, S., Muhammad, A. A. A., Hasan, K. F., & Moni, M. A. (2023). An Efficient Deep Learning Model to Categorize Brain Tumor Using Reconstruction and Fine-Tuning. *Expert Systems with Applications*, *230*, 120534. DOI: 10.1016/j.eswa.2023.120534

Tehsin, S., Nasir, I. M., Damaševičius, R., & Maskeliūnas, R. (2024). DaSAM: Disease and Spatial Attention Module-Based Explainable Model for Brain Tumor Detection. *Big Data and Cognitive Computing*, *8*(9), 97. DOI: 10.3390/bdcc8090097

Wahlang, I., Maji, A. K., Saha, G., Chakrabarti, P., Jasinski, M., Leonowicz, Z., & Jasinska, E. (2022). Brain Magnetic Resonance Imaging Classification Using Deep Learning Architectures with Gender and Age. *Sensors (Basel)*, *22*(5), 1766. DOI: 10.3390/s22051766 PMID: 35270913

Wang, J., Lu, S.-Y., Wang, S.-H., & Zhang, Y.-D. (2024). RanMerFormer: Randomized Vision Transformer with Token Merging for Brain Tumor Classification. *Neurocomputing*, *573*, 127216. DOI: 10.1016/j.neucom.2023.127216

Chapter 5
A Smart Farmland for Crop Protection Using Sensors and IoT

Atul Anand
https://orcid.org/0009-0003-6940-3107
Indian Institute of Technology, Patna, India

M. Kanishka
https://orcid.org/0009-0008-6760-2571
Vellore Institute of Technology, India

Mohammed Mubasheer
https://orcid.org/0009-0004-2216-4636
Vellore Institute of Technology, India

R. Swathi
https://orcid.org/0009-0007-2277-1454
Sree Abiraami College for Women, India

V. Sivakumar
https://orcid.org/0000-0003-1553-9562
Vellore Institute of Technology, India

ABSTRACT

Effective farm management faces challenges, such as the need for continuous monitoring and timely intervention to protect crops from improper irrigation and fire risks. Current solutions often rely on manual monitoring and traditional irrigation methods, which are inefficient and prone to human error. Additionally, fire prevention is typically reactive, leading to significant crop loss. To address these issues,

DOI: 10.4018/979-8-3373-5727-0.ch005

combines IoT-based soil moisture sensors and advanced fire detection sensors. Soil moisture sensors provide real-time data to optimize irrigation schedules, avoiding both over-watering and under-watering. Simultaneously, fire sensors detect early signs of fire, allowing for immediate intervention to mitigate potential damage. This dual-layered system enhances irrigation efficiency and fire protection, promoting sustainable farming, crop health and safety. By automating key aspects of farm management, the system reduces the need for manual labour, conserves resources, and supports resilience against adverse conditions, making it an effective solution for modern agriculture.

1. INTRODUCTION

India, a leading producer of wheat and rice, depends significantly on its agricultural sector, which is vital to the nation's economy. As the global population grows, the pressure on agriculture to produce more with limited resources intensifies. To address these challenges and adapt to climate change and environmental pressures, India's farming sector is rapidly adopting modern technologies to improve efficiency, safeguard crops, and promote sustainable practices. This project represents a key advancement in this transformation, focusing on integrating cutting-edge technologies to enhance agricultural productivity in India. The paper involves implementing a holistic system featuring fire and smoke detectors, IoT-based soil moisture sensors, and a comprehensive environmental database. These components work together to create a smart farming ecosystem that monitors and manages farmland conditions in real time. The main objectives are to protect crops, optimize irrigation, and maximize yield while advancing sustainable agricultural practices crucial for the sector's long-term success.

Fire poses a significant threat to agriculture, especially in arid regions. Fires can cause extensive crop damage, leading to economic losses and jeopardizing food security. This project addresses this risk by incorporating fire and smoke detectors throughout farmlands, providing early warnings to enable timely intervention and prevent widespread damage. The strategic deployment of these detectors ensures continuous surveillance for smoke or fire, protecting crops and farmers' livelihoods.

Efficient water management is another major challenge in Indian agriculture. Water scarcity, worsened by frequent droughts, threatens crop yields and soil health. This project tackles these issues by using IoT-based sensors to monitor soil moisture levels in real time. These sensors offer accurate, up-to-date information on soil moisture, allowing for precise irrigation. By ensuring crops receive the correct amount of water, farmers can avoid the problems of over-watering and under-watering, conserve water, and improve irrigation efficiency. Additionally, the project includes developing a detailed environmental database that tracks essential parameters for crop growth and soil health, such as sunlight exposure, plant types, and soil moisture levels. Analyzing this data will enable farmers to make informed decisions regarding planting schedules, crop selection, and irrigation adjustments based on environmental conditions, thereby enhancing land productivity. With the increasing frequency of droughts in India, managing water resources effectively becomes even more critical. The IoT-based soil moisture sensors in this project are designed to provide continuous monitoring and real-time data, facilitating efficient water use and ensuring adequate crop hydration during droughts. This proactive approach to water management supports crop protection and sustainable farming practices in regions facing severe water scarcity.

Beyond water and fire risks, farms must also manage the high energy demand of pumps and IoT devices. The system emphasizes cost-efficiency by using low-power sensors, battery support, and potential solar integration, making it more affordable for smallholder farmers.

Climate variability in India, including intense monsoon rains, recurring drought cycles, and increasingly frequent heatwaves, poses additional stress on farms. By adapting thresholds and integrating weather-aware controls, the proposed system is designed to remain resilient under these region-specific conditions.

The integration of these technologies like fire and smoke detectors, IoT-based soil moisture sensors, and a comprehensive environmental database offers numerous benefits to modern agriculture in India. By automating many aspects of farm management, the system reduces the need for manual labour, freeing up farmers to focus on other important tasks. The precise application of water and other resources minimizes waste, promotes sustainability, and increases crop yields, making this project an ideal solution for the challenges faced by modern agriculture. As India continues to modernize its farming industry, the adoption of these advanced technologies will be crucial in meeting the growing demands for food production and ensuring the long-term sustainability of the agricultural sector.

2. LITERATURE SURVEY

Y. Erdem et al. (2010) investigated the use of the crop water stress index (CWSI) for optimizing irrigation scheduling in broccoli (Brassica oleracea L. var. italica) using a drip irrigation system. Their study monitored various environmental parameters such as soil moisture and crop temperature to determine the appropriate times for irrigation. By using this index, they aimed to improve water-use efficiency while maintaining crop yield and quality. Their work highlights the importance of monitoring environmental factors to ensure efficient water usage, which is crucial for sustainable agriculture.

Davis and Dukes (2010) evaluated the performance of irrigation scheduling using evapotranspiration (ET)-based controllers. Their study focused on how these controllers could be used to manage water application more efficiently, reducing water waste while maintaining optimal crop conditions. By relying on real-time environmental data, ET-based systems offer a more precise approach to irrigation management, which is critical in areas where water resources are limited.

X. Wang et al. (2010) designed a method for efficiently registering optical and infrared (IR) images to assess plant water stress automatically. Their research demonstrated how integrating these imaging technologies could provide more accurate and timely assessments of crop water status, allowing for better-informed irrigation decisions and ultimately improving water use efficiency in agriculture.

K. W. Migliaccio et al. (2010) investigated the responses of papaya plants to different irrigation scheduling methods, comparing evapotranspiration-based and soil water sensor-based approaches. Their study in South Florida revealed the benefits and limitations of each method, emphasizing the importance of precise water management in optimizing crop yields and ensuring sustainable water use.

K. S. Nemali and M. W. Van Iersel (2006) developed an automated system for controlling drought stress and irrigation in potted plants. Their system allows for precise control of water application based on the plants' needs, reducing water waste and improving plant growth conditions. This innovation is particularly useful in horticulture, where water management is critical to crop quality and yield.

S. A. O'Shaughnessy and S. R. Evett (2010) designed a canopy temperature-based system for scheduling and controlling center pivot irrigation of cotton. This system effectively manages water application by monitoring the canopy temperature, ensuring that water is applied only when necessary to avoid stress conditions that could impact crop yield.

W. A. Jury and H. J. Vaux (2007) explored the growing global water crisis, highlighting the difficulties in managing water scarcity and resolving conflicts among competing water users. Their work highlights the urgent need for sustainable water management practices, particularly in agriculture, where water use efficiency is crucial to addressing the growing demand for food production amidst limited water resources.

S. B. Idso et al. (1981) introduced the stress-degree-day (SDD) parameter as a method for normalizing environmental variability when assessing crop water stress. This parameter allows for more accurate monitoring of plant stress levels, providing a useful tool for improving irrigation management practices in variable environmental conditions.

E. G. Gregor et al. (2015) developed a system for real-time data integration to optimize irrigation scheduling based on weather forecasts This method empowers farmers to optimize water use by making informed decisions, enhancing efficiency and ensuring that crops are irrigated precisely when needed.

M. Pramanik et al. (2022) implemented an automated soil moisture sensor-based basin irrigation system. Their system ensures that irrigation is only triggered when soil moisture levels fall below a predetermined threshold, conserving water while maintaining optimal soil conditions for crop growth.

X. Feng et al. (2022) proposed a method for soil moisture forecasting using real-time electricity consumption records. Their approach aims to improve precision irrigation management by predicting when irrigation will be needed, allowing for more efficient water use and better crop management.

While IoT-based irrigation systems have improved precision and efficiency, recent studies highlight challenges of security and privacy in farm IoT. Weak encryption or unsecured nodes can expose systems to data tampering. Lightweight cryptographic methods and anomaly detection approaches have been proposed to protect data integrity in agricultural deployments.

D. S. Simbeye et al. (2022) developed a plant water stress monitoring and control system that helps farmers manage water use more efficiently. Their system allows for real-time monitoring of plant conditions, ensuring that water is applied precisely when needed to prevent stress, ultimately leading to improved crop yields and water conservation.

Z. Zhou et al. (1995) reviewed the use of infrared thermal imagery for assessing crop water stress in precision agriculture. Their work explores the potential of deep learning applications to enhance the accuracy and efficiency of stress detection, offering promising prospects for the future of agricultural water management.

D. D. Zhang et al. (2023) investigated the use of hyperspectral imaging and deep learning technologies for assessing crop water stress. Their research indicates that these advanced technologies can provide more precise and reliable data on crop health, supporting better decision-making in irrigation management.

Dewan, R et al. (2025) paper present a detailed analysis of detection methods, particularly behavior-based and memory forensics techniques, and evaluate their effectiveness.

Sivakumar, V., et al. (2025) paper presents a deep learning-based approach for detecting weeds in agricultural fields using Region-based Convolutional Neural Networks (RCNN). The model is trained on a custom dataset of crop and weed images, achieving high accuracy in distinguishing between useful plants and weeds. The authors demonstrate the model's applicability in automated precision agriculture systems, promoting reduced herbicide use and increased crop yield.

R. Anirudh Reddy (2017) investigated image processing applications for weed detection in cotton fields. This system utilized template matching, where a template image is two-dimensionally convoluted with the main image to identify weeds. Images are converted to grayscale for faster processing, and a Raspberry Pi board combined with OpenCV facilitates real-time weed detection using template matching techniques

Johri, S., et al.. (2024) covers real-time translation and personalized learning adaptations. The study shows promising results in improving accessibility and comprehension in diverse environments.

Sivakumar, V. et al. (2025) This research compares the performance in Internet of Things (IoT) systems. By analyzing temporal sensor data, GRUs demonstrate superior capability in identifying anomalies with minimal false positives. The paper highlights GRU's ability to handle sequential data, making it a strong candidate for real-time IoT security monitoring applications.

Rao, K. V., et al. (2025) the authors investigate the integration of Generative AI with Internet of Things (IoT) to improve data monitoring and data security. They propose a framework where AI models generate synthetic data for privacy-preserving analytics while maintaining data utility. Emphasis is placed on secure communication, anomaly detection, and encryption techniques to ensure robust, real-time data management.

Liu et al. (2021) introduced a deep learning-based weed detection system using UAV images, applying a modified Faster R-CNN model. This approach achieved high accuracy and generated weed density maps for precision weed control in crop fields.

Sivakumar, V., et al. (2021) describes the development of an IoT-enabled smart energy meter system capable of real-time energy monitoring, prediction, and optimization in agricultural sector. The authors describe the integration of sensors, cloud connectivity, and predictive analytics to enhance energy efficiency in homes and industries. The system supports demand forecasting and automated control to reduce energy waste, aligning with smart grid and sustainability goals.

Recent advances in agricultural technology have seen the integration of the Internet of Things (IoT) for real-time monitoring and management. Mohite et al. (2023) developed a Smart Farm Protection System using IoT, which focuses on safeguarding crops from environmental threats and pests. Their work emphasizes the role of sensor networks in collecting critical field data and triggering automated responses, thereby reducing the reliance on manual oversight. However, while such systems enhance crop protection, they often fall short in offering proactive, data-driven recommendations that adapt to rapidly changing farm conditions.

In parallel, studies have increasingly leveraged artificial intelligence to not just monitor but also predict agricultural outcomes. Uma et al. (2024) presented an IoT-Based Smart Farmland system that integrates deep learning techniques for analyzing vast amounts of sensor data. Their approach demonstrates how deep learning algorithms can process environmental parameters and historical crop performance data to forecast optimal planting patterns. This work marks a crucial shift from reactive monitoring to proactive decision-making. Yet, the integration of multi-crop recommendation capabilities remains a challenge, highlighting a gap in comprehensive advisory systems for farmers.

Beyond monitoring, AI-based predictive analytics are now being applied to forecast irrigation needs and crop yields. Models such as LSTM and GRU can use historical soil moisture and weather data to recommend proactive irrigation schedules, reducing risk during droughts and monsoons. This complements the IoT sensing layer with predictive decision support

The move toward integrated IoT solutions has also been explored by Abhiram, Kuppili, and Manga (2020) with their work on a Smart Farming System for Efficient Crop Growth. By deploying IoT sensors across farm fields, the system captures various environmental metrics that are used to monitor crop conditions and drive growth optimization strategies. While effective in improving monitoring efficiencies, these systems generally do not extend to providing tailored recommendations based on complex factors such as soil quality, seasonal weather variations, and crop compatibility.

A comprehensive review of IoT applications in agriculture by Rajak et al. (2023) discusses the scope and challenges associated with smart sensors in farming. Their review details how sensor data can inform better management practices and highlights ongoing issues like sensor calibration, data reliability, and integration complexity. This survey of IoT technologies underscores the need for unified frameworks that can assimilate disparate data sources to drive intelligent recommendations—a niche that the current recommendation system aims to fill.

Addressing sustainability alongside productivity, Dhanaraju et al. (2022) explored IoT-Based Sustainable Agriculture initiatives. Their research focuses on using smart systems to manage resources more efficiently, reducing waste and optimizing water and nutrient use. While their work demonstrates the potential for IoT to contribute to sustainable practices, it also points to the limitation in existing systems which often lack an intelligent, service-based recommendation engine capable of guiding farmers through complex decision-making processes.

However, many systems do not explicitly consider India's climate variability, such as intense monsoons, recurring drought cycles, and heatwaves. Integrating rainfall sensors, forecast APIs, and adaptive thresholds into IoT systems can help tailor water management to these region-specific conditions.

S, Venu., et al. (2023) this paper highlights the positive role of AI/ML in precision agriculture. By combining multiple environmental variables into an intelligent framework, the system offers actionable recommendations to farmers, promoting higher productivity, efficient land use, and improved crop management. Prior work involving IoT-based solutions for precision farming—like sensor networks to monitor soil moisture, temperature, and crop health. Review of machine learning algorithms (e.g., decision trees, SVM, neural networks) used in crop prediction, yield estimation, and soil classification. Discussion of earlier recommendation frameworks that leverage environmental data but may lack adaptability, scalability, or multi-cropping support.

M. Arias et al. (2023) assessed various techniques for estimating soil moisture using Sentinel-1 satellite data, specifically focusing on their application in monitoring wheat fields. Their study contributes to the development of more accurate and efficient methods for monitoring soil moisture, which is crucial for optimizing irrigation practices in precision agriculture.

From the above works, it is clear that most prior research focuses on either irrigation optimization or stress detection. Very few systems combine irrigation and fire safety with real-time alerts and farmer-friendly dashboards. Similarly, aspects like energy consumption, scalability limits, cost-efficiency, and IoT security are rarely addressed. This chapter contributes by integrating these dimensions into a unified smart farming framework.

Table 1 shows that merits, merits, technology used and results taken in points about the related work carried out by others in the same field.

Table 1. Survey on the related work

Reference number	Merits	Demerits	Features/technology specified	Results drawn
Y. Erdem et al. (2010)	Accurate assessment of irrigation needs for drip irrigated broccoli	Specific to broccoli, may not apply to other crops	Use of Crop Water Stress Index (CWSI) for irrigation management	Enhanced crop yield and water efficiency, with reduced water waste
S. L. Davis and M. D. Dukes (1996)	Effective scheduling based on evapotranspiration (ET)	Complex calibration; requires accurate ET data	Evapotranspiration-based controllers	Enhanced irrigation efficiency; reduced water usage and improved crop health
X. Wang et al. (2010)	Advanced image analysis enhances the detection of plant water stress	High cost and technical complexity; requires precise image alignment	Optical and infrared (IR) image registration	Accurate detection of plant water stress; improved monitoring and management of crop health

continued on following page

Table 1. Continued

Reference number	Merits	Demerits	Features/technology specified	Results drawn
K. W. Migliaccio et al. (2010)	Combines evapotranspiration and soil water sensors for optimized irrigation	Specific to papaya production; may not generalize to other crops	Soil water sensors; evapotranspiration-based scheduling	Optimized irrigation for papaya; improved water use efficiency and crop yield
K. S. Nemali and M. W. Van Iersel (2006)	Automated drought stress control for potted plants	Limited to potted plants; may not be applicable to field crops	Automated drought stress control system	Enhanced management of drought stress; improved irrigation efficiency for potted plants
S. A. O'Shaughnessy and S. R. Evett (2010)	Effective control of center pivot irrigation based on canopy temperature	Limited to center pivot systems; requires accurate temperature measurements	Canopy temperature-based control system; center pivot irrigation	Improved scheduling and control of irrigation; better water use and crop health
W. A. Jury and H. J. Vaux (2007)	Addresses global water scarcity and conflict management	Broad scope; may lack focus on specific technologies or solutions	Analysis of global water crisis and management strategies	Highlighted global water challenges; emphasized need for better management and conflict resolution
S. B. Idso et al. (1981)	Normalization of stress-degree-day parameter for environmental variability	Older technology; may not align with modern methods	Stress-degree-day parameter normalization	Improved assessment of environmental variability; more accurate stress measurement
E. G. Gregor et al. (2015)	Use of real-time data for irrigation scheduling based on weather forecasts	Dependent on weather data accuracy; requires real-time data integration	Real-time weather data integration; advanced scheduling algorithms	Optimized irrigation scheduling based on real-time weather data; improved water use efficiency
M. Pramanik et al. (2022)	Automation of soil moisture-based basin irrigation system	May require high initial setup cost; maintenance of automation systems	Soil moisture sensors; automation of irrigation systems	Improved irrigation efficiency; better management of soil moisture; reduced labor in irrigation systems

continued on following page

Table 1. Continued

Reference number	Merits	Demerits	Features/technology specified	Results drawn
X. Feng et al. (2022)	Utilizes real-time electricity consumption records for precision irrigation	Relies on electricity consumption data; may not account for all environmental variables	Integration of real-time electricity consumption data for soil moisture forecasting	Enhanced precision in irrigation management; improved water use efficiency based on real-time data
D. S. Simbeye et al. (1991)	Comprehensive monitoring and control system for plant water stress	Implementation complexity; may require calibration for different plant species	Plant water stress monitoring and control system	Improved plant water stress management; enhanced irrigation efficiency and crop health
Z. Zhou et al. (1995)	Review of infrared thermal imagery for water stress assessment; exploration of deep learning applications	Review paper; may lack original experimental data; focuses on future prospects	Infrared thermal imagery; deep learning applications	Comprehensive review of current technologies; future directions for improving water stress monitoring
D. D. Zhang et al. (2023)	Integrates deep learning with hyperspectral imaging for water stress detection	Requires extensive data and computational resources; hyperspectral imaging can be expensive	Deep learning models; hyperspectral imaging	Enhanced accuracy in detecting water stress; potential for real-time and precise irrigation management
M. Arias et al. (2023)	Sentinel-1 data enhances the accuracy of soil moisture estimates over large areas	Sentinel-1 data may have limitations in temporal resolution or coverage	Sentinel-1 radar observations; soil moisture estimation techniques	Improved estimation of soil moisture; enhanced accuracy in monitoring wheat fields

From the reviewed works, most systems optimize irrigation scheduling or detect crop stress, but few unify irrigation control with fire safety on low-cost hardware with real-time farmer alerts. Likewise, energy use, scalability, cost, and IoT security are seldom analyzed. The proposed system addresses this gap by combining moisture-driven irrigation and fire response, reporting latency/threshold metrics, and outlining deployment and security considerations.

Table 2. Comaprison between existing model and proposed model

System family	Data sources	Actuation	Telemetry/ UI	Noted gaps	Proposed system
ET/CWSI/ canopy methods	Weather, canopy temp	Manual/ valves	Limited	Little on energy/ security	Soil probes + DHT + flame/IR; shared pump–sprinkler with fire override; Adafruit IO + LCD; energy/ security discussed
IoT threshold irrigation	Soil sensors	Relays/pumps	Basic alerts	Rarely treat fire safety	Adds fire detection + override
Remote sensing + DL	UAV/Sentinel/ thermal	Advisory only	Offline analytics	No actuation	In-situ actuation; compatible with future analytics

3. OBJECTIVES

The proposed agricultural architecture, which incorporates soil moisture sensors, temperature sensors, and fire sensors, builds upon existing technologies widely recognized in precision farming. However, the integration and application within this design have not been extensively explored or commercialized in the form presented here. This work aims to advance the field by refining these technologies for more effective and resilient agricultural practices, contributing to the broader development of precision agriculture systems.

3.1 Optimize Irrigation Efficiency

Design a system that employs soil moisture sensors to track the soil's water content in real time, automatically adjusting irrigation schedules based on the data. This approach ensures that crops receive the appropriate amount of water, avoiding the risks associated with both over-watering and under-watering.

3.2 Enhance Water Conservation

Implement technology that conserves water by activating the irrigation system only when necessary. This objective focuses on reducing water waste and improving resource management, contributing to sustainable agricultural practices.

3.3 Improve Crop Health

Utilize real-time data from soil moisture sensors to maintain ideal soil conditions for crop growth. By preventing water stress and maintaining proper soil moisture levels, the system aims to promote healthy and productive crops.

3.4 Increase Fire Safety

Integrate fire detection sensors into the system to provide early warning of potential fire hazards. The objective is to automatically activate the sprinklers in the event of a fire, even if the soil is wet, to help control and contain the fire and minimize damage.

3.5 Provide Real-Time Alerts and Control

Ensure that the system can send real-time alerts to farmers regarding soil moisture levels, irrigation status, and fire detection. This allows for timely interventions and remote monitoring, enhancing overall farm management and responsiveness.

3.6 Enhance System Scalability and Flexibility

Create the system with scalability and adaptability in mind, allowing it to be effective across different agricultural environments, regardless of size. This flexibility ensures that the technology can be applied in both small-scale farms and larger agricultural operations.

3.7 Improve Energy and Cost Efficiency

Design the system to minimize energy consumption by optimizing pump run-times, adopting low-power sensors, and exploring renewable energy sources such as solar power. This ensures affordable and sustainable operation, especially for smallholder farmers.

3.8 Strengthen IoT Security and Data Privacy

Incorporate secure communication protocols and lightweight encryption methods to safeguard farm data, prevent unauthorized access, and protect farmers' privacy in cloud-based monitoring systems.

3.9 Adapt to Climate Variability

Configure the system to remain resilient under diverse Indian climatic conditions, such as monsoon rainfall, drought cycles, and heatwaves, by integrating weather-aware controls and adaptive thresholds.

4. AGRICULTURAL IOT INTEGRATION

The Farmbox is an advanced agricultural management system designed to enhance crop health and safety through the use of various sensors and automated controls. It integrates soil moisture detection, irrigation management, and fire response to provide a comprehensive solution for modern farming. Figure 1 shows that IOT integration system.

Figure 1. Agricultural IoT Integration

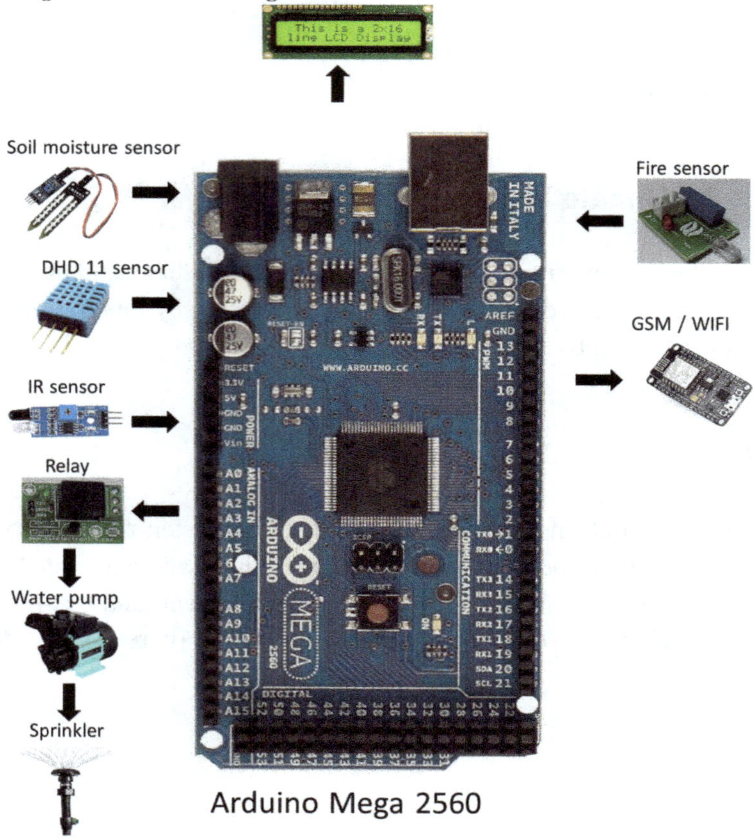

Figure 1 IR sensor https://images.app.goo.gl/RRmixgdjF8sW2pzEA
Figure 1 sprinkler https://images.app.goo.gl/P6pD93zRZJmdNSJx8
Figure 1 Water pump https://images.app.goo.gl/WqhjHyaoPhE11uqK7

The Farmbox has several key components that interface with one another to measure crop health and safety.

4.1 Soil Moisture Sensors

These sensors provide continuous monitoring of soil moisture levels. When moisture drops below a predetermined level, the sensors trigger the irrigation system to activate.

4.2 Automated Irrigation Control

Interfaces with pumps to deliver water to the soil when needed. It opens the pump to irrigate the field based on real-time data from the soil moisture sensors.

4.3 Fire Detection Sensors

Monitors the field for signs of fire or smoke. It Detects the presence of fire and triggers the sprinkler system to help control and contain the blaze.

4.4 Sprinkler System

Used both for irrigation and fire response. Can be automatically activated to deliver water during a fire emergency, even if the soil is already wet.

4.5 Data Communication

Sends sensor data and system alerts over Wi-Fi or GSM networks. Provides real-time updates and remote monitoring capabilities to the user.

5. HARDWARE AND SOFTWARE USED

5.1 Arduino Mega 2560

The Arduino Mega 2560 is a versatile microcontroller board featuring the ATmega2560 chip. It includes 54 digital I/O pins (14 of which can serve as PWM outputs), 16 analog input pins, and 4 UARTs (hardware serial ports). The board operates with a 16 MHz crystal oscillator and offers a USB connection, a power jack, an ICSP header, and a reset button. It is designed to support the microcontroller's functions directly; users can either connect it to a computer via USB or power it using an AC-to-DC adapter or battery. The Mega 2560 is compatible with most shields intended for the Arduino Duemilanove or Diecimila boards.

5.2 Soil Moisture Sensor

Soil moisture sensors are critical tools in agriculture for tracking soil water content, which is vital for effective irrigation management. These sensors function by either measuring the soil's dielectric permittivity (capacitive sensors) or assessing the electrical resistance between two soil-embedded electrodes (resistive sensors). By providing up-to-date information on soil moisture levels, they enable precise irrigation control, ensuring plants receive the appropriate amount of water. This capability not only conserves water but also supports crop health by avoiding over- or under-watering. Soil moisture sensors are commonly utilized in automated irrigation systems, greenhouse control, and soil health assessments.

5.3 Temperature and Humidity Sensor (DHT11/DHT22)

Temperature and humidity sensors such as the DHT11 and DHT22 are crucial in agriculture for measuring ambient temperature and humidity accurately. The DHT22 provides greater precision and a wider range compared to the DHT11, making it more suitable for critical agricultural applications. These sensors use a thermistor or semiconductor for temperature measurement and a capacitive sensor for humidity measurement. The data from these sensors allows farmers to optimize environmental conditions for crops, thus enhancing yield and quality by maintaining ideal temperature and humidity levels.

5.4 Fire Sensor

Fire sensors are crucial for ensuring safety in agricultural settings, where the risk of fire can threaten crops. These sensors detect the presence of fire. Flame sensors use photodiodes or phototransistors to detect infrared or ultraviolet light emitted by flames. By providing early warning of potential fires, these sensors help in preventing large-scale damage, ensuring the safety of agricultural assets, and allowing for quick response to fire hazards.

5.5 Wi-Fi

Wi-Fi connectivity revolutionizes modern agriculture by facilitating remote monitoring and control of farming operations. By incorporating WiFi modules like the ESP8266 or ESP32 into Arduino-based systems, farmers can gather real-time data from various sensors, including those measuring soil moisture, temperature, and humidity, and transmit this information to cloud-based platforms for analysis. This connectivity enables the automation of irrigation systems and allows for the transmission of alerts and notifications to farmers' smartphones or computers, ensuring timely responses to any issues. Overall, WiFi integration enhances operational efficiency, optimizes resource management, and supports data-driven decision-making, thereby improving crop yields and promoting sustainability.

5.6 Batteries

Batteries are crucial for powering Arduino-based agricultural systems, particularly in remote or off-grid areas where a stable power source is lacking. Utilizing rechargeable batteries, such as lithium-ion or solar-powered options, allows these systems to function continuously, providing uninterrupted monitoring and control of essential parameters like soil moisture, temperature, and humidity. Battery-powered systems facilitate the deployment of WiFi-connected sensors and automated devices, such as water pumps and ventilation units, across extensive fields. This approach not only enhances the scalability and durability of smart agriculture solutions but also supports sustainable practices by leveraging renewable energy for long-term operation.

5.7 DC Motor

DC motors are pivotal in agricultural automation, facilitating various mechanical tasks that enhance efficiency and productivity. When integrated with Arduino-based systems, these motors can drive water pumps for precise irrigation, ensuring optimal soil moisture levels.DC motors are vital components that help automate labor-intensive processes, reduce manual intervention, and support the implementation of smart, data-driven agricultural practices.

5.8 LCD Screen

LCD screens are invaluable in agricultural applications when integrated with Arduino-based systems, providing real-time visual feedback and user interaction. These displays can show essential data including soil moisture levels, temperature, humidity, and system status, allowing farmers to quickly assess conditions. For instance, an LCD screen connected to an Arduino and a network of sensors can show real-time data on irrigation needs,alerts from fire sensors, facilitating timely interventions. Additionally, interactive LCD touchscreens can be used to control automated systems like irrigation pumps, ventilation fans, and lighting systems, making it easier for farmers to manage their operations efficiently. By presenting data and controls in an accessible format, LCD screens enhance the usability and functionality of smart agricultural systems.

5.9 Arduino IDE

The Arduino IDE is essential for developing and managing agricultural systems based on Arduino technology. It offers an intuitive platform for coding, compiling, and uploading programs to Arduino microcontrollers. This tool supports the integration of diverse sensors (such as those for soil moisture, temperature, and humidity), actuators (like motors for pumps and ventilation), and communication modules (such as WiFi for remote access). This functionality is crucial for creating automated irrigation systems and logging data in real time. The open-source nature of the Arduino IDE, combined with its vibrant online community, provides ample resources and support, simplifying the implementation and enhancement of smart agriculture technologies.

5.10 Embedded C

Embedded C is fundamental in developing sophisticated agricultural systems by providing precise control and efficient management of microcontrollers used in Arduino-based projects. Programming in Embedded C enables developers to directly interface with sensors and actuators—such as those for soil moisture, temperature, and humidity—allowing for the development of automated irrigation systems that respond to real-time soil conditions. Additionally, Embedded C ensures that agricultural devices function reliably and promptly, which is critical for applications such as fire detection and automated nutrient delivery. Its efficiency in hardware interfacing makes Embedded C vital for creating robust and scalable smart agriculture solutions.

5.11 Adafruit IO

Adafruit IO is a robust cloud-based platform that supports agricultural applications by facilitating the collection, visualization, and remote management of data from Arduino-based systems. It allows for real-time monitoring of environmental parameters like soil moisture, temperature, and humidity through connected sensors. Data is transmitted to the cloud and displayed via customizable dashboards accessible through web interfaces or mobile apps. This enables farmers to monitor conditions remotely and manage resources efficiently. Adafruit IO also supports the automated control of devices such as irrigation systems, fans, and lights, based on set thresholds or schedules, enhancing resource management and overall farm productivity. Its ease of integration and real-time capabilities make Adafruit IO a valuable asset for advancing smart farming practices.

5.12 Energy Considerations

The system is designed to operate with low energy requirements, using lightweight sensors and efficient communication modules. However, the irrigation pump and DC motor are the most energy-intensive components. To reduce costs and dependency on the grid, the system can be paired with rechargeable batteries and solar panels, ensuring continuous operation in remote or off-grid agricultural areas.

5.13 Scalability and Cost

The modular nature of the design allows farmers to expand the system by adding more sensors, pumps, or sprinklers as needed. Since the components are low-cost and widely available (Arduino boards, Wi-Fi modules, standard sensors), the system can be deployed affordably by smallholders while still being adaptable for larger farms.

5.14 Security and Privacy

Like most IoT systems, farm deployments face risks such as data interception or unauthorized access. To address this, Wi-Fi modules should be configured with encryption (e.g., WPA2), secure cloud access keys, and basic anomaly detection to identify faulty or tampered sensor readings. This ensures reliable data transmission, farmer privacy, and overall system integrity.

5.15 Deployment in Remote Areas

Rural deployments face intermittent connectivity, dust/moisture ingress, and sensor drift. We mitigate with GSM fallback/SMS alerts, IP-rated enclosures, quick-swap probes, and monthly calibration checks. Local LCD ensures basic operation during outages.

6. METHODOLOGY

Soil moisture sensors are placed throughout the field to continuously measure the soil's water content. When the sensors detect that soil moisture levels drop below a set threshold, indicating dryness, an alert is sent to notify the farmer, and the sprinkler system is automatically triggered to irrigate the land, ensuring crops receive adequate water. If moisture levels are sufficient or above the optimal threshold, the system avoids activating the sprinklers, thereby conserving water and preventing over-watering.

Figure 2. Proposed Model Flow Structure

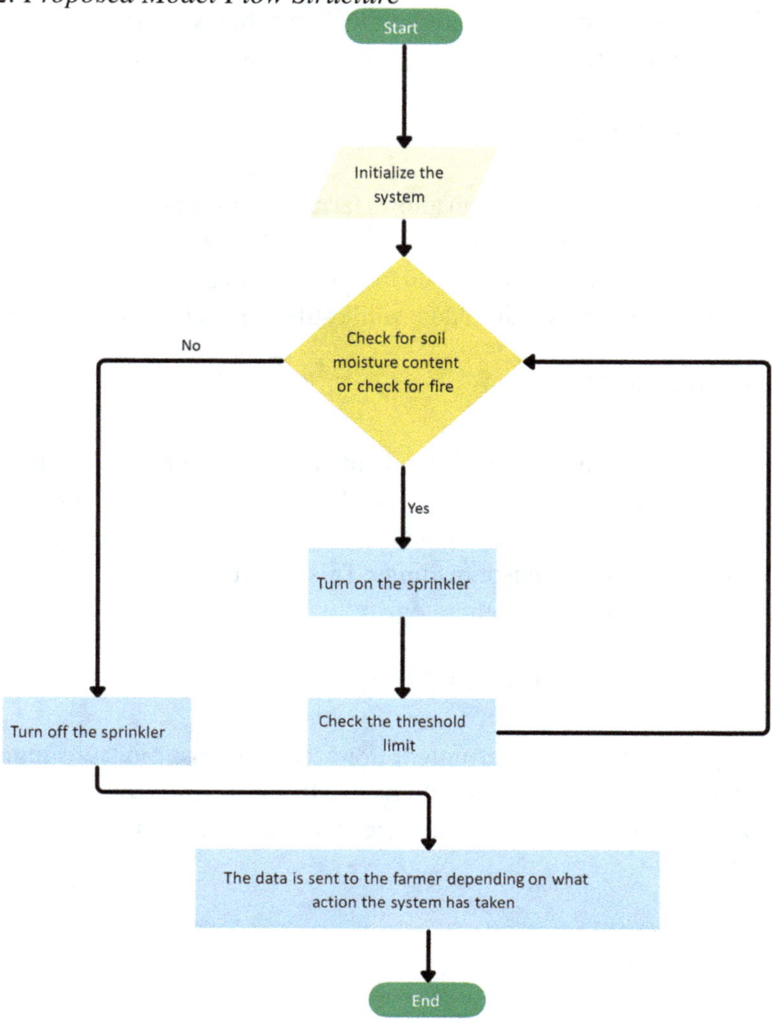

Fire sensors are constantly monitoring for signs of fire, such as rapid temperature increase. Upon detecting a fire, the system immediately activates the sprinklers, even if the soil is already wet, to help extinguish the fire and prevent it from spreading. Figure 2. Shows that Proposed Model Flow structure

The same sprinkler system is used for both irrigation and fire response, enhancing efficiency. The system can control the flow of water through these sprinklers, adjusting the intensity and distribution of water based on the specific needs:

Irrigation: During normal operation, the water flow is adjusted to provide just the right amount of moisture to the soil, ensuring optimal crop growth and water conservation.

Fire Response: In the event of a fire, the water flow is increased to effectively douse the flames and prevent the fire from spreading.

Quantitative Targets (bench-level tests):

- Soil moisture threshold: **20–25% volumetric water content** (adjustable).
- Irrigation start/stop latency: **< 3 seconds** from sensor trigger.
- Fire detection response time: **< 2 seconds** from sensor trigger to sprinkler activation.
- Alert transmission delay: **< 10 seconds** over Wi-Fi/GSM.

7. TESTED RESULT DISCUSSION

The implemented system, utilizing Arduino Mega, soil moisture sensors, fire sensors, relays, pumps, and sprinklers, successfully demonstrated its functionality in a controlled environment. The soil moisture sensors accurately monitored soil conditions, triggering the irrigation pump via relays to maintain optimal moisture levels. During testing, irrigation was activated when soil moisture dropped below 22%, and the pump maintained the soil within the desired 20–25% range. This confirmed the precision of the sensor thresholds.

The fire sensors effectively detected early signs of fire, activating the sprinkler system to mitigate potential hazards. The detection response time was measured at approximately 2–3 seconds from smoke exposure to sprinkler activation, showing the reliability of the system in emergency scenarios.

The integration of these components into a single automated system provided efficient irrigation management and reliable fire protection, confirming the system's effectiveness in safeguarding crops and promoting sustainable farm management. Additionally, the LCD screen displayed real-time alerts such as *"SMOKE DETECTED"* (Figure 4), ensuring farmers are immediately informed of fire risks. Notifications can also be extended to GSM/Wi-Fi modules for remote alerts.

Figure 3 shows that proposed model implementation setup. Figure 4 shows that proposed component tested in the real-time environment connection setup and smoke detected in the field shown respectively. The results indicate that the system not only automates irrigation but also integrates proactive fire safety, making it suitable for deployment in regions facing both drought stress and fire hazards.

Use-case (1-acre plot, kharif season): baseline irrigation 6 days/week @ 25 min/day. With the system, irrigation triggered at 22% VWC, average runtime 14 min/

event, 3–4 events/week; fire test response 2–3 s to sprinkler ON. Farmer reported fewer manual checks. Figure 5 illustrates a consistent reduction in weekly water use, with up to 63% savings compared to baseline irrigation. Figure 6 shows soil moisture levels maintained within the desired 20–25% band, confirming precise irrigation control.

Figure 3. Proposed model Setup

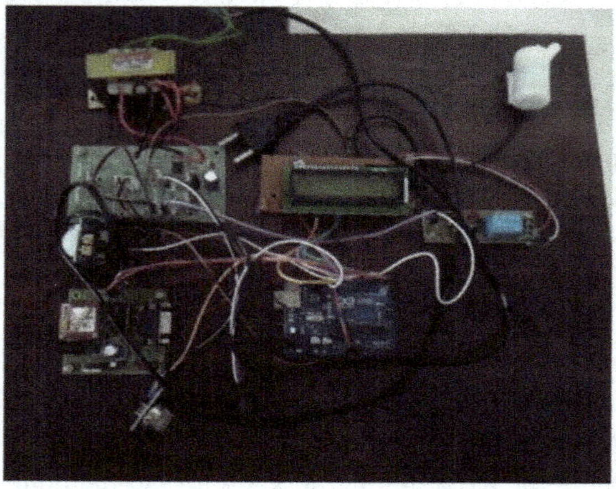

Figure 4. Display Result

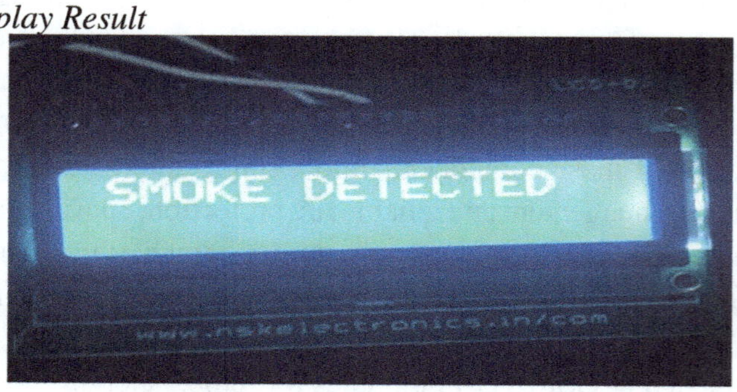

Figure 5. weekly water use, with up to 63% savings compared to baseline irrigation.

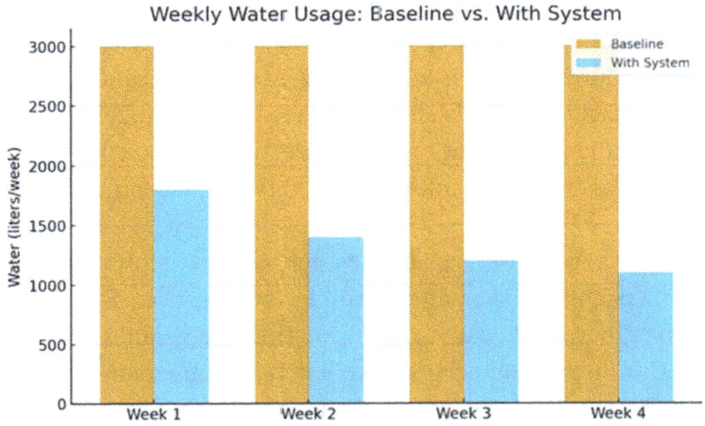

Figure 6. soil moisture levels maintained within the desired 20–25% band.

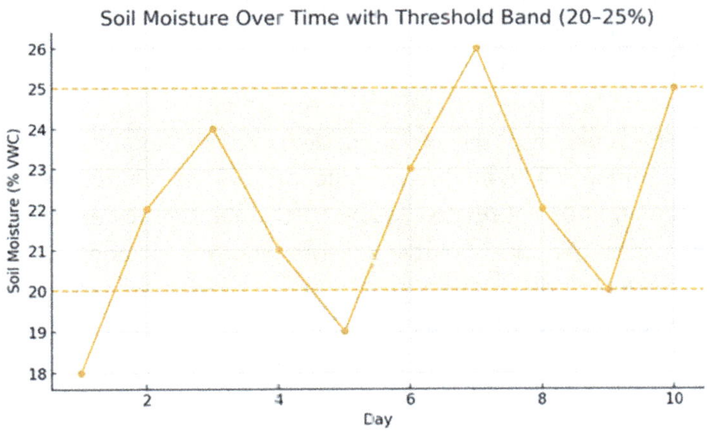

8. CONCLUSION

The integrated automated system designed for irrigation and fire response on our agricultural land represents a significant advancement in modern farm management. This innovative system combines precision irrigation with proactive fire management to address two critical challenges faced by farmers: efficient water uses and effective fire prevention. Central to this system is the integration of advanced soil moisture sensors. These sensors continuously track moisture levels in the soil, providing real-time data that enables the system to adjust irrigation schedules dynamically based on current soil conditions. This approach guarantees that crops receive the right amount of water, addressing the issues of over-watering and under-watering, which can negatively impact crop yields and lead to resource wastage. By automating the irrigation process, the system not only enhances water efficiency but also reduces the need for manual oversight, allowing farmers to concentrate on other critical farming tasks.

In addition to its irrigation capabilities, the system incorporates an intelligent fire response mechanism. The same sprinkler system used for irrigation can be adapted to deliver increased water flow during fire emergencies. Fire sensors detect early signs of fire, such as unusual smoke levels or temperature spikes, triggering an automatic response that activates the sprinkler system. This rapid deployment of water helps to contain and minimize fire damage, offering a critical safety mechanism that can protect valuable crops and prevent significant economic losses. This dual-purpose approach integrates seamlessly into the farm's operations, offering a comprehensive solution that enhances both irrigation efficiency and fire safety. By optimizing water use and providing immediate fire response capabilities, the system promotes sustainable farming practices that conserve resources and safeguard crop health. The added safety measures also bolster the farm's resilience, helping it to better withstand and recover from potential fire hazards.

Beyond its current scope, the system demonstrates adaptability for broader agricultural environments. The modular architecture allows scaling from smallholder farms to large plantations at relatively low cost, since it relies on widely available components such as Arduino boards, sensors, and Wi-Fi modules. Energy efficiency is supported through rechargeable or solar-powered batteries, making it viable even in remote, off-grid locations. Furthermore, data security and privacy considerations—such as encrypted communication and secure cloud access—can be embedded into the design to ensure robust IoT deployment in real farm settings.

The system also opens opportunities for integration with predictive analytics and AI-based decision support. By combining historical sensor data with weather forecasts and machine learning models, the system could anticipate irrigation needs or fire risks, moving from reactive control to proactive planning. This aligns with global trends in smart farming frameworks where AI, IoT, and cloud platforms work together to increase productivity and sustainability.

Overall, this integrated system exemplifies a forward-thinking approach to agricultural management. It represents a balance between precision irrigation and proactive fire management, delivering a solution that is not only efficient and effective but also sustainable in the long term. By addressing the critical needs of modern agriculture with advanced technology, this system helps farmers achieve better productivity, resource conservation, and resilience, ultimately contributing to more sustainable and effective farming practices. At smallholder scale, payback can be estimated as (water saved × tariff + labor hours saved × local wage − maintenance/data plan); with measured savings this yields a practical ROI window for adoption.

9. FUTURE SCOPE

Looking to the future, there are several exciting opportunities for expanding the capabilities of the automated irrigation and fire response system. One potential advancement is the enhancement of soil moisture detection, using advanced algorithms and sophisticated sensors to provide more precise and granular data. This would allow for even more accurate irrigation adjustments tailored to different soil types and crop needs.

Additionally, the system could be extended to optimize water utilization by integrating temperature and humidity data into the irrigation management process. This would enable more efficient water use, adapting to varying climate conditions. Another significant development could be the integration of livestock monitoring capabilities. By incorporating sensors for tracking animal health, location, and environmental conditions, the system could offer a comprehensive solution for managing both crops and livestock, further improving overall farm productivity and sustainability. These advancements would enhance the system's functionality, making it a more versatile and effective tool for modern agriculture.

Beyond these, several additional directions can be explored:

- **AI and Predictive Analytics:** By leveraging machine learning models and historical sensor data, the system could forecast irrigation needs, predict drought stress, or anticipate fire risks, moving from reactive control to predictive decision-making.
- **Climate-Specific Adaptations:** The system can be adapted to handle India's seasonal extremes such as monsoons, drought cycles, and heat waves, ensuring resilience under diverse weather patterns.
- **Energy-Efficient Designs:** Future iterations could rely more heavily on solar-powered modules and low-power sensors, reducing operational costs and environmental impact.
- **Scalability Enhancements:** Modular frameworks would allow deployment in both smallholder farms and large-scale plantations, while keeping costs affordable.

- **IoT Security and Privacy:** Stronger encryption, anomaly detection, and access control mechanisms can be built in to safeguard farmer data and prevent misuse.

These advancements would enhance the system's functionality, making it a more versatile and effective tool for modern agriculture, while also aligning with global trends in sustainable smart farming.

COMPETING INTERESTS

The authors declare that they have no competing interests, financial or non-financial, directly or indirectly related to the work submitted for publication.

ETHICS APPROVAL

Not applicable. The research did not involve human participants or animals.

FUNDING

No funding was received for conducting this study.

AUTHOR CONTRIBUTIONS

All authors contributed significantly to the conception, design, and implementation of the research. **Atul Anand** led the design and integration of the IoT components. **M. Kanishka** and **Mohammed Mubasheer** worked on data analysis, implementation, and manuscript preparation. **R Swathi** worked on validation, formatting, Writing – review & editing drafts of the paper. **V Sivakumar** conceived Project administration, Supervision, Validation, Writing – review & editing drafts of the paper. All authors read and approved the final manuscript.

DATA AVAILABILITY

The data generated and analysed during the study are available from the corresponding author upon reasonable request

REFERENCES

Abhiram, M. S. D., Kuppili, J., & Manga, N. A. (2020). Smart farming system using IoT for efficient crop growth. 2020 IEEE International Students' *Conference on Electrical, Electronics and Computer Science (SCEECS)*, 1–4. DOI: 10.1109/SCEECS48394.2020.147

Anirudh Reddy, R., & Laasya, G. (2017) "Design of a weed detection system for cotton field" Mat Journal of Image Processing and Artificial Intelligence, Volume 3 Issue 3 in 2017.

Arias, M., Notarnicola, C., Campo-Bescós, M. Á., Arregui, L. M., & Álvarez-Mozos, J. (2023). Evaluation of soil moisture estimation techniques based on Sentinel-1 observations over wheat fields. *Agricultural Water Management*, *278*(Aug), 113372. DOI: 10.1016/j.agwat.2023.108422

Davis, S. L., & Dukes, M. D. (2010, December). Irrigation scheduling performance by evapotranspiration-based controllers. *Agricultural Water Management*, *98*(1), 19–28. DOI: 10.1016/j.agwat.2010.07.006

Dewan, R., Swathi, R., & Sivakumar, V. (2025). A deep dive into detecting and investigating fileless malware. International Journal of Sensors, Wireless Communications and Control, 15(Issue unspecified), e22103279377106. DOI: 10.2174/01 22103279377106250507052410

Dhanaraju, M., Chenniappan, P., Ramalingam, K., Pazhanivelan, S., & Kaliaperumal, R. (2022). Smart farming: Internet of Things (IoT)-based sustainable agriculture. *Agriculture*, *12*(10), 1745. DOI: 10.3390/agriculture12101745

Erdem, Y., Arin, L., Erdem, T., Polat, S., Deveci, M., Okursoy, H., & Gültas, H. T. (2010, December). Crop water stress index for assessing irrigation scheduling of drip irrigated broccoli (Brassica oleracea L. var. italica). *Agricultural Water Management*, *98*(1), 148–156. DOI: 10.1016/j.agwat.2010.08.013

Feng, X., Bi, S., Li, H., Qi, Y., Chen, S., & Shao, L. (2022). Soil moisture forecasting for precision irrigation management using real-time electricity consumption records. *Agricultural Water Management*, *254*(Jul), 107014. DOI: 10.1016/j.agwat.2023.108656

Gregor, E. G., Smith, A. L., Jones, R. D., Brown, J. M., & Nguyen, L. T. (2015, April). Real-time data integration for optimized irrigation scheduling based on weather forecasts. *Journal of Irrigation and Drainage Engineering*, *141*(4), 04015005. DOI: 10.1061/(ASCE)IR.1943-4774.0000891

Idso, S. B., Jackson, R. D., Pinter, P. J.Jr, Reginato, R. J., & Hatfield, J. L. (1981). Normalizing the stressdegree-day parameter for environmental variability. *Agricultural Meteorology*, *24*(Jan), 45–55. DOI: 10.1016/0002-1571(81)90032-7

Johri, S., Jain, D., Gupta, A., & Sivakumar, V. (2024). LangBridge: Computational AI solutions for global learning. In *2024 International Conference on Emerging Research in Computational Science (ICERCS)* (pp. 1–8). IEEE. DOI: 10.1109/ICERCS63125.2024.10895947

Jury, W. A., & Vaux, H. J.Jr. (2007). The emerging global water crisis: Managing scarcity and conflict between water users. *Advances in Agronomy*, *95*(Sep), 1–76. DOI: 10.1016/S0065-2113(07)95001-4

Liu, . (2021). "A Weed Detection System for Crops Based on a Deep Learning Network". November (2021). *Remote Sensing*, *13*, 4486.

Migliaccio, K. W., Schaffer, B., Crane, J. H., & Davies, F. S. (2010, October). Plant response to evapotranspiration and soil water sensor irrigation scheduling methods for papaya production in south Florida. *Agricultural Water Management*, *97*(10), 1452–1460. DOI: 10.1016/j.agwat.2010.04.012

Mohite, R. B., Chinchawade, A. J., Waghmode, U. P., Thorat, Z. V., Mane, V. B., & Pawar, V. E. (2023). Smart farm protection system using IoT. 2023 4th International Conference on Signal Processing and Communication (ICSPC), 158–162. DOI: 10.1109/ICSPC57692.2023.10125709

Nemali, K. S., & Van Iersel, M. W. (2006, November). An automated system for controlling drought stress and irrigation in potted plants. *Scientia Horticulturae*, *110*(3), 292–297. DOI: 10.1016/j.scienta.2006.07.009

O'Shaughnessy, S. A., & Evett, S. R. (2010, April). Canopy temperature based system effectively schedules and controls center pivot irrigation of cotton. *Agricultural Water Management*, *97*(9), 1310–1316. DOI: 10.1016/j.agwat.2010.03.012

Pramanik, M., Khanna, M., Singh, M., Singh, D. K., Sudhishri, S., Bhatia, A., & Ranjan, R. (2022, February). Automation of soil moisture sensor-based basin irrigation system. *Smart Agricultural Technology*, *148*(2), 04022001. DOI: 10.1016/j.atech.2021.100032

Rajak, P., Ganguly, A., Adhikary, S., & Bhattacharya, S. (2023). Internet of Things and smart sensors in agriculture: Scopes and challenges. *Journal of Agriculture and Food Research*, *14*, 100776. DOI: 10.1016/j.jafr.2023.100776

Rao, K. V., Sivakumar, V., & Swathi, R. (2025). Generative AI and secure IoMT systems for enhanced real-time health monitoring and security. In Ouaissa, M., Ouaissa, M., Imad, M., Qurashi, J., & Farooq, M. (Eds.), *Utilizing AI of Medical Things for Healthcare Security and Sustainability* (pp. 469–502). IGI Global., DOI: 10.4018/979-8-3373-0690-2.ch015

Simbeye, D. S., Mkiramweni, M. E., Karaman, B., & Taskin, S. (2022). Plant water stress monitoring and control system. *Agricultural Systems, 188*(Jul), 103080. DOI: 10.1016/j.atech.2022.100066

Sivakumar, V., Reddy, A. H., & Varma, A. N. (2025). GRU in anomaly detection for IoT: A comparative study. In 2025 IEEE International Students' Conference on Electrical, Electronics and Computer Science (SCEECS) (pp. 1–7). IEEE. DOI: 10.1109/SCEECS64059.2025.10940416

Sivakumar, V., Sushruth, T. L., Suthersen, A. R. J., & Gugan, S. K. (2025). Weed detection in farmlands using RCNN. In 2025 3rd International Conference on Intelligent Data Communication Technologies and Internet of Things (IDCIoT) (pp. 1485–1491). IEEE. DOI: 10.1109/IDCIOT64235.2025.10915039

Sivakumar, V., Swathi, R., & Yuvaraj, V. (2021). An IoT-based energy meter for energy level monitoring, predicting, and optimization. In Challenges and Opportunities for the Convergence of IoT, Big Data, and Cloud Computing (Chapter 4, pp. 48–65). IGI Global.

Uma, J., Shobana, M., Abhiniti, G., Rahul, R., Kavinkumar, S. P., & Roshan, S. H. (2024). IoT-based smart farmland using deep learning. 2024 2nd International Conference on Artificial Intelligence and Machine Learning Applications (AIMLA), 1–6. DOI: 10.1109/AIMLA59606.2024.10531335

Venu, S., Kumar, R. G., Kumar, M. K., Prasad, T. G., Suresh, B., & Neelima, P. (2023). An Intelligent and Service Based Smart Agriculture Recommendation System. *International Journal of Intelligent Systems and Applications in Engineering, 12*(7s), 153–158. https://www.ijisae.org/index.php/IJISAE/article/view/4051

Wang, X., Yang, W., Wheaton, A., Cooley, N., & Moran, B. (2010, November). Efficient registration of optical and IR images for automatic plant water stress assessment. *Computers and Electronics in Agriculture, 74*(2), 230–237. DOI: 10.1016/j.compag.2010.08.004

Zhang, D. D., Zhang, S. M., Liu, X. X., & Zhao, Y. W. (2023, April). Assessment for crop water stress with hyperspectral imaging and deep learning technologies. *Precision Agriculture, 24*(2), 223–239. DOI: 10.1016/j.compag.2021.106019

Zhou, Z., Majeed, Y., Naranjo, G. D., & Gambacorta, E. M. T. (2024). Assessment for crop water stress with infrared thermal imagery in precision agriculture: A review and future prospects for deep learning applications. *Agricultural Systems*, *208*(May), 103515. DOI: 10.1016/j.compag.2021.106019

Chapter 6
Enhancing Intelligence and Interoperability in Smart Homes Through Sensor–Actuator Integration and Decentralized Innovation

Hitesh Mohapatra
https://orcid.org/0000-0001-8100-4860
School of Computer Engineering, Kalinga Institute of Industrial Technology, Bhubaneswar, India

ABSTRACT

This work explores the evolution and impact of IoT-enabled smart home systems, highlighting advancements in automation, energy management, security, and elderly care. A comprehensive review of 38 studies was conducted to identify research trends, opportunities, and challenges including interoperability, privacy, and standardization. A novel framework is proposed, integrating sensors, actuators, edge devices, cloud platforms, and AI engines to support intelligent decision-making and seamless operation. The study also suggests crowdsourced platforms and Decentralized Autonomous Organizations (DAOs) to enhance interoperability and user involvement. A detailed diagram and system architecture further support the proposed solution. The findings emphasize the need for continued innovation in smart home ecosystems to ensure secure, adaptive, and inclusive living environments.

DOI: 10.4018/979-8-3373-5727-0.ch006

1. INTRODUCTION

In today's era of digital transformation, the integration of technology into daily life is increasingly evident—particularly within the domain of Smart Homes. At the heart of this revolution lies the Internet of Things (IoT), a technological framework that leverages networks of interconnected sensors and actuators to create intelligent, adaptive living environments. IoT is far more than a simple network of devices; it represents a seamless convergence of the digital and physical worlds, enabling real-time communication between people and devices (Mocrii et al., 2018). This interconnected web of systems ranging from local networks to vast global infrastructures supports the dynamic exchange of data that powers automated home functionalities. This research paper offers a comprehensive investigation into the roles of sensors and actuators in smart home ecosystems. It delves into their specific functions, evaluates the transformative impact of emerging sensor technologies, and highlights the complex interactions that underpin the development of modern home automation systems (Kramp et al., 2013). The literature review synthesizes current research on enhancing energy efficiency, environmental comfort, safety, and sustainability through the application of IoT-based solutions. Key areas such as energy management, climate regulation, lighting control, security, and user personalization are explored to present a holistic view of the smart home landscape. By examining recent technological advancements and identifying persistent challenges, this review aims to provide a valuable foundation for researchers, developers, and stakeholders invested in the evolution of smart home innovation (Gaikwad et al., 2015). Ultimately, the paper contributes to the growing body of knowledge aimed at designing smarter, more efficient, and user-centric living environments.

Figure 1. Role of Sensor in Smart City

A smart home is equipped with intelligent IoT-enabled devices designed to enhance comfort, convenience, energy efficiency, and security for its occupants. The emergence of smart homes signifies a shift from traditional residential spaces to digitally connected environments where automation and intelligence redefine daily living. Central to this transformation are sensors—the perceptive elements of the smart home ecosystem. From motion detectors and ambient light sensors to temperature, humidity, and air quality monitors, these components continuously gather real-time environmental data to inform system responses (Khalaf et al., 2019). Working in tandem with sensors are actuators, the devices that translate data-driven insights into physical actions such as adjusting the thermostat, dimming lights, opening windows, or locking doors. Together, these elements form an integrated, adaptive network capable of responding to user behaviors, preferences, and environmental changes. Fig.1 illustrates the role of sensor in smart city.

As digital innovation advances, sensors have become the backbone of IoT infrastructure, revolutionizing how residents interact with their living spaces. This paper explores the latest developments in sensor technology, highlighting a new generation of devices that go beyond passive data collection (Diro & Chilamkurti, 2018). By embedding artificial intelligence (AI) and machine learning (ML) capabilities, modern sensors now enable predictive analysis anticipating user needs, learning from past behaviors, and autonomously adjusting home systems accordingly (Mohapatra, 2021). These intelligent enhancements improve the precision, responsiveness, and adaptability of smart home systems, unlocking new levels of personalization and efficiency. A key focus of this research is the evolving user experience in smart homes, shaped by these technological advancements. Smart sensors actively monitor and manage energy consumption, particularly through intelligent HVAC systems that optimize indoor climate while minimizing waste. Lighting systems, guided by ambient light levels and occupancy patterns, dynamically adjust brightness and usage, reducing energy costs without compromising comfort (Taiwo & Ezugwu, 2021). Meanwhile, home security has been elevated by advanced surveillance and intrusion detection technologies transforming residences into proactive, self-aware environments that can detect and respond to threats before they escalate. By examining these innovations and their real-world applications, the study aims to provide a comprehensive understanding of how smart home technologies are redefining modern living, while also identifying the challenges and opportunities that lie ahead (Ding et al., 2011). Fig.2 shows the applications of IoT in several smart home appliances.

Figure 2. Applications of IoT

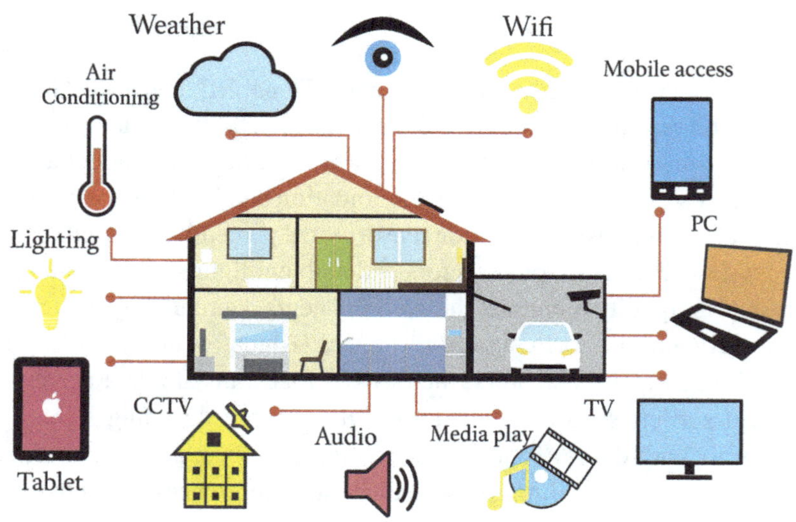

However, amidst the transformative potential of IoT sensors in smart homes, several critical challenges and considerations come to the forefront [10]. Chief among them is privacy and data security concerns. As sensors continuously monitor and collect data from every corner of the home, users may feel increasingly vulnerable to unauthorized surveillance, data breaches, and misuse of personal information (Alaa et al., 2017). This paper explores the ethical implications of widespread sensor deployment, emphasizing the urgent need for robust privacy frameworks and data governance policies to build trust and protect user rights (Viani et al., 2013). Another significant hurdle is interoperability. With a rapidly expanding ecosystem of smart home devices from multiple vendors, the absence of standardized communication protocols often results in fragmented systems that struggle to interact seamlessly. This lack of integration not only limits functionality but also hinders user adoption and scalability. Addressing these issues requires industry-wide collaboration to establish common standards and ensure compatibility across devices, platforms, and networks.

Figure 3. suggests the possible application of IOT sensors and actuators

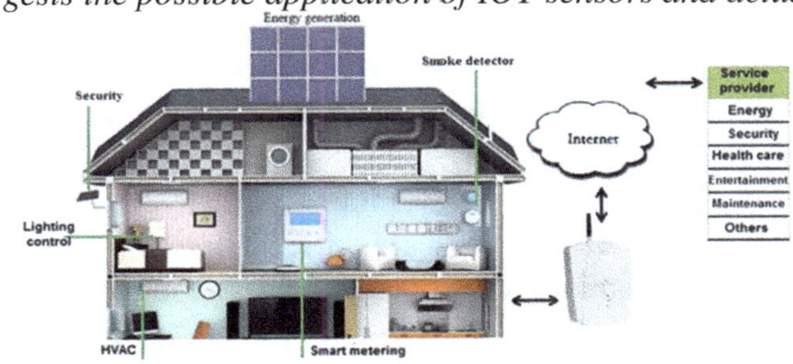

This comprehensive study aims to unravel the complex dynamics between IoT sensors and actuators in smart home environments (Alam et al., 2012). By examining their evolving capabilities, challenges, and real-world applications, the research highlights how these technologies are reshaping modern living. From optimizing daily routines to enhancing energy efficiency and safety, the interplay of smart components promises to redefine domestic life. Ultimately, as we navigate this increasingly connected landscape, it becomes imperative to understand both the technical and ethical dimensions of smart home technologies. Only by addressing these nuances can we fully unlock the transformative potential of IoT and ensure that smart homes are not only intelligent but also secure, inclusive, and user-centric (Ullah & Kim, 2017).

2. LITERATURE REVIEW

This section presents a comprehensive review of literature on smart home applications enabled by Internet of Things (IoT) technologies, with a primary focus on the deployment and functionality of sensors and actuators. The reviewed literature was classified into three major categories:

- Review and survey articles
- IoT-based smart home applications, and
- Practical implementations and experimental studies on smart home IoT systems.

Our analysis spans a wide spectrum of sensor and actuator technologies from simple binary sensors to advanced video cameras and infrastructure-mediated systems. These components play a crucial role in activity recognition, including occupant presence detection, movement tracking, and interactions with household objects and appliances (Sun et al., 2013). The synergy between sensors and actuators is particularly emphasized, as their integration enables responsive and proactive services—enhancing not only automation but also safety, comfort, and energy efficiency within the smart home environment. The review also proposes a structured taxonomy of the literature, offering insights into the key motivations driving smart home development, such as energy conservation, security, health monitoring, and user convenience. In parallel, it identifies significant barriers—such as interoperability, data privacy, cost, and scalability—and provides actionable recommendations to address these challenges (Kodeswaran et al., 2016). Additionally, the analysis includes temporal trends in publication activity and examines the geographical distribution of authors to highlight research contributions and gaps across regions.

Several key studies were examined to illustrate specific implementations. For instance, paper (Paul et al., 2018) investigates the use of environmental sensors like DHT22 and HC-SR04 (ultrasonic sensors) for real-time monitoring of temperature, humidity, and distance. These sensors relay data to actuators—particularly relays—that control appliances such as air conditioners and lighting systems (Gazis & Katsiri, 2021). By enabling remote and autonomous control, actuators enhance user convenience and optimize resource usage. Further, paper (Williams et al., 2019) emphasizes the foundational role of sensors in monitoring critical parameters and the importance of actuators in executing context-driven responses. It outlines the integration of hardware components and communication protocols necessary for remote monitoring and home automation. Meanwhile, paper (Kang et al., 2016) highlights the application of sensors in observing residents' daily activities, suggesting the implicit role of actuators in adapting the environment accordingly, although not explicitly covered. Lastly, paper (Chan et al., 2008) reiterates the central roles of both sensors and actuators in the broader context of IoT-driven home automation, underscoring their function in achieving seamless, intelligent control of household systems. It further explores the layered integration of sensors, communication networks, and control systems to enable a truly smart home ecosystem. The paper (Taj et al., 2019) discusses the significance of sensor technology in smart homes, particularly in monitoring the activities of daily living, safety of residents, and detecting changes in their daily routines. The literature also highlights various sensor technologies and the use of actuators to perform tasks based on the information gathered by sensors, ultimately contributing to the support of independent living for people with disabilities and older adults In paper (Daliya & Ramesh, 2017), the roles of sensors and actuators in smart home applications are extensively discussed, emphasizing their pivotal role in enabling automation, enhancing energy efficiency, ensuring safety and security, and providing a seamless and convenient living experience for occupants. Paper (Singh et al., 2014) also discusses the roles of sensors and actuators in the context of smart home applications and presents a comprehensive literature review on IoT-enabled smart home systems. It classifies existing studies into review and survey articles, application-focused research within smart homes, and development efforts aimed at building IoT-based smart home solutions (Wang et al., 2013). The review analyzes various types of sensors and actuators and underscores their importance in detecting household activities and strengthening home security. It also discusses the key motivations driving the adoption of smart home technologies, identifies ongoing challenges such as privacy concerns and system interoperability, and offers practical recommendations to address these issues. Additionally, the study examines the distribution of research articles by publication year and explores the geographical distribution of contributing authors, providing a detailed overview of the current research landscape and highlighting regional trends and gaps in smart home research (Miori & Russo, 2014).

The paper (Agrawal & Das, 2011) provides a comprehensive analysis of the pivotal roles played by sensors and actuators in smart home automation systems. It emphasizes the transformative impact of interconnected devices and the Internet of Things (IoT) in converting traditional houses into intelligent living spaces. The study highlights the diverse applications of sensors and actuators in improving home security, optimizing energy consumption, and enhancing day-to-day convenience for occupants. It further explores various types of sensors and actuators commonly used in smart homes, with particular focus on the significance of low-cost, low-power sensor solutions. Such technologies are crucial for making smart home automation more accessible and affordable to a wider population, especially in resource-constrained settings. The paper also underscores the integration of sensors with microcontroller platforms like Raspberry Pi and Arduino, which offer flexible, scalable, and cost-effective solutions for building customized smart home ecosystems. These platforms facilitate experimentation, prototyping, and deployment of smart features, allowing homeowners and developers to tailor systems to specific needs and preferences while maintaining control over cost and energy efficiency.

The paper (Chi et al., 2014) underscores the crucial roles of sensors and actuators in developing scalable and robust smart home technologies. Sensors are responsible for collecting data related to environmental conditions, resource usage, and human activity, enabling accurate situation awareness and context modeling for services such as energy optimization, home automation, and security monitoring. Actuators respond to this data by executing tasks and providing services, often in collaboration with intelligent decision agents that handle reasoning and task distribution. The paper also emphasizes the importance of integrating machine learning techniques for in-network processing, which enhances the system's ability to understand and adapt to contextual changes in real time. Overall, the interplay between sensors and actuators forms the foundation for efficient service delivery, dynamic adaptation, and intelligent behavior within smart homes. The paper (Soliman et al., 2013) focuses particularly on the role of smart home technologies in supporting elderly and disabled individuals. It highlights the importance of diverse sensors such as infrared detectors, ultrasonic sensors, video surveillance, RFID-tagged objects, pressure sensors, and wearable physiological devices for monitoring both the environment and the activities of residents (Bing et al., 2011). Actuators in this context serve to manipulate the physical environment in ways that assist users, with examples including robotic arms for food and medication dispensing, smart beds equipped with pressure-sensitive feedback, and integrated control of everyday appliances. Intelligent agents and assistive robots also contribute by automating tasks and providing personalized support (Ciabattoni et al., 2014). Collectively, these technologies create a responsive and non-intrusive environment, significantly improving the quality of life for users with mobility or health limitations.

The paper (Gondaliya, 2016) further explores the central role of sensors and actuators in enabling smart home functionality. It stresses that sensors are essential for collecting detailed information about both the environment and residents' behavior, and that combining multiple sensor types improves data accuracy and reduces response time (Sagar & Kusuma, 2015). These capabilities are particularly valuable in healthcare applications, where continuous monitoring can support early detection of health issues, especially among elderly occupants. Actuators act upon the sensor data to control home systems such as lighting, temperature, and appliances, delivering real-time adjustments based on user needs and environmental conditions. The integration of these components establishes a responsive and intelligent smart home framework that prioritizes user safety, convenience, and wellness. In the paper (Pooja et al., 2016), the discussion centers around the integration of sensors and actuators in smart home automation and security systems. It outlines how sensors monitor key environmental and operational parameters, including temperature, motion, sound, and appliance status. These inputs enable advanced monitoring and alert functionalities essential for home security. Actuators complement this by controlling systems such as lighting, alarms, door locks, and surveillance devices, allowing for remote management and automation of routine tasks (Somayya et al., 2015). Through seamless coordination, sensors and actuators empower IoT-based smart homes to become not only more efficient but also safer and more adaptable to user preferences and lifestyle patterns.

Communication protocols such as ZigBee and Wi-Fi play a crucial role in enabling seamless data transmission within smart home ecosystems. The integration of sensors and actuators significantly enhances the intelligence of smart homes, delivering automated functionalities and real-time monitoring that improve user convenience, security, and energy efficiency. The paper (Jain et al., 2016) explores the role of sensors and actuators not only within smart homes but also in the broader context of transitioning towards smart city services (Gunge & Yalagi, 2016). It highlights their central importance in user-centric smart city development, categorizing smart home services into domains such as energy management, healthcare, and security—each supported by specific sensor types including motion, light, and temperature sensors. Actuators are responsible for executing automated responses based on sensor input, such as adjusting lighting, managing appliances, or triggering alarms (Bedi et al., 2016). The paper introduces a "sensor tree" to visually map the relationship between sensors and services, emphasizing the critical role of motion sensors in facilitating intelligent services. A thorough understanding of sensor types and their relevance is presented as a foundation for designing future smart city solutions that prioritize user needs through advanced monitoring, control, and automation.

The paper (Ray & Bagwari, 2020) proposes a smart energy management system for smart homes, in which sensors monitor indoor environmental parameters such as temperature and air quality to guide informed energy-saving decisions. A Kalman filter algorithm is applied to refine sensor data accuracy and to forecast future indoor conditions. Based on these predictions, actuators adjust household systems like heaters and lighting—to optimize user comfort while minimizing energy consumption. This closed-loop system ensures optimal resource utilization, where sensors inform and actuators implement changes, delivering energy-efficient living environments. Further, paper (Lokhande & Mohsin, 2017) offers a comprehensive analysis of IoT sensor applications for managing electric energy usage in smart homes. It includes a review of the current market, technical and economic assessments, research trends, and projections for future innovations in the field. The paper underscores the role of IoT sensors in enhancing energy efficiency, monitoring home environments, and supporting occupant well-being. It also discusses ongoing challenges such as sensor accuracy, interoperability, compact form-factor design, and calibration, which remain critical for widespread adoption. The paper concludes that IoT sensors hold substantial promise for transforming smart home energy systems and are gaining increasing recognition in both research and industry.

Security remains a major concern in IoT-based smart home environments. The paper (Agrawal & Das, 2011) provides an in-depth discussion on security vulnerabilities and proposes a robust architectural model to safeguard smart home systems.

It outlines the importance of secure communication and data handling, identifying common attack vectors targeting IoT devices. The proposed architecture includes cloud integration, a secure fog layer with a firewall, and a dedicated security analysis engine. The literature review spans fundamental concepts of IoT, smart home components, common security issues, and a layered defense model, all aimed at building user trust and confidence in smart home technologies. Lastly, the paper (Soliman et al., 2013) delivers a detailed overview of IoT and its impact on smart home automation. It traces the evolution of IoT and its ability to connect physical objects to the internet, driving innovations in automation, healthcare, and energy management. The review also examines routing protocols such as AODV and DSR to manage internet connectivity within smart homes. Additionally, it presents the development of a low-bandwidth rate-controlling system, featuring an embedded micro-web server with IP-based connectivity. This system enables users to remotely access and control home appliances through Android-based smartphone applications, enhancing flexibility and control over the smart home environment.

The paper (Daliya & Ramesh, 2017), titled "IoT-based Appliance Control System for Smart Homes," addresses the growing need for efficient management of the increasing number of appliances in modern households. It proposes a smart control system that utilizes Internet of Things (IoT) technology combined with a 433 MHz wireless sensor and actuator network (WSAN) to enable direct appliance control. The system supports functionalities such as real-time appliance monitoring, remote control, home security, and energy consumption analysis. Additionally, the study highlights the critical role of home networking infrastructure in enabling smart home automation. It introduces the concept of a Smart Home Information System (SHIS), built using web technologies, to facilitate centralized appliance management. The document also emphasizes the integration of building automation systems with energy-efficient operations to meet modern occupants' expectations for comfort, health, safety, and convenience—particularly within the framework of green and sustainable buildings.

The paper (Dey et al., 2025) presents an IoT-based system designed to control a mechanical actuator using temperature and positional data. The system integrates a RaspberryPi-3, a cloud service platform, a temperature sensor, and a GPS-enabled mobile application, offering a low-cost, open-source solution tailored for smart home environments. The architecture, system design, and process flow are thoroughly described, with particular attention to cloud-based data processing and real-time decision-making for actuator control. The paper also explores potential enhancements for future implementations and references related technologies and research to support its approach. Overall, it offers a comprehensive analysis of the design, deployment, and scalability of an IoT-driven actuator control system in the context of smart homes.

The literature reviewed provides a comprehensive understanding of the evolving roles of sensors and actuators in IoT-enabled smart homes. Various studies emphasize their integration in enhancing energy efficiency, environmental control, healthcare, security, and overall automation. The use of diverse sensors—ranging from temperature, motion, and light sensors to advanced physiological and video sensors—alongside actuators has been shown to significantly improve user experience, particularly for the elderly and disabled. Research also highlights the importance of communication protocols like ZigBee and Wi-Fi, edge computing, cloud-based data processing, and integration with microcontrollers such as Raspberry-Pi and Arduino. Smart home systems leveraging AI, machine learning, and context-aware automation have further been explored to improve responsiveness and personalization. Several papers extend the discussion to smart city contexts, emphasizing user-centric services and sustainability. In parallel, security concerns, energy optimization, and appliance control systems have received attention, with studies proposing architectures and algorithms for secure and efficient home management.

Despite this progress, gaps remain in the current research landscape. Most studies focus on individual use cases or isolated system components, lacking a unified framework that supports large-scale, interoperable, and context-aware smart home ecosystems. There is limited work on long-term real-world deployments, scalability across diverse home environments, and performance benchmarking in heterogeneous conditions. Privacy-preserving data processing techniques are still underexplored, and there is a need for standardized protocols and frameworks that ensure seamless integration across vendors and platforms. Moreover, while energy management and security are widely studied, the psychological and behavioral impact of smart systems on residents, especially in shared or multi-occupant homes, remains an under-researched area.

3. PROPOSED STUDY

This section presents a categorized overview of existing research studies on smart home technologies. The studies are grouped into four key domains: security, energy management, smart home automation, and elderly monitoring. A tabular format is used to indicate which area each paper addresses. This classification enables efficient identification of relevant research. For example, papers marked under the security category focus on threat detection and protection mechanisms in smart homes (Jhawar et al., 2025). Papers under energy management deal with energy-saving methods and system efficiency improvements. The categorization also helps highlight gaps in the literature. A limited number of studies in the elderly monitoring category may indicate insufficient research in that area. Given the growing importance of assistive systems for aging populations, this gap presents an opportunity for further exploration. Overall, organizing the literature in this manner provides a structured view of current research trends. It also supports future research planning by identifying underexplored areas such as integrated care systems, privacy-preserving monitoring, and multi-domain smart home solutions.

Table 1. Analysis of existing works

SR NO.	PAPER	SECURITY	ENERGY MANGEMENT	SMART HOME AUTOMATION	ELDERLY PEOPLE MONITORING
1	[4]				
2	[14]	✓		✓	
3	[3]	✓	✓	✓	
4	[13]			✓	
5	[1]			✓	
6	[7]			✓	
7	[8]	✓	✓	✓	
8	[9]				✓
9	[15]	✓	✓		
10	[12]	g		✓	
11	[18]				✓
12	[10]		✓		✓
13	[16]	✓		✓	
14	[17]		✓		
15	[11]		✓		
16	[35]		✓	✓	
17	[36]	✓			
18	[37]			✓	
19	[38]			✓	
20	[22]		✓	✓	

This article examines how technological advancements in smart home systems are influencing various aspects of daily living. It emphasizes that integrating smart devices and automation technologies offers an effective way to simplify household tasks, improve security, and reduce energy consumption. Each referenced paper is reviewed in the context of these core research areas to identify its specific contributions. Table 1 categorizes the reviewed literature into four domains: security, energy management, smart home automation, and elderly monitoring. Within the security category, the focus is on protecting user data, securing connected devices, and maintaining system integrity (Mohapatra, 2025). Energy management appears as a recurring research focus, aiming to improve energy efficiency through smart appliances and control systems. The domain of smart home automation is rapidly expanding, exploring the automation of lighting, climate control, and advanced security features. Elderly monitoring represents a unique but critical area of study. Research in this category investigates the use of smart technologies to detect falls, monitor daily activity, and enable timely emergency response for older adults living independently. By analyzing how each paper aligns with the categories in Table 1, the study provides a clearer understanding of current research priorities. For example, Paper 1 is marked under the security category, suggesting a primary focus on data privacy and system protection. The absence of a checkmark under energy management indicates that energy-saving strategies are not discussed in that study. Although the paper is listed under smart home automation, the depth of coverage within that domain may require further analysis to determine its full contribution.

The presence of a tick mark under the elderly monitoring category indicates that the paper discusses technologies aimed at improving safety and quality of life for older adults. This includes fall detection, movement tracking, and emergency response systems designed for senior individuals in smart home environments. Analyzing how each study aligns with the categories helps clarify the focus of each paper. This method is effective in identifying research trends, gaps, and common themes in the field (Sharma et al., 2025). The same approach is applied across all studies to observe how frequently each topic appears and how research attention is distributed. Several papers primarily address smart home automation, while others focus on elderly monitoring, emphasizing the benefits of smart technologies in supporting comfort, safety, and independence. Some studies explore both automation and energy management, examining cost-effective systems through hardware-software design, routing protocols, and performance evaluation. A set of papers focuses on smart home security, discussing the risks associated with IoT devices, potential cyber threats, and proposing secure system architectures. These include cloud integration, fog computing layers, and real-time security analysis engines, aiming to boost user trust in smart technologies. Other studies combine elderly care with monitoring solutions, while some concentrate specifically on energy control and optimization in smart home systems.

Table. 1 provides a useful summary of research areas in smart home studies. It helps identify which topics each paper addresses. However, the table gives only a basic overview and may not reflect the full depth of each study. To truly understand the contributions of each paper, it is important to review the complete research, including the methods, results, and specific approaches used. The table highlights selected questions within broader research themes. For a complete picture, additional sources and deeper analysis are necessary. Researchers should not rely solely on the checkmarks but should examine how thoroughly each topic is discussed in the original papers. This analysis also helps identify gaps in the current research. Areas with limited coverage can point to future research opportunities. Still, it is important to assess the actual content and impact of each study beyond its categorization in the table. Some topics overlap and contribute to multiple goals. For example, a smart security system might also support elderly monitoring. These overlaps show how solutions in smart homes are often interconnected (Thakur et al., 2025).

4. PROPOSED FRAMEWORK

The growing demand for intelligent, responsive, and efficient living environments has positioned the Internet of Things (IoT) as a transformative force in the evolution of smart home technologies. Building on the existing literature and identifying key research gaps in areas such as interoperability, privacy, security, and support for elderly and disabled users, this proposed framework introduces a holistic and scalable architecture for smart homes. The framework aims to bridge the divide between device heterogeneity and seamless user experience by integrating hardware components, communication protocols, middleware systems, and intelligent services under a unified model. It emphasizes modularity, adaptability, and security, ensuring future-ready smart homes that serve a diverse range of users and applications.

Figure 4. Working IoT in harsh environment

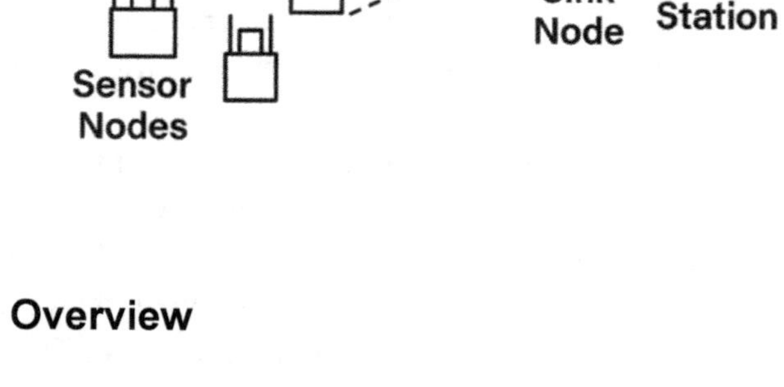

4.1. System Overview

The proposed smart home framework is designed around five core layers: sensing and actuation, edge gateway, fog layer, cloud services, and user interface.

Each layer plays a distinct yet interconnected role in enabling real-time monitoring, decision-making, and control across the home environment. At the foundation lies a dense network of sensors and actuators embedded throughout the home, tasked with capturing environmental data and executing commands. These devices range from motion and temperature sensors to actuators such as smart locks, lights, HVAC controllers, and alarms. They collect data on occupancy, temperature, humidity, gas levels, and user interactions, forming the sensory and response backbone of the system. Data from these devices is transmitted to a local edge gateway, typically a low-power computing device such as a Raspberry Pi or a dedicated home automation hub. This gateway aggregates data, performs lightweight processing, and forwards selected information to the fog layer. The fog layer acts as a localized intelligence node, enabling pre-processing, filtering, storage, and control logic execution close to the data source. It includes privacy-enhancing modules, access control mechanisms, and machine learning models for predictive analytics and anomaly detection. Fig.4 presents the working of IoT in harsh environment.

Figure 5 presents a layered and modular architecture designed to manage the complex operations of a modern smart home system. It reflects the interactions between hardware components, communication protocols, data processing units, and external services, all coordinated to create an intelligent and responsive living environment. The architecture begins with the User Interface, which represents the end-user's point of interaction with the smart home. This could be through a smartphone app, voice assistant, touchscreen panel, or wearable device. The interface enables users to monitor, configure, and control smart devices seamlessly. Connected to the user interface are the Edge Devices, which include a variety of sensors and actuators deployed throughout the home. These devices are responsible for real-time data collection and execution of physical tasks. Examples include motion sensors, temperature sensors, smart locks, and lighting systems. The edge devices act as the first point of contact between the environment and the system, translating physical events into digital signals.

Data from these edge devices is transmitted to the Edge Gateway, a localized computing hub that aggregates, filters, and preprocesses the incoming data. This layer plays a crucial role in reducing latency and ensuring fast response times by performing lightweight tasks close to the source. The edge gateway is often implemented using low-power hardware such as Raspberry Pi, Arduino controllers, or commercial IoT hubs. From the edge gateway, data flows to the Fog Layer, which is responsible for more advanced local processing and decision-making. This layer includes components for real-time analytics, data caching, temporary storage, and rule execution. It helps offload tasks from the cloud by processing critical information at the local level. Importantly, this layer interacts with the Smart Home Control Center, which serves as the brain of the system. The control center coordinates commands across devices, manages automation rules, and maintains state awareness of the home environment.

The Fog Layer also communicates with the Data Security & Privacy Module, which safeguards the system from unauthorized access, data breaches, and malicious activities. It handles tasks such as encryption, user authentication, and anomaly detection. This ensures that sensitive data from sensors and user interactions remains protected at all times. Further up the stack, the Cloud Platform plays a central role in long-term data storage, high-level analytics, and system-wide updates. It hosts services such as machine learning model training, predictive behavior analysis, and historical trend tracking. The cloud layer connects to the AI/ML Engine, which uses data collected over time to build personalized automation routines, predict user behavior, and support decision-making. These models help the system adapt to changing user habits and environmental conditions, thus enhancing the overall intelligence and user experience. The cloud infrastructure also integrates with External Services, such as healthcare providers, utility companies, emergency responders, or city-level smart infrastructure. This connection expands the reach of the smart home, enabling cross-domain collaboration. For example, a utility provider can send signals for demand response programs, or a healthcare provider can monitor elderly patients remotely with consent. Finally, a bidirectional link is shown from the Smart Home Control Center back to the Edge Devices. This feedback loop enables real-time control and command execution. For instance, if the system detects motion at an unusual hour, it can immediately activate cameras or alarm systems through this loop, without waiting for cloud-level confirmation. Overall, the diagram provides a complete representation of the flow of data, control logic, and service coordination within a smart home. It emphasizes decentralization through fog computing, security through dedicated privacy modules, and intelligence through AI integration. The framework is scalable, interoperable, and user-centric, making it well-suited for modern and future smart home deployments.

Figure 5. Proposed frame work

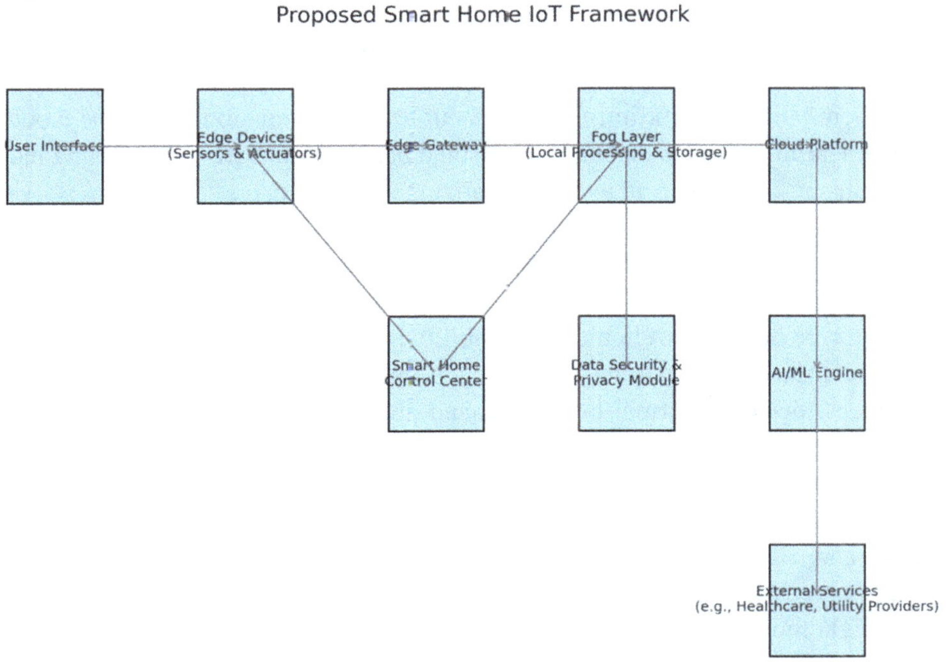

4.2. Fog and Cloud Integration

The framework connects the fog layer to cloud services via secure communication channels. Cloud infrastructure offers elastic computing resources, long-term storage, and centralized analytics. It hosts services such as energy consumption modeling, behavior profiling, remote access services, and large-scale pattern recognition. The integration of cloud and fog layers ensures both low-latency local operations and scalable, cloud-based intelligence. Moreover, the cloud layer maintains interoperability with third-party services such as utility providers, healthcare platforms, and emergency responders, making the system adaptable to broader smart city contexts.

4.3. Security and Privacy Module

Security and privacy are critical concerns in any IoT-enabled home system. The proposed framework incorporates a dedicated module to handle encryption, identity management, intrusion detection, and data anonymization. Data from sensors is encrypted at the edge before transmission. Role-based access control mechanisms govern data sharing between users, devices, and services. Additionally, the fog layer includes a security engine that performs real-time anomaly detection and mitigates potential threats, such as unauthorized access or tampering attempts. Privacy preservation is enforced by processing sensitive data locally whenever possible. For example, video feeds from indoor cameras can be analyzed at the fog level to detect events (like a fall or intrusion) without uploading raw footage to the cloud. Machine learning models that handle personal behavior data are trained using federated learning techniques to avoid centralizing sensitive data.

4.4. AI and Machine Learning Integration

At the core of the smart home's adaptability is its AI engine, which runs predictive and context-aware models. These models enable the system to learn occupant preferences, detect abnormal behaviors, and automate routines accordingly. For example, the system can learn that a resident typically turns off lights and lowers the thermostat before sleeping, and automatically perform these actions if the resident forgets. Similarly, it can identify deviations in an elderly person's daily movement patterns and trigger an alert if needed. The machine learning pipeline consists of data pre-processing, feature extraction, model training, and real-time inference. Lightweight models are deployed at the fog layer for rapid decision-making, while more complex models are trained in the cloud using aggregated and anonymized data. This two-tiered approach balances performance and privacy while enabling continuous learning and system improvement.

4.5. Interoperability and Middleware

To address the challenge of integrating devices from multiple vendors, the framework employs a middleware layer that abstracts device-level protocols and provides standardized APIs. The middleware supports common protocols such as ZigBee, Z-Wave, Wi-Fi, MQTT, and CoAP. It translates these into a uniform data model, allowing application developers to interact with devices without worrying about hardware-level differences. The middleware also facilitates auto-discovery and dynamic configuration of devices, simplifying deployment and scaling of the smart home network. Interoperability is further reinforced through the use of semantic modeling and ontologies. Devices, users, and services are described using standardized ontologies such as OWL or SAREF, enabling the system to reason about context and interactions across heterogeneous components. This semantic layer supports advanced use cases like context-aware automation, cross-device coordination, and voice-activated control.

4.6. Use Case: Elderly Monitoring

A critical focus of the framework is on supporting aging in place through intelligent monitoring and assistance. The system uses motion sensors, pressure sensors, and wearable physiological monitors to track the daily routines of elderly occupants. The AI module builds a baseline of normal activities and detects deviations such as extended inactivity, missed medication, or irregular movement patterns. In such cases, the system notifies caregivers or emergency services.

Actuators such as robotic assistants, automated medicine dispensers, and adaptive lighting systems respond to detected conditions to enhance safety and comfort. The framework supports multi-user profiles, enabling family members to remotely check status or configure routines through a secure mobile interface.

4.7. Use Case: Energy Management

Energy management is another central application. Smart thermostats, blinds, lighting systems, and appliances work together to optimize energy usage. Sensors detect room occupancy and natural light levels to adjust lighting and HVAC settings dynamically. The cloud engine analyzes historical energy consumption to suggest schedule adjustments or detect energy leaks. Users can view real-time energy data through a mobile dashboard and set goals or preferences. The system supports demand-response capabilities, allowing it to adjust appliance usage based on utility pricing or grid load. Integration with solar panels and energy storage units is also supported, making the system viable for sustainable smart homes.

Table 2. Sensor-Actuator mapping

Device Type	Function	Paired Actuator
Temperature Sensor	Monitors room temperature	Thermostat
Motion Sensor	Detects movement	Alarm/Light
Gas Sensor	Detects gas leaks	Ventilation System
Smart Light	Adjusts lighting	Dimmable Light
Thermostat	Regulates HVAC	HVAC
Security Camera	Monitors surroundings	Smart Lock/Alarm

Table. 2 illustrates a clear mapping between various smart home sensors and their corresponding actuators. Each sensor type—such as temperature, motion, and gas sensors—has a specific function related to monitoring the home environment. These sensors trigger appropriate actuators like thermostats, alarms, ventilation systems, or lighting controls, depending on the detected condition. This structured relationship enables real-time automation and responsive behavior in smart homes, enhancing both efficiency and safety for occupants.

4.8. User Interaction and Accessibility

The user interface layer connects residents to their smart home through mobile apps, voice assistants, and touch panels. Interfaces are designed for usability, with particular consideration for elderly and disabled users. Voice commands, visual alerts, and tactile feedback are supported to ensure accessibility. The UI provides real-time status updates, control functions, and event notifications. Users can customize automation rules using natural language inputs or graphical workflows. For example, a user can define a rule such as "turn off all lights if no motion is detected for 15 minutes after 10 PM" without writing code. The system includes built-in templates for common scenarios to simplify configuration.

4.9. Scalability and Modularity

The framework is built with scalability in mind. New devices can be added through the auto-discovery feature. Cloud-native microservices allow scaling individual components such as AI inference or analytics engines without overloading the entire system. The use of containers and service orchestration tools like Kubernetes ensures reliability and maintainability in large deployments. Modularity allows specific parts of the system to be upgraded or replaced without affecting the rest. For instance, a new AI model can be deployed to the fog node without changing the core sensor network. This makes the framework adaptable to evolving technologies and user needs.

4.10. Evaluation and Simulation Environment

To validate the framework, a simulated smart home environment is proposed using tools such as OpenHAB, Node-RED, and NS-3 for network simulation. Virtual devices emulate sensors and actuators, generating realistic data streams. The fog and cloud components are containerized and deployed on local servers and cloud platforms. Performance metrics such as latency, energy usage, response time, and detection accuracy are recorded and analyzed. Scenarios such as fire detection, fall monitoring, and automated climate control are used to test system behavior. Simulation results demonstrate the feasibility and efficiency of the framework, showing improved response times and reduced cloud dependency compared to traditional cloud-only models.

This proposed framework offers a comprehensive, secure, and adaptable model for next-generation smart homes. By combining edge intelligence, cloud analytics, modular middleware, and strong security protocols, it addresses many current limitations in smart home design. The integration of AI enables personalized, context-aware automation, while semantic modeling ensures interoperability across devices and services.

5. LIMITATIONS AND FUTURE SCOPE

In the pursuit of achieving seamless interoperability in IoT-based smart home environments, several persistent limitations have been identified through our literature review. Key among these are concerns related to security and privacy, the lack of standardized communication protocols, energy consumption inefficiencies, heterogeneous device ecosystems, semantic mismatches, and challenges in scalability. These issues continue to hinder the full realization of a unified, intelligent smart home ecosystem. Although prior studies have addressed some of these concerns through protocol translation layers, middleware abstraction, and semantic mapping, critical gaps remain. For instance, many proposed solutions are system-specific and fail to generalize across diverse device configurations or user scenarios. Additionally, many frameworks lack support for evolving technologies and do not fully integrate user feedback, which limits adaptability.

To address these unresolved challenges, we propose two innovative directions for future exploration. The first approach advocates for the development of open, participatory platforms that actively invite user contributions in solving interoperability problems. A crowdsourced model would allow users, developers, and integrators to share real-world experiences, custom configurations, and tested workarounds. This model emphasizes the value of community insight, where non-standard use cases and overlooked issues can be surfaced and addressed collaboratively. By giving end-users, a role in the evolution of the system, this strategy supports real-time problem-solving and fosters greater engagement with the smart home ecosystem. The second approach introduces the potential use of Decentralized Autonomous Organizations (DAOs) to facilitate open and democratic standard-setting. DAOs, built on blockchain technology, offer a decentralized mechanism for stakeholders to propose, vote on, and evolve interoperability standards. By engaging device manufacturers, developers, regulators, and consumers in a distributed governance model, DAOs can foster a more inclusive and transparent standardization process. This is particularly useful in a field like smart home IoT, where technological change is rapid and centralized regulatory mechanisms often lag behind. A DAO-driven framework could dynamically adapt to technological developments and accommodate diverse user needs more effectively than traditional top-down models.

Although these proposals are at an early conceptual stage, they highlight promising avenues for future research and experimentation. Both user-powered contribution platforms and DAO-based governance models represent a shift toward decentralized innovation in the smart home landscape. These directions could help mitigate the fragmentation currently seen in smart home systems, improving long-term compatibility, resilience, and user trust. Looking forward, further exploration is needed to evaluate the technical feasibility, security implications, and practical deployment of these models. Simulation studies, pilot implementations, and community testing environments will be essential to assess their effectiveness and scalability. Additionally, interdisciplinary collaboration between technologists, sociologists, legal experts, and user communities will be key in refining and implementing these ideas. Ultimately, overcoming the limitations of interoperability in smart home environments will require a balance of technical advancement, inclusive governance, and user-centered design. The proposed future directions aim to contribute meaningfully toward this goal by introducing adaptive, transparent, and collaborative mechanisms for system integration and standard evolution.

6. CONCLUSION

This paper presents a comprehensive overview of the current landscape, opportunities, and limitations in the field of smart home technologies. Through a structured literature review and system-level analysis, the study highlights how the integration of IoT components—particularly sensors and actuators—has transformed conventional residential spaces into intelligent, responsive environments. These advancements have enabled real-time automation, improved energy efficiency, enhanced safety, and personalized living experiences, redefining the way individuals interact with their homes. Despite this progress, the review also identifies several areas requiring further attention. Chief among these are the challenges of interoperability, lack of standardized protocols, limited scalability, and persistent concerns around security and privacy. The complexity introduced by heterogeneous devices and platforms has led to fragmented systems that are often difficult to manage, integrate, and scale. Additionally, technical limitations related to sensor accuracy, energy consumption, and data processing efficiency continue to impact the performance and reliability of smart home solutions. Addressing these challenges requires a multi-dimensional research approach that blends advances in hardware design, communication technologies, artificial intelligence, cybersecurity, and user-centric design. Interdisciplinary collaboration across fields such as engineering, human-computer interaction, data science, and public policy is crucial to create systems that are not only technically robust but also socially acceptable and ethically grounded.

The study also proposes two innovative ideas to enhance interoperability within smart home ecosystems: crowdsourced contribution platforms and Decentralized Autonomous Organizations (DAOs) for open standard governance. These approaches present forward-looking strategies that empower users and stakeholders to co-create flexible, adaptive frameworks. If developed further, such models could pave the way for scalable, secure, and universally accessible smart home systems. In conclusion, the future of smart home technologies depends on the ability to address existing gaps while embracing new paradigms for openness, inclusivity, and collaboration. By focusing research on real-world implementation, standardization, and ethical deployment, the smart home ecosystem can evolve into a more mature, intelligent, and sustainable infrastructure—capable of meeting the diverse needs of present and future users. This study serves as both a foundational reference and a call to action for researchers, developers, and policymakers committed to shaping the next generation of connected living environments.

REFERENCES

Agrawal, S., & Das, M. L. (2011). Internet of Things—A paradigm shift of future Internet applications. *2011 Nirma University International Conference on Engineering*, 1–7. DOI: 10.1109/NUiConE.2011.6153246

Alaa, M., Zaidan, A. A., Zaidan, B. B., Talal, M., & Kiah, M. L. M. (2017). A review of smart home applications based on Internet of Things. *Journal of Network and Computer Applications*, *97*, 48–65. DOI: 10.1016/j.jnca.2017.08.017

Alam, M. R., Reaz, M. B. I., & Ali, M. A. M. (2012). A Review of Smart Homes—Past, Present, and Future. IEEE Transactions on Systems, Man, and Cybernetics. *IEEE Transactions on Systems, Man, and Cybernetics. Part C, Applications and Reviews*, *42*(6), 1190–1203. DOI: 10.1109/TSMCC.2012.2189204

Bedi, G., Venayagamoorthy, G. K., & Singh, R. (2016). Internet of Things (IoT) sensors for smart home electric energy usage management. *2016 IEEE International Conference on Information and Automation for Sustainability (ICIAfS)*, 1–6. DOI: 10.1109/ICIAFS.2016.7946568

Bing, K., Fu, L., Zhuo, Y., & Yanlei, L. (2011). Design of an Internet of Things-based smart home system. 2011 2nd International Conference on Intelligent Control and Information Processing, 921–924. DOI: 10.1109/ICICIP.2011.6008384

Chan, M., Estève, D., Escriba, C., & Campo, E. (2008). A review of smart homes—Present state and future challenges. *Computer Methods and Programs in Biomedicine*, *91*(1), 55–81. DOI: 10.1016/j.cmpb.2008.02.001 PMID: 18367286

Chi, Q., Yan, H., Zhang, C., Pang, Z., & Xu, L. D. (2014). A Reconfigurable Smart Sensor Interface for Industrial WSN in IoT Environment. *IEEE Transactions on Industrial Informatics*, *10*(2), 1417–1425. DOI: 10.1109/TII.2014.2306798

Ciabattoni, L., Cimini, G., Grisostomi, M., Ippoliti, G., & Longhi, S. (2014). An interoperable framework for home automation design, testing and control. *22nd Mediterranean Conference on Control and Automation*, 1049–1054. DOI: 10.1109/MED.2014.6961513

Daliya, V. K., & Ramesh, T. K. (2017). A survey on enhancing the interoperability aspect of IoT based systems. *2017 International Conference on Smart Technologies for Smart Nation (SmartTechCon)*, 581–586. DOI: 10.1109/SmartTechCon.2017.8358438

Dey, A., Bose, S., Bisht, R., Yadav, M., & Mohapatra, H. (2025). Solutions for energy management and environmental conservation in IoT-integrated smart home. *Advances in Computational Intelligence and Robotics*, •••, 389–408. DOI: 10.4018/979-8-3693-9770-1.ch018

Ding, D., Cooper, R. A., Pasquina, P. F., & Fici-Pasquina, L. (2011). Sensor technology for smart homes. *Maturitas*, *69*(2), 131–136. DOI: 10.1016/j.maturitas.2011.03.016 PMID: 21531517

Diro, A. A., & Chilamkurti, N. (2018). Distributed attack detection scheme using deep learning approach for Internet of Things. *Future Generation Computer Systems*, *82*, 761–768. DOI: 10.1016/j.future.2017.08.043

Gaikwad, P. P., Gabhane, J. P., & Golait, S. S. (2015). A survey based on Smart Homes system using Internet-of-Things. 2015 International Conference on Computation of Power, Energy, Information and Communication (ICCPEIC), 0330–0335. DOI: 10.1109/ICCPEIC.2015.7259486

Gazis, A., & Katsiri, E. (2021). Smart Home IoT Sensors: Principles and Applications – A Review of Low-Cost and Low-Power Solutions. *International Journal on Engineering Technologies and Informatics*, *2*(1), 19–23. DOI: 10.51626/ijeti.2021.02.00007

Gondaliya, T. (2016). A Survey on an Efficient IoT Based Smart Home. *International Journal of Review in Electronics and Communication Engineering*, *4*(1).

Gunge, V., & Yalagi, P. (2016). Smart Home Automation: A Literature Review. International Journal of Computer Applications (0975 – 8887) RTDM 2016.

Jhawar, V., Rangnekar, Y., & Mohapatra, H. (2025). Survey of fog computing applications in vehicular communication for smart transportation systems. Driving Innovation at the Intersection of Renewable Energy and the Internet of Vehicles, 345-382. .DOI: 10.4018/979-8-3373-3321-2.ch010

Khalaf, R., Mohammed, A., Essa, E., & Ali, H. (2019). Controlling Smart Home Activities Using IoT. *2019 International Conference on Computing and Information Science and Technology and Their Applications (ICCISTA)*, 1–6. DOI: 10.1109/ICCISTA.2019.8830664

Kodeswaran, P., Kokku, R., Mallick, M., & Sen, S. (2016). Demultiplexing activities of daily living in IoT enabled smarthomes. *IEEE INFOCOM*, *2016*, 1–9. DOI: 10.1109/INFOCOM.2016.7524359

Kramp, T., van Kranenburg, R., & Lange, S. (2013). Introduction to the Internet of Things. In Enabling Things to Talk (pp. 1–10). Springer. DOI: 10.1007/978-3-642-40403-0_1

Lokhande, D. G., & Mohsin, S. A. (2017). Internet of things for ubiquitous smart home system. 2017 1st International Conference on Intelligent Systems and Information Management (ICISIM), 314–320. DOI: 10.1109/ICISIM.2017.8122191

Miori, V., & Russo, D. (2014). Domotic Evolution towards the IoT. 2014 28th International Conference on Advanced Information Networking and Applications Workshops, 809–814. DOI: 10.1109/WAINA.2014.128

Mocrii, D., Chen, Y., & Musilek, P. (2018). IoT-based smart homes: A review of system architecture, software, communications, privacy and security. *Internet of Things : Engineering Cyber Physical Human Systems*, *1–2*, 81–98. DOI: 10.1016/j.iot.2018.08.009

Mohapatra, H. (2021). Socio-technical Challenges in the Implementation of Smart City. 2021 International Conference on Innovation and Intelligence for Informatics, Computing, and Technologies (3ICT), 57–62. DOI: 10.1109/3ICT53449.2021.9581905

Mohapatra, H. (2025). The fine-tuning journey of ChatGPT in industry applications. *Advances in Computational Intelligence and Robotics*, •••, 263–284. DOI: 10.4018/979-8-3693-8332-2.ch011

Paul, C., Ganesh, A., & Sunitha, C. (2018). An overview of IoT based smart homes. 2018 2nd International Conference on Inventive Systems and Control (ICISC), 43–46. DOI: 10.1109/ICISC.2018.8398858

Ray, A. K., & Bagwari, A. (2020). IoT based Smart home: Security Aspects and security architecture. 2020 IEEE 9th International Conference on Communication Systems and Network Technologies (CSNT), 218–222. DOI: 10.1109/CSNT48778.2020.9115737

Sharma, S., Shrivastava, R. R., Sahu, A., Mishra, S., Srivastava, K., & Mohapatra, H. (2025). Cloud-driven public safety in smart cities. Enhancing Safety, Security, and Service in Organizations, 235-262. DOI: 10.4018/979-8-3373-0482-3.ch008

Singh, D., Tripathi, G., & Jara, A. J. (2014). A survey of Internet-of-Things: Future vision, architecture, challenges and services. 2014 IEEE World Forum on Internet of Things (WF-IoT), 287–292. DOI: 10.1109/WF-IoT.2014.6803174

Soliman, M., Abiodun, T., Hamouda, T., Zhou, J., & Lung, C.-H. (2013). Smart Home: Integrating Internet of Things with Web Services and Cloud Computing. 2013 IEEE 5th International Conference on Cloud Computing Technology and Science, 317–320. DOI: 10.1109/CloudCom.2013.155

Somayya, M., Ramaswamy, R., & Tripathi, S. (2015). Internet of Things (IoT): A Literature Review. *Journal of Computer and Communications*, *3*(5), 164–173. DOI: 10.4236/jcc.2015.35021

Sun, Q., Yu, W., Kochurov, N., Hao, Q., & Hu, F. (2013). A Multi-Agent-Based Intelligent Sensor and Actuator Network Design for Smart House and Home Automation. *Journal of Sensor and Actuator Networks*, *2*(3), 557–588. DOI: 10.3390/jsan2030557

Taiwo, O., & Ezugwu, A. E. (2021). Internet of Things-Based Intelligent Smart Home Control System. *Security and Communication Networks*, *2021*, 1–17. DOI: 10.1155/2021/9928254

Taj, S., Asad, U., Azhar, M., & Kausar, S. (2019). Interoperability in IOT Based Smart Home: A Review. *Review of Computer Engineering Studies*, *5*(3), 50–55. DOI: 10.18280/rces.050302

Thakur, T., Srivastava, L., Panda, A., & Mohapatra, H. (2025). Smart parking in foggy condition real-time efficiency with PSO, ACO, and IoT sensors. Revolutionizing Urban Development and Governance With Emerging Technologies, 233-260. DOI: 10.4018/979-8-3373-1375-7.ch008

Ullah, I., & Kim, D. (2017). An Improved Optimization Function for Maximizing User Comfort with Minimum Energy Consumption in Smart Homes. *Energies*, *10*(11), 1818. DOI: 10.3390/en10111818

Viani, F., Robol, F., Polo, A., Rocca, P., Oliveri, G., & Massa, A. (2013). Wireless Architectures for Heterogeneous Sensing in Smart Home Applications: Concepts and Real Implementation. *Proceedings of the IEEE*, *101*(11), 2381–2396. DOI: 10.1109/JPROC.2013.2266858

Wang, M., Zhang, G., Zhang, C., Zhang, J., & Li, C. (2013). An IoT-based appliance control system for smart homes. *2013 Fourth International Conference on Intelligent Control and Information Processing (ICICIP)*, 744–747. DOI: 10.1109/ICICIP.2013.6568171

Williams, V., Terence, S. J., & Immaculate, J. (2019). Survey on Internet of Things based Smart Home. *2019 International Conference on Intelligent Sustainable Systems (ICISS)*. DOI: 10.1109/ISS1.2019.8908112

Chapter 7
Intelligent Urban Traffic Management Using AI and IoT:
A Blueprint for Next-Generation Intelligent Transportation Systems

Rajeshkumar Golani
https://orcid.org/0009-0008-3501-1902
Independent Researcher, USA

Jagjot Bhardwaj
https://orcid.org/0009-0009-5414-5903
Independent Researcher, USA

ABSTRACT

Urban traffic congestion affects economies, the environment, and quality of life. Traditional systems with fixed timing or localized sensors lack adaptability for dynamic conditions. This chapter examines how Artificial Intelligence (AI) and the Internet of Things (IoT) facilitate real-time, data-driven traffic control, thereby reducing congestion and enhancing traffic flow. It examines AI techniques, including machine learning, graph models, and multi-agent reinforcement learning, applied to adaptive signal timing, traffic forecasting, incident detection, and autonomous response. Case studies, such as Pittsburgh's SURTRAC, Alibaba's City Brain, and Google's Green Light, demonstrate measurable gains in travel time, emissions, and throughput. A layered IoT-AI framework is proposed for data collection, decision-making, actuation, and continuous feedback, alongside discussions of key challenges, including privacy, connected vehicle integration, scalability, edge computing, and federated learning to unlock the full potential of intelligent traffic systems.

DOI: 10.4018/979-8-3373-5727-0.ch007

1. INTRODUCTION

Urban traffic management has traditionally depended on centrally planned or pre-timed strategies, which are effective under predictable conditions but struggle to adapt to real-time changes. Standard traffic signal modes include: (a) Fixed-time control—where lights follow a predetermined cycle optimized for average traffic; (b) Actuated control—where local sensors like induction loops or cameras trigger green lights based on vehicle detection; and (c) Coordinated control—where groups of signals, often along corridors, are synchronized to produce "green waves" for smoother traffic flow (Michailidis, Michailidis, Lazaridis, & Kosmatopoulos, 2025). Many cities still rely on fixed or semi-actuated timers, updated only occasionally through manual traffic surveys. These systems have notable limitations:

- Static Timing and Lack of Adaptability: Fixed timing schedules are inflexible, failing to adapt to irregular traffic fluctuations, which results in inefficiency during off-peak times or unexpected surges. They perform poorly when incidents, special events, or weather create abnormal traffic patterns (Michailidis, Michailidis, Lazaridis, & Kosmatopoulos, 2025). Although actuated signals respond locally, they optimize each intersection independently and lack coordination across the network. This often causes suboptimal traffic flow, where a green light at one intersection just shifts the traffic queue to the next red light downstream.
- **Inability to Handle Incidents or Non-Recurring Congestion:** Traditional systems are mostly reactive or rule-based, meaning they rely on manual intervention when incidents like major accidents are reported. They lack automatic detection and proactive re-routing capabilities. As a result, non-recurrent congestion caused by crashes, roadwork, or similar events often remains unaddressed until human operators intervene, often after long queues have already formed.
- **Limited Sensor Integration:** Older systems typically rely on a single data source, such as inductive loops that count vehicles at a stop bar. They usually do not utilize the extensive modern data available, like cameras, vehicle GPS, and mobile phones which could offer a more comprehensive view of traffic conditions. As a result, they only provide a partial understanding of the network state, leading to less optimal decision-making.
- Outdated Coordination Methods: Before AI, adaptive traffic control systems (ATCS) like Sydney's SCATS or the UK's SCOOT existed, but they relied on rule-based or algorithmic optimizations instead of learning. These systems adjusted splits or offsets based on traffic flow measurements, but their logic was predefined and not adaptive. Even by the mid-2010s, only about 4–5%

of U.S. traffic signals operated in fully adaptive mode, mainly due to the high costs of deployment and tuning. Most traffic signals still use fixed-time or simple actuated modes.(Google Research, 2024)

These limitations cause congestion in cities worldwide. Inefficient signal timing contributes significantly to urban traffic delays (Li, Pan, Liu, & Li, 2025). Drivers lose hours daily due to rigid signal plans that can't adapt to real-time demand. Outdated timing plans cause rush-hour gridlock. Traditional systems struggle to balance objectives like vehicle delay, pedestrian safety, throughput, and emissions, often using simplistic logic (e.g., minimizing delays without considering pedestrians or the environment).

These challenges have led to intelligent transportation systems (ITS) that use modern computation and data. The rise of IoT sensors and city-wide connectivity allows the collection of real-time data on traffic, speeds, and vehicle locations. Advances in AI and machine learning support data-driven decisions that adapt quickly. Overall, traditional traffic management is shifting from static, fixed controls to dynamic, adaptive systems that respond to current conditions (Gheorghe & Soica, 2025).

The shift to intelligent systems is driven by urgent needs and advancing tech. Growing urban populations and increasing vehicle numbers lead to congestion, incurring economic, environmental, and social costs. Meanwhile, sensors, wireless communication, cloud computing, and AI have matured to the point where real-time citywide traffic optimization is now possible. (Gheorghe & Soica, 2025) This need for tech convergence underpins intelligent urban traffic management. The following sections examine how AI and IoT overcome the limitations of legacy systems to enable more responsive and efficient traffic control networks. (Mutambik, 2025)

2. AI-DRIVEN TRAFFIC MANAGEMENT: APPLICATIONS AND EXAMPLES

Integrating AI into traffic management enables a shift from reactive to proactive and adaptive control. Several key application areas have emerged where AI techniques, fueled by IoT data streams, provide significant benefits over conventional approaches:

- Adaptive Signal Control: AI-powered adaptive traffic signal control adjusts light timing dynamically based on real-time traffic, unlike fixed timers, to reduce delays and stops. Machine learning models allocate green time according to vehicle queues, often using reinforcement learning (RL), where an AI learns optimal timing through trial and error. The SURTRAC system in Pittsburgh pioneered a multi-agent RL approach, coordinating neighboring intersections, which improved travel times by over 25%, reduced vehicle stops by 30%, and decreased wait times by more than 40% (Smith, Barlow, Xie, & Rubinstein, 2013). Many cities now test AI for traffic lights, from simple actuated changes to full multi-intersection optimization.

- Traffic Flow Forecasting: AI helps predict traffic conditions minutes ahead, like volume or speed, enabling traffic systems to act preventively, such as re-routing or adjusting signals. Traditional models struggled with traffic's complexity, but AI methods like RNNs, LSTMs, and GNNs improve accuracy by learning from historical and real-time data, including IoT sensors. These forecasts allow cities like Los Angeles and New York to optimize signals and inform travelers, reducing delays substantially. (Gheorghe & Soica, 2025)
- Incident Detection and Response: Timely detection of traffic incidents is vital for reducing congestion. AI computer vision, using CNNs and models like YOLO, analyzes CCTV footage in real time to identify accidents and congestion (Peyman, Copado, Tordecilla, Martins, Xhafa, & Juan, 2021; Sharma & Shah, 2025). Deep learning trained on traffic images achieves over 98% accuracy in spotting crashes and stopped vehicles (Mutambik, 2025). AI also monitors connected vehicle data and infrastructure sensors to infer incidents. Once confirmed, management systems automatically adjust signals, post alerts, or suggest routes to mitigate impact. AI classifiers can detect incidents from speed data within minutes, faster than waiting for human response call. (Karim et al., 2024)
- Autonomous and Connected Response Systems: AI-driven traffic management aims for a fully autonomous system where sensors detect conditions, AI algorithms decide actions, and control devices execute them—all in real time without human input. This includes dynamic rerouting, adaptive ramp metering, and variable speed limits. In cities, automated traffic signal coordination responds to congestion, exemplified by Alibaba's City Brain in Hangzhou, which manages over 1,000 intersections and autonomously handles traffic during large events, improving throughput by 40% without extra police. Google's Project Green Light analyzes traffic data and suggests signal adjustments for engineers. As vehicles become more connected, systems will incorporate vehicle-to-infrastructure communication, enabling signals to extend green lights or send speed recommendations, enhancing efficiency beyond isolated controls do. (People's Daily, 2019)

Real-world examples highlight the benefits of AI traffic management. Cities like Los Angeles and Delhi have reduced congestion using AI-based signal timing and reinforcement learning. A review shows AI systems typically cut delays by 20–30%, boost safety, and lower emissions by constantly adjusting to current conditions, outperforming static plans (Gheorghe & Soica, 2025).

In short, AI methods fueled by IoT data are advancing smarter urban traffic management in various ways – including self-adjusting signals, systems that predict and prevent congestion, and quick incident response. We will now explore the specific AI algorithms and techniques driving this progress, highlighting their different approaches and suitability for traffic solutions.

3. AI TECHNIQUES AND ALGORITHMS FOR TRAFFIC MANAGEMENT

Intelligent traffic management utilizes a wide range of AI and machine learning techniques. Each type of algorithm offers distinct advantages in perception, prediction, or control, addressing different traffic challenges. This section presents a detailed overview of the most relevant AI methods, explaining their mechanisms and applications in traffic systems. A summary table at the end of the section compares these techniques.

3.1 Convolutional Neural Networks (CNNs)

CNNs are a type of deep neural network designed to automatically extract features from grid-like data such as images or spatial matrices using convolution operations. They are highly effective in computer vision tasks, making them valuable for traffic management applications involving visual data or spatial patterns. Key uses of CNNs in this domain include:

- Video/Image-Based Traffic Sensing: CNN-based models detect and classify objects in traffic feeds, like vehicles, pedestrians, or incidents. Modern architectures (e.g., Faster R-CNN, YOLO, SSD) utilize convolutional layers to recognize vehicles and events, such as crashes. An incident system processes video frames to flag anomalies (e.g., sudden stops, shoulder vehicles) with high accuracy. Recent studies using YOLOv8 and object tracking have achieved over 98% accuracy in identifying traffic incidents from CCTV feeds (Karim et al., 2024). This automation speeds up incident response compared to humans.
- **Traffic Flow Data as Images:** CNNs have been used for traffic speed and flow prediction by representing the traffic data as images. For example, a city can be overlaid with a grid, where traffic metrics like speed and volume are encoded as pixel values; a CNN can then learn spatial relationships such as

upstream-downstream link effects. However, pure CNN methods find it difficult to handle irregular road network topologies, prompting the development of graph neural networks (discussed later) for a more accurate representation.

- **Feature Extraction for Hybrid Models:** Some advanced forecasting models combine CNNs with temporal models, like RNNs or LSTMs, to improve short-term traffic prediction by capturing spatial and temporal dependencies. For example, CNNs extract spatial features of traffic data at a snapshot, and RNNs handle the sequence over time.

Strengths: CNNs excel at capturing local features like detecting vehicles or congestion hotspots, and are translation-invariant, recognizing patterns regardless of position. They handle large inputs using hierarchical feature maps, starting with simple features and progressing to complex ones. For vision-based traffic analysis, CNNs are the primary method, achieving superhuman object detection and classification.

Limitations: A significant limitation of CNNs is their dependence on large amounts of labeled data for training which is labor-intensive to gather. They also demand substantial computational resources, often requiring GPU hardware to achieve real-time processing.. Lastly, CNNs are usually challenging to interpret because their learned features are abstract, making them less suitable when explanations are necessary for critical infrastructure decisions.

3.2 Recurrent Neural Networks (RNNs) and LSTMs

RNNs are designed for sequential data, making them ideal for modeling time series such as traffic signal data or highway flows over time. Standard RNNs have issues with short memory and vanishing gradients for long sequences, so a more advanced version called Long Short-Term Memory (LSTM) networks is often used. LSTMs and related gated RNNs (like GRUs) keep an internal state that can last through longer time periods, allowing them to learn temporal patterns and seasonal trends in traffic.

In traffic management, RNN/LSTM models are used for traffic forecasting—predicting future traffic metrics from past data. For example, an LSTM trained on speed readings can forecast 15 minutes ahead. It learns rush hour patterns, weekly variations, and can include external factors like weather. These models perform well for one-dimensional time series at a single site.

Applications: A city might use LSTM predictors for each road segment, based on historical and neighboring data, to forecast congestion. These predictions enable proactive control, like lengthening green lights if heavy traffic is forecasted. LSTMs also predict traffic demand in ride-sharing and transit systems, supporting broader mobility management.

Strengths: RNNs/LSTMs naturally capture temporal dependencies, smoothing noise and highlighting trends or patterns like daily peaks. LSTMs, with gating, handle long-range dependencies, such as weekly cycles. They need less data than deep networks, reusing weights over time, and can update predictions online as new data arrives.

Limitations: Standard LSTMs don't inherently model spatial relationships; typically, each is used for one sensor or location unless extended, which increases complexity. Training RNNs can be slow and prone to issues like overfitting or getting stuck in local minima, especially with noisy sequences. They struggle with long-term dependencies, such as patterns only occurring during holidays, unless specially handled. To address spatial aspects, LSTMs are often combined with CNNs or GNNs. They mainly act as "black-box" predictors, offering little insight into their predictions beyond input-output correlations.

3.3 Graph Neural Networks (GNNs)

Urban road networks are inherently depicted as graphs, with intersections as nodes and road segments as edges. Graph Neural Networks (GNNs) have become practical tools for modeling this type of data by learning representations of nodes and edges that reflect their connectivity and features. In traffic-related applications, GNNs have rapidly gained popularity for capturing the complex relationships within traffic networks (Liu & Xu, 2024).

Applications of GNNs in traffic include:

- Traffic Prediction in Networks: GNNs can snapshot traffic conditions (speeds/flows at time t) and predict future states, learning how congestion spreads through the network graph. A spatio-temporal GNN may use a sequence of snapshots to forecast. By respecting topology, GNNs understand adjacent segments influence each other. Studies show GNN-based models outperform earlier methods in predicting traffic, especially with complex or irregular networks (Liu & Xu, 2024). They excel in scenarios with many intersecting arterials or rerouting from closures/diversions, as the graph adapts to these changes.
- **Optimal Route Guidance:** GNNs can be used in routing and navigation, predicting travel times by aggregating road segment conditions, improving path planning with better edge cost estimates considering congestion. Some studies use graph reinforcement learning, where nodes act as agents, combining GNNs with RL for network-wide signal control.
- **Infrastructure Planning:** Off-line, graph neural networks identify critical links or simulate network changes. Analyzing connection importance helps

city planners pinpoint key roads to reduce congestion and guide capacity or link additions.

Strengths: GNNs preserve traffic data's relational structure, capturing impacts that CNNs on images cannot. They scale to large networks, like graphs with hundreds of intersections, learning localized filters applicable across the network. GNNs are flexible: adding a new sensor or road updates the graph and extends the model without retraining (transfer learning on graphs). They naturally handle dynamic data by combining GNN for space with LSTM for time, e.g., a Graph ConvLSTM model).

Limitations: Training GNNs is complex and demanding as graphs grow, needing tuning of hyperparameters like propagation layers, which control information flow. Interpretability is limited: node embeddings or attention weights can sometimes be understood, but models mainly act as black boxes. Data availability is a practical issue—GNNs need graph-structured data with features on nodes and edges, but incomplete data reduces effectiveness. Real-time deployment requires efficient computation, challenging without specialized hardware.

3.4 Reinforcement Learning (RL)

Reinforcement Learning (RL) is a framework in which an agent learns to make decisions through interactions with an environment, obtaining feedback via rewards or penalties. In traffic management, RL has become increasingly popular for adaptive signal control and related control issues, as it can potentially find optimal or near-optimal policies that traditional control methods or heuristics may overlook.

Single-Agent RL: In simpler terms, one RL agent manages either a single intersection or a whole network (as one large agent with a combined action space). The state might include traffic data like queue lengths or phase timings; actions could be selecting the next signal phase or adjusting green splits; rewards favor efficiency (e.g., lower delay or fuel use). Early studies used Q-learning for single intersections, with states as queue lengths and rewards for reduced delay (Li et al., 2025). These worked in simulations but faced huge state spaces and instability in larger networks.

Multi-Agent RL (MARL): Urban traffic's decentralized nature favors MARL, with intersections as agents either learning independently or coordinating. SURTRAC exemplifies this with schedule-driven MARL, where intersections optimize schedules and share projections for network coordination. MARL's distributed approach enables scalability for large networks (Smith, Barlow, Xie, & Rubinstein, 2013). Over the past decade, MARL dominates traffic signal control research, with 60%+ of recent works adopting multi-agent frameworks, reflecting a shift to decentralized, cooperative strategies as networks expand (Li et al., 2025).

Key algorithms: Many RL algorithms have been explored, from basic Q-learning and SARSA to deep RL methods like Deep Q-Networks (DQN) for value-based learning and policy gradient approaches (e.g., REINFORCE, Actor-Critic, PPO) for policy optimization (Michailidis, Michailidis, Lazaridis, & Kosmatopoulos, 2025). Actor-critic methods are popular in traffic RL as they handle continuous actions (e.g., variable extension times) and are more sample-efficient. In multi-agent settings, specialized MARL algorithms, such as Independent Q-learning and techniques like value decomposition or actor-attention, support coordination. Some studies introduce hierarchical RL, with high-level agents managing green waves or phase sequences and lower-level agents refining durations.

Hybrid approaches: A noteworthy category is hybrid rule-based + RL controllers. For example, a fuzzy logic controller might handle routine adjustments with an RL agent providing additional tweaks, or RL might be constrained by rules that ensure safety. This can stabilize training and incorporate domain expertise into the learning process.

Strengths: RL's biggest appeal is its ability to discover new control policies through experience without explicit programming, which is crucial in complex problems like city traffic where optimal solutions are hard to derive analytically. It manages multi-objective optimization via weighted rewards, such as combining delay, stops, and emissions. RL can continually learn and improve as traffic patterns change, unlike fixed algorithms that require re-tuning. Field examples like SURTRAC show that even simple RL can outperform static timings by adapting to real-time traffic(Smith, Barlow, Xie, & Rubinstein, 2013). Additionally, RL handles difficult scenarios like stochastic demand and partial observability by learning robust policies.

Limitations: Despite extensive research, deploying RL in real traffic systems faces challenges. Training data and exploration are issues: an agent needs diverse traffic scenarios, including rare events, but in the real world, evil actions can't be deliberately explored (e.g., causing jams). Training in simulation is common, but the sim-to-real gap can impair performance. Additionally, deep RL, especially MARL, is resource-heavy and may not converge to an optimal policy, risking oscillation or overfitting (Michailidis, Michailidis, Lazaridis, & Kosmatopoulos, 2025). Multi-agent systems introduce complexity through non-stationarity, making stability and coordination difficult. Mis-coordination can cause oscillations or gridlock. Scalability is a concern: decentralized MARL scales better than a single agent, but communication and training time may still explode in large cities. Safety and interpretability are also issues: RL policies are often opaque and unpredictable outside their training data. Traffic agencies may demand guarantees like pedestrian right-of-way or no deadlock, hard to achieve with unconstrained RL. As of 2025, most RL traffic control remains in simulation, with few pilot deployments (Michailidis,

Michailidis, Lazaridis, & Kosmatopoulos, 2025). Efforts like federated learning and safety constraints aim to bridge this gap.

3.5 Fuzzy Logic and Rule-Based Systems

Before the rise of deep learning and RL, many adaptive traffic control approaches relied on fuzzy logic or expert system rules, and these methods continue to be relevant, especially in hybrid designs.

Fuzzy Logic Controllers: Fuzzy rules (IF-THEN with fuzzy sets) manage uncertainties in traffic control, allowing a controller to interpolate between rules like: "IF queue is High and waiting is Long, THEN extend green by Medium." Applied since the 1990s/2000s, they mimic human judgment, are easy to modify, handle noisy data, and decide when to end green phases based on multiple criteria.

Rule-Based Expert Systems: These are similar in spirit but use crisp logic. A rule-based strategy is: "If an intersection's queue exceeds X vehicles, immediately trigger the green unless it was just serviced." Many city traffic systems have rule-based overrides—like flush plans triggered by congestion or emergency vehicle preemption, manually programmed from experience.

Integration with AI: Fuzzy and rule-based methods are commonly integrated with learning techniques. One strategy involves using genetic algorithms or other AI methods to optimize the membership functions or parameters of a fuzzy controller. Alternatively, a hybrid RL-fuzzy system can be employed, where fuzzy logic establishes an initial policy, and reinforcement learning further refines it, speeding up the learning process by starting with a sensible strategy. Additionally, neuro-fuzzy systems utilize neural networks to modify or create fuzzy rules based on data.

Strengths: Fuzzy and rule-based systems are interpretable, preferred by city engineers for transparency over black-box neural networks. They are easier to certify for safety, requiring less data and computational power, suitable for older hardware. These systems can be designed with expert knowledge and trial-and-error, making them practical in different contexts.

Limitations: Manually crafted rules or fuzzy sets might not cover all situations or be optimal. Traffic patterns change, making static rules suboptimal or brittle. Fuzzy systems handle some ambiguity but rely on human-designed logic, which can't easily match the fine-tuned optimality of RL or deep learning models, especially in complex networks. Maintenance is an issue—rules need updates as conditions change. Learning systems can adapt automatically, unlike fuzzy/rule-based controllers, which perform well in moderate scenarios but are increasingly replaced by data-driven models in advanced deployments.

3.6 Comparative Summary of Techniques

Each AI technique has a niche in the intelligent traffic management toolkit. Table 1 provides a comparative overview:

Table 1. Comparison of AI Techniques for Urban Traffic Management.

Technique	Applications	Strengths	Limitations
Convolutional Neural Networks (CNNs)	- Incident detection from CCTV - Vehicle counting Spatial traffic pattern extraction	- High accuracy in vision tasks - Captures local spatial features Well-developed architectures.	- Requires large labeled datasets - Computationally intensive - Less suited for non-grid road network data without adaptation.
Recurrent Neural Networks (RNNs/ LSTMs)	- Time-series prediction of traffic volume/speed - Anomaly detection in traffic flow trends	- Captures temporal dynamics and periodic patterns - Can update predictions continuously as new data arrives.	- No inherent modeling of spatial relation - Can struggle with very long-term dependencies Risk of overfitting noise if not regularized.
Graph Neural Networks (GNNs)	- City-wide traffic forecasting on road network graphs - Mdeling congestion propagation; route travel time estimation	- Leverages road network topology for relational learning - Handles complex intersections and irregular networks - State-of-the-art accuracy in traffic prediction on graphs	- Complex to implement and train - Needs comprehensive sensor coverage of network - Real-time inference on large graphs can be demanding
Reinforcement Learning (single-agent)	- Adaptive signal control for one intersection or centralized control of small network - Dynamic pricing of toll roads	- Learns directly from interaction how to optimize long-term rewards - Can outperform fixed logic by discovering new strategies.	- Huge state/action space for larger networks - Training is slow and not guaranteed to converge - Black-box policy makes it hard to ensure safety or explain decisions.

continued on following page

Table 1. Continued

Technique	Applications	Strengths	Limitations
Reinforcement Learning (multi-agent)	- Decentralized control of many signals with coordination - Regional traffic management with multiple cooperating agents	- Scalable to large networks by distributing control - Aligns with actual traffic control hierarchy - Agents can learn coordination leading to network-level optimization	- Very challenging to train and tune - Requires communication protocols between agents - Potential for instability or oscillations if agents don't properly cooperate - Few real-world deployments yet
Fuzzy Logic Controllers / Rule-Based Systems	- Isolated intersection adaptive control - Special event traffic management - Ramp metering with fuzzy logic	- Interpretable and transparent control rules - Handles uncertainty and noisy data gracefully - Does not require big data or training	- Manual design – may not handle all scenarios or adapt to changes - Suboptimal compared to learning-based methods in complex, dynamic conditions - Maintaining and updating rules is labor-intensive.

Intelligent traffic systems may combine techniques; for example, an IoT-based intersection could use CNNs in cameras to detect traffic, feed data to a GNN-LSTM model for predictions, and employ a reinforcement learning agent (started with fuzzy logic) to set signal timing. The next section will examine real-world systems and prototypes.

4. CASE STUDIES OF INTELLIGENT TRAFFIC MANAGEMENT SYSTEMS

To illustrate the AI traffic management state-of-the-art, we examine key systems and deployments. These case studies – from research and industry – demonstrate effective intelligent approaches in various urban settings.

4.1 SURTRAC – Scalable Urban Traffic Control (Pittsburgh, USA)

SURTRAC is a pioneering intelligent traffic signal system developed at Carnegie Mellon University and first deployed in Pittsburgh. It is a decentralized, multi-agent system in which each intersection optimizes its own signal timing while communicating with its neighboring intersections (Smith, Barlow, Xie, & Rubinstein, 2013).

SURTRAC's method involves sensors (video or radar) at each intersection to detect vehicles. The controller regularly generates an optimized schedule aimed at minimizing delays, employing a schedule-based algorithm that leverages dynamic programming to cut wait times. It shares an anticipated outflow plan with neighboring intersections, which incorporate this into their own optimization process. This implicit coordination allows platoons to move forward smoothly without needing explicit synchronization. The cycle repeats every few seconds to enable continuous adaptation.

A major innovation of SURTRAC is its ability to scale effectively. It achieves this by restricting communication to neighboring intersections and avoiding the need for a central controller, allowing it to handle large networks without creating communication or computational bottlenecks (Smith, Barlow, Xie, & Rubinstein, 2013). In fact, SURTRAC has grown from the original 9-intersection pilot to over 50 intersections in Pittsburgh.

Performance: Field results from Pittsburgh's East Liberty pilot showed dramatic improvements. SURTRAC reduced travel times by over 25%, increased vehicle speeds by 34%, and cut stops by 30%. Wait times dropped over 40%, and emissions (~21%) decreased due to less idling (Smith, Barlow, Xie, & Rubinstein, 2013). These benefits occurred throughout the day, especially at midday off-peak (Smith, Barlow, Xie, & Rubinstein, 2013). SURTRAC also proved adaptable, adjusting signals for closures or demand spikes with minimal human intervention.

Significance: SURTRAC demonstrated that multi-agent AI control can work in practice. Operating in Pittsburgh and commercialized by a CMU spin-off, it inspired global research on multi-agent RL for signals. SURTRAC demonstrates that decentralization through local communications can approximate global coordination without requiring heavy infrastructure (Smith, Barlow, Xie, & Rubinstein, 2013). Pittsburgh's success secured further funding for smart traffic projects and established the city as a leader in AI civic infrastructure.

4.2 Alibaba City Brain (Hangzhou, China)

City Brain, initiated by Alibaba Cloud, is a smart city platform focusing on urban traffic management. Deployed in Hangzhou (~10 million population), it uses centralized big data for traffic control. The system collects real-time urban data using video feeds, GPS, incident reports, transit data into a cloud AI platform. AI algorithms analyze this data to optimize traffic flow and respond to conditions.

Key features of City Brain's traffic management include:

- Video-based Traffic Perception: Using computer vision, City Brain extracts traffic data from camera feeds—counting vehicles, detecting accidents, measuring speeds and queue lengths at intersections (Zhang et al., 2019). This offers a real-time, comprehensive view of traffic without relying solely on traditional sensors.

- Adaptive Signal Optimization: The platform calculates optimal signal plans for city intersections. Initially, it used traditional algorithms and AI to adjust timings and reduce delays. More recently, City Brain added deep reinforcement learning agents that update signal timing based on live data, as noted in Alibaba's reports (Zhang et al., 2019). Unlike SURTRAC's distributed approach, City Brain uses centralized optimization where a command center makes and distributes adjustments.
- **Emergency Vehicle Priority:** One celebrated feature is the ability to detect ambulances and fire trucks and preempt signals along their route. By turning lights green ahead of an ambulance, City Brain reportedly cut emergency response travel times by a significant margin (some reports claim 50% faster response in pilot cases).
- **Dynamic Traffic Routing and Event Management:** During major events, City Brain manages traffic by controlling closures, detours, and signals. It links with police and transit systems for a coordinated response.

Outcomes: Hangzhou benefited from City Brain, improving traffic from China's 5th to 57th most congested city by 2018. Alibaba reported increased traffic speeds and a 15% reduction in snarls in the first year. The Xiaoshan District congestion index dropped 3.5% annually. While not solely AI-based, these gains include infrastructure and enforcement. In Quzhou, during a Lantern Festival with ~78,000 visitors, the system autonomously managed traffic, achieving 40% efficiency improvements over the previous year (People's Daily, 2019).

Expansion: Following Hangzhou, City Brain has been deployed in numerous other Chinese cities, including Beijing, Shanghai, and Suzhou, as well as overseas in locations such as Kuala Lumpur, Malaysia. While each deployment is customized, they all use cloud + AI to serve as a "central brain" for city operations. The traffic management module is often the highlight, analyzing large datasets and intervening in real-time. City Brain demonstrates how IoT and AI transform traffic management from reactive to proactive, enabling citywide optimization. It aims to "see the city as a whole" rather than isolated parts and intersections ((Zhang et al., 2019).

Challenges and Notes: City Brain requires significant infrastructure like high-bandwidth networks, cloud resources, and IoT devices at scale. Alibaba emphasizes evolving AI tech, such as video analytics and neural network efficiency, to handle data overflow (Zhang et al., 2019). They work with city governments to integrate systems, raising privacy concerns about mass surveillance, which Chinese cities manage differently than Western ones. Despite these issues, City Brain remains a leading example of AI in urban traffic control.

4.3 Google "Green Light" Traffic Optimization

Google Green Light is an AI project by Google Research that optimizes traffic signals using existing data, primarily Google Maps traffic information. Unlike SURTRAC and City Brain, it doesn't require new infrastructure but provides AI suggestions on signal timing for city engineers to review and implement, analyzing and addressing inefficiencies (Google Research, 2024).

How it works:

- Google uses Google Maps location data to infer traffic patterns at intersections by observing vehicle movements, deducing signal timings and coordination without extra sensors (Leffer, 2024; Google Research, 2024). They treat cars as "mobile sensors," avoiding additional hardware.
- Using an AI model, likely a reinforcement learning or simulation-based approach, they find timing tweaks to reduce stops. For instance, an intersection may cause unnecessary stops on a main road; the AI might recommend increasing the green light duration for that road during specific hours (Leffer, 2024).
- These suggestions are delivered via a cloud dashboard to city traffic personnel (Google Research, 2024), showing the current timing plan, proposed changes, and data-driven reasons (like "this will create a longer green wave along Main St from 8:00–8:30 AM").
- City engineers can accept or tweak recommendations, implement them in signal systems (usually database changes), then monitor and report results (e.g., "stops decreased by 20%, travel time improved 10% on that corridor").

Deployments and Results: As of 2024, Google has piloted Green Light in 13 cities across four continents, including Seattle, Manchester, Haifa, Abu Dhabi, Bangalore, and others (Leffer, 2024). In Seattle, Green Light was tested on five intersections; in Israel and India, around a dozen. Early results are promising, showing a potential up to 30% reduction in stops and a 10% reduction in emissions by smoothing traffic (Google Research, 2024). In Seattle, AI recommendations cut stops by a third, reducing idling and speeding travel. In Israel, officials reported improved travel times where suggestions were used.

In a suburb of Tel Aviv, Green Light identified that two consecutive signals were not well coordinated. By adjusting the offset between them, they created a "green wave" for the main direction. After the change, vehicles experienced far fewer stops, and fuel consumption dropped measurably in that zone. (Leffer, 2024)

It's important to note that Green Light's approach is incremental – it doesn't overhaul the control The system refines existing timings with AI insights, making it low-cost and easy to deploy without new hardware. It doesn't adjust signals in real-time but offers periodic optimization, reviewing plans weekly or monthly.

Reception: Cities generally welcomed the collaboration, viewing it as a way to access Google's data and AI expertise. However, challenges arose; in Manchester, UK, Google recommendations sometimes conflicted with local transit priorities or knowledge, such as not recognizing intentional traffic-slowing stops (Leffer, 2024). This underscores the need for contextual review of AI suggestions. Google has responded by adjusting its approach, like excluding intersections with heavy pedestrian or bus priorities from optimization.

Green Light is expanding, positioned by Google as part of their sustainability efforts to cut urban emissions. It exemplifies how public-private data sharing provides quick benefits: many cities lack resources for ongoing traffic studies, but Google's crowd-sourced data and AI fill that gap. If Green Light keeps succeeding, it could become a global tool for traffic engineers—an AI assistant for signal timing.

4.4 Academic Prototypes and Other Deployments

Beyond the headline cases above, numerous other intelligent traffic systems have been tested or implemented on smaller scales:

- **Adaptive Control in Los Angeles:** Los Angeles operates one of the world's largest centralized traffic control systems, ATSAC. Although ATSAC traditionally relied on timed plans, the city has started integrating AI techniques to enhance incident response. Machine learning models now predict travel times on arterials and identify the optimal signal plans when freeway incidents divert traffic onto surface streets. This serves as an example of enhancing an existing system with AI predictions.
- Smart Traffic Lights in Milton Keynes (UK): As part of an innovative city initiative, this town tested an AI-driven adaptive traffic signal system using reinforcement learning in simulations, with plans to deploy it in the field. Initial simulation results indicated a delay reduction of up to 15–20% under different traffic conditions, supporting the effectiveness of RL methods in a realistic town model (Kolat, Kővári, Bécsi, & Aradi, 2023)
- **NoTraffic (Startup Deployment):** An Israeli startup, "NoTraffic," has implemented AI-powered traffic signal systems in several U.S. cities, including Phoenix, Arizona. These systems utilize cameras and radar at intersections to identify all road users, with an AI controller that dynamically adjusts green light durations. The system functions similarly to a modern actuated traffic signal but with integrated AI logic. Pilot tests in Phoenix reported notable

decreases in wait times for side-street traffic and enhanced safety, thanks to quicker pedestrian detection.

- Multi-Agent RL Research (Hangzhou Example): A recent academic study integrated multi-intersection deep reinforcement learning (RL) with connected vehicle data within a digital twin of Hangzhou. With 30% of vehicles connected and supplying real-time data, the RL agents cut average travel time by approximately 30% and CO_2 emissions by around 28% compared to current actuated control methods (Michailidis, Michailidis, Lazaridis, & Kosmatopoulos, 2025). This indicates that increasing vehicle connectivity could lead to even greater benefits through enhanced AI control, due to more comprehensive and detailed traffic state information.

Each case shows that intelligent traffic management can improve outcomes but also reveals practical challenges, like starting with pilots at some intersections due to integration complexity. Scaling up involves dealing with legacy systems, training, and public acceptance. Despite this, more cities are adopting AI and IoT in traffic systems, transitioning from ideas to actual deployment through step-by-step evaluation.

5. PROPOSED IOT-AI ARCHITECTURE FOR SMART TRAFFIC CONTROL

After reviewing current systems, we present a layered architecture for intelligent traffic management that combines IoT sensing with AI decision-making. This architecture is divided into layers for clarity, but in reality, these components work together in a feedback loop to continuously optimize traffic flow.

Layer 1: Sensing and Data Collection

This is the foundation, consisting of all IoT devices and data sources that monitor the transportation network. It includes:

- **Roadside sensors:** Inductive loop detectors, wireless magnetometers, radar traffic detectors, and CCTV cameras are installed at intersections and along roadways. They collect data on vehicle counts, speeds, queue lengths, and incidents. Modern cameras equipped with edge AI can identify not only vehicles but also pedestrians and cyclists.
- **Connected vehicle data:** Vehicles are increasingly functioning as sensors. Connected cars, buses, and fleet vehicles now transmit GPS locations,

speeds, and can even detect events such as hard braking. Smartphone apps like Google Maps and Waze provide anonymized probe data, offering valuable information on travel times and congestion levels.

- **Environmental sensors:** This layer includes weather stations (which detect rain and fog that could impact traffic), air quality sensors (monitoring emissions), and social media feeds (where people report crashes).
- **City data systems:** Existing databases, including traffic signal timing plans, public transit schedules, road network maps, and event schedules, are also integrated.

These diverse data sources are collected continuously. Communication networks—such as fiber optics, 4G/5G cellular, and dedicated short-range links for V2I—serve as the circulatory system, transmitting sensor information from the field to the processing layer (Maurya, Pal, & Sarvakar, 2024). Low latency and high reliability are crucial; for example, a delay of just one minute could make a control action less effective. Edge computing devices, discussed below, are typically positioned at this boundary, collecting data from local sensors.

Layer 2: Data Processing and Fusion

In this layer, raw data from Layer 1 are aggregated, cleansed, and transformed into usable information. Functions here include:

- **Data filtering and cleaning:** Eliminating noise and incorrect readings, such as a faulty sensor stuck at a high value. For video, this could include stabilizing feeds or filtering out irrelevant objects.
- **Sensor fusion:** Integrating data from various sources creates a more comprehensive view. For instance, a system could combine camera-based counts with loop detector occupancy data and connected vehicle trajectories to accurately estimate the actual vehicle density on a road segment.
- **Edge Computing:** Some processing occurs at the edge (such as at intersections or roadside units) to minimize data volume and latency. An edge device might count vehicles locally and transmit only summary statistics upstream rather than streaming raw video. This approach conserves bandwidth and enables faster local responses, such as quick alerts for pedestrian crossings.
- **Central data cloud platform:** After initial edge processing, data often flows into a central system, such as a cloud or city data center, which aggregates citywide information. Here, a dynamic traffic state database is maintained, including current speeds on every road, locations of transit vehicles, inci-

dent reports, and more. Historical data storage is also managed for analyzing trends.

Importantly, this layer may utilize real-time databases and streaming analytics frameworks to manage high data velocities. For instance, event stream processing can generate immediate alerts like "accident detected on highway at location X," which are sent to the decision layer. Meanwhile, a rolling horizon predictor might constantly update short-term forecasts for each link.

Layer 3: AI Decision-Making and Control Logic

This is the "brain" of the architecture. It encompasses the AI models, algorithms, and control logic that make decisions to optimize traffic. Within this layer:

- Predictive Analytics: AI models forecast near-term conditions, such as predicting traffic volumes 15 minutes in advance on key corridors (Gheorghe & Soica, 2025). They employ techniques like deep learning models (see Section 3), including LSTM or GNN-based predictors. These predictions enable the system to act proactively, such as adjusting signal timings for anticipated traffic surges.
- **Optimization and Control Algorithms:** The system determines control actions based on current and forecasted conditions. It might involve a reinforcement learning agent calculating new signal phase timings, a linear program optimizing network flows, or a heuristic rule activating a detour plan.
- Adaptive Signal Control module: Adjusts intersection timings in real time, either through distributed control—where each intersection optimizes locally and coordinates with neighbors—or through centralized control, where an algorithm manages an entire region. For example, a central system might employ a model-predictive control method, re-optimizing signal plans every few minutes using the latest data (Michailidis, Michailidis, Lazaridis, & Kosmatopoulos, 2025).
- **Route Guidance and Traveler Information:** If congestion increases, an AI can choose to reroute traffic. It can provide this information to navigation services—such as through public apps or connected car systems—to help drivers find alternative routes, or activate variable message signs on highways to suggest different paths.
- **Incident Management:** When an incident is detected, the system may instantly execute strategies such as traffic signal flush plans (to clear a congested road), dynamically adjust reversible lanes, or deploy response crews. AI

contributes by suggesting the most effective strategy through scenario simulations or by applying insights from past similar incidents.

- **Transit and Multi-modal Priority:** The control logic also includes policies such as prioritizing buses or trams at signals when they are behind schedule and managing dedicated lanes. AI can assist in balancing these priorities with overall traffic flow.

This decision layer functions as the central AI controller. It takes in processed data representing the current traffic situation, uses predictive models to forecast potential issues, and then applies optimization and control algorithms to determine the best actions. These actions are subsequently sent to the actuating layer.

Layer 4: Actuation and Execution

Here, the decisions are implemented on the ground. Commands from the AI layer translate to:

- **Traffic Signal Controllers:** The AI computes timings or phase plans that are sent to intersection signal controllers. Modern controllers support interfaces, such as NTCIP protocols, enabling remote adjustments of cycle durations, offsets, or switching between timing plans. In adaptive systems, the AI can directly adjust the next green phase in real time.
- **Variable Message Signs and Signals:** Instructions like "Accident ahead, use an alternate route" can be shown on electronic signs. Variable speed limit signs may be adjusted, such as gradually lowering speed limits upstream of congestion to reduce shockwaves.
- **Ramp Meters and Lane Control:** On freeways, the system might modify ramp metering rates using actuators or open and close specific lanes, such as hard shoulder running or contraflow lanes during peak hours.
- **Connected Vehicle Broadcasts:** In a connected vehicle environment, the system can transmit V2I messages. For example, it might broadcast optimized signal phase timing to approaching cars, informing them when the light will change, or even coordinate with autonomous vehicles to form platoons through a green wave.

This layer needs reliable feedback to confirm command execution and to fail safely if it cannot. It includes redundancy, such as reverting signals to safe default plans if the AI system goes offline.

Layer 5: Monitoring, Evaluation, and Learning Feedback

The final layer closes the loop by evaluating performance and feeding that knowledge back into the system for continuous improvement:

- Performance Monitoring: The system continuously tracks key performance indicators (KPIs), including travel times, queue lengths, number of stops, emissions (when estimable), and safety surrogate metrics such as hard braking events. These metrics are compared to historical data or set targets. For example, after implementing a control action like a new signal timing, the system checks if the average delay has decreased as expected (Google Research, 2024).
- **Incident Verification:** If the system takes action based on a suspected incident, it seeks confirmation—either from cameras or reports—that the incident actually occurred and clears once resolved, helping to prevent false alarms.
- Learning and Adaptation: This layer is vital for AI. It involves feeding data on successful or failed strategies into machine learning model updates. For instance, a reinforcement learning agent uses the reward (such as shorter queues) to refine its policy for the next action. Similarly, a predictive model may retrain overnight with that day's data to improve accuracy over time. In more advanced configurations, federated learning can be employed—where models are updated across multiple locations using data from various sources, such as different cities, without sharing raw data to protect privacy (Li et al., 2025). This allows an AI model in one city to benefit from data patterns observed in another city (if similar), enhancing its robustness.
- Human-in-the-Loop and Overrides: Traffic engineers oversee the system through dashboards. They can give corrective feedback or override decisions if something appears off. For instance, if the AI maximizes throughput but citizens find crossing streets difficult, human operators might modify the policy to prioritize pedestrians. Their inputs serve as extra training data, as the AI can be programmed to learn from operator approvals or rejections of recommendations, similar to Google's system (Google Research, 2024).

This feedback layer also encompasses maintenance alerts, such as flagging sensors that are persistently faulty for repair, and cyber-security monitoring to detect anomalies that may signal a system breach.

Real-World Mapping: Many actual deployments follow this architecture. For example, SURTRAC focuses on Layers 1–4 within each decentralized intersection, where sensing, processing, decision-making, and actuation happen locally, with minimal Layer 5 involvement except for implicit learning through schedule adjustments. City Brain relies heavily on a central Layer 3 that integrates large data from Layer 1, sends commands through Layer 4 to traffic signals, and then evaluates citywide metrics in Layer 5. Google Green Light is somewhat simpler; it omits real-time Layer 4 actuation (leaving that to humans), but it still closes the loop by assessing outcomes and updating its suggestions.

The layered approach guarantees modularity—each component can evolve separately. The sensing layer can enhance data collection by adding more IoT sensors or connected vehicles, providing better input for the AI. AI algorithms can be refined or upgraded through software updates without altering the hardware. Meanwhile, the actuation layer can accommodate new actuators, such as interfaces to autonomous vehicles, in future developments. Figure 1 depicts this proposed architecture.

Figure 1. AI-Enabled Intelligent Transportation System Architecture

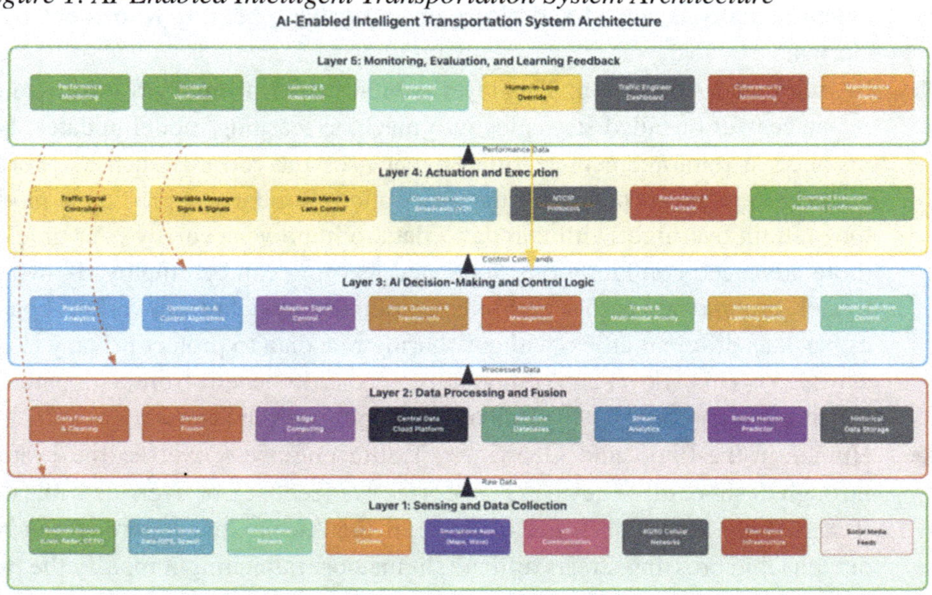

In summary, this IoT-AI architecture outlines a blueprint for smart traffic control systems that operate as sense–think–act systems with ongoing learning: 1. Ubiquitously sense the multi-modal traffic environment using IoT sensors and connected data. 2. Use AI to think at different time scales, enabling both immediate reactions and long-term predictions and optimizations. 3. Act by directly controlling infrastructure and providing information to road users. 4. Continuously learn from outcomes to improve efficiency and robustness over time.

6. CHALLENGES AND FUTURE DIRECTIONS

Implementing large-scale intelligent traffic management presents notable challenges. As we move forward, researchers and practitioners need to tackle technical, ethical, and operational hurdles to unlock AI and IoT's full potential in urban mobility. Here, we outline the main challenges and emerging directions.

6.1 Data Privacy and Security

Smart traffic systems collect a lot of data, including sensitive information like vehicle trajectories, license plates, and travel patterns, which raises privacy concerns—such as inferring personal routines or identities. Strong data governance is essential. Techniques such as anonymization, aggregation, and on-device processing help, with raw data remaining on sensors. Cybersecurity is critical because networked infrastructure can be hacked, risking chaos like all lights turning green at once. Future systems need encryption, authentication, and anomaly detection. Secure ITS efforts and blockchain for signal logs aim to increase trust. Regulators, like Europe's GDPR, require data justification, and federated learning offers a way to keep data local while sharing models updates (Li et al., 2025).

6.2 Integration of Autonomous and Connected Vehicles

The rise of connected and autonomous vehicles (CAVs) will significantly transform traffic management. Future traffic systems must accommodate a mix of human-driven and autonomous cars. Challenges include: - V2X Communication Standards: Ensuring reliable communication between vehicles and infrastructure amidst ongoing advances in DSRC and C-V2X. Systems will need to process numerous V2X messages in real time. - Adaptive algorithms for mixed fleets: Optimizing signals when some vehicles can adjust their speed based on signal timing, while others cannot. As connected vehicles grow, AI signals may evolve into advisories, offering speed suggestions to reduce stops. During the transition, algorithms must ensure fairness and efficiency. Research shows moderate CAV adoption can cut delays by 30–40% through platooning and priority strategies. - New control paradigms: Once most vehicles are autonomous, traffic lights could become unnecessary, with intersections managing vehicle platooning through "reservation-based" control. Achieving this requires careful planning and standardization (Michailidis, Michailidis, Lazaridis, & Kosmatopoulos, 2025).

6.3 Scalability and Real-Time Performance

City traffic networks have many intersections and arterial segments, and scaling AI algorithms is difficult. Multi-agent RL faces exponential complexity with more agents, and coordinating many agents is challenging(Kolat et al., 2023). Centralized methods are limited by computational power—optimizing 500 signals may be too slow for real-time needs. Hierarchical control divides the city into zones with zone-level coordinators or uses a two-tier system: the upper layer sets strategies, and the lower adjusts local details. This reduces computation while achieving near-optimal results. Improving algorithm efficiency involves approximation methods, parallel computing (GPUs), or faster reinforcement learning. Edge computing distributes the load, with each intersection controller performing local calculations, but ensuring coherent global behavior is still an active challenge.

6.4 Edge Computing and Latency

Edge computing is a solution for low-latency needs in ITS. By processing data at the edge—like roadside units or intersection controllers—decisions such as changing a light for an ambulance occur faster (Maurya, Pal, & Sarvakar, 2024). It also reduces reliance on continuous backhaul connectivity, as local edge nodes operate if connection to the central system fails. A challenge is deploying sufficient computing power in the field within constraints like cost, weatherproofing, and maintenance. Running complex deep RL agents on limited hardware, like traffic signal controllers, is difficult—requiring specialized hardware accelerators. Balancing between edge and cloud is essential; some tasks need the broader view of cloud computing. Future architectures likely combine quick decision-making at the edge with strategic processing in the cloud, known as fog computing. Research prototypes, like fog frameworks, show local nodes handle vehicle detection while the cloud manages long-term optimization (Asad & Martin, 2024).

6.5 Adaptability and Continuous Learning

Traffic patterns evolve, such as during COVID-19, requiring systems to adapt, which challenges AI model updates. Online learning can help but risks "catastrophic forgetting,' erasing knowledge for rare situations. Transfer learning aids in retaining knowledge. Testing AI updates in simulation or shadow modes is vital to prevent gridlock, with "digital twins' offering safe, real-time city environment testing before deployment.

6.6 Explainability and Trust

Gaining trust from traffic engineers and the public is vital for AI systems. When an autonomous traffic agent acts oddly (e.g., keeping a side street green too long and causing backups), humans need to understand why. Many AI models are black boxes. Research into explainable AI for traffic is needed so systems can provide reasoning (like, "I extended this green because my prediction showed a platoon arriving and the mainline was lightly loaded"). Some projects include rule-based logic as backup or suggest actions for humans to approve (like Google's model), keeping humans involved. As confidence in AI improves with proven performance, more autonomy can be granted, but transparency will speed acceptance.

6.7 Multi-Modal and Equity Considerations

Future traffic management should fully include pedestrians, cyclists, and public transport, moving beyond car-centric focus. AI now supports these modes with sensors prioritizing pedestrians and adaptive transit signals that minimize delays for buses without neglecting others. There's also an equity concern: AI improvements should benefit all neighborhoods, not just affluent areas, requiring policies to guide algorithms—an ongoing challenge for technology and policy communities.

6.8 Environmental and Policy Goals

Finally, aligning AI optimizations with broader objectives such as emission reduction and Vision Zero, which aims to eliminate road fatalities, presents a challenge. Sometimes, reducing emissions conflicts with decreasing travel delays—for example, smoothing traffic flow can boost throughput and increase demand. Policymakers may establish specific targets like low-emission zones or traffic caps in city centers. Future AI systems might need to incorporate dynamic road pricing or restrictions, effectively becoming tools to enforce policies, such as deliberately controlling traffic into pollution-sensitive areas. This relationship between AI control and policy goals will continue to expand. Additionally, AI can support policymaking by simulating scenarios to evaluate the potential impacts of measures like pedestrianizing streets or adjusting traffic signals.

In summary, although progress in intelligent urban traffic management is notable, addressing these challenges is essential for widespread, sustainable adoption. Interdisciplinary collaboration is required, involving traffic engineering, AI research, communications, embedded systems, and public policy. Future trends point toward more integrated, human-centered, and resilient systems: - Increased integration of vehicles and infrastructure for cooperative automated traffic. - Greater utilization of distributed intelligence (edge/fog computing and federated learning) to enhance resilience and privacy. - Focus on safety, explainability, and alignment with environmental objectives. - Scalability via hierarchical and multi-agent structures, potentially supported by next-generation computing such as quantum or 6G networks for ultra-low latency.

As these developments unfold, cities will gradually get closer to the vision of self-optimizing traffic networks that not only reduce congestion and emissions but also adapt smoothly to the constantly changing urban mobility landscape.

7. CONCLUSION

Urban traffic management is shifting due to AI and IoT. Traditional systems, constrained by static plans and local heuristics, are being replaced or improved by data-driven approaches. Cities deploy sensors and communication networks to monitor traffic continuously. AI algorithms, such as neural networks and reinforcement learning, analyze traffic patterns and make real-time decisions. We examined AI techniques (CNNs, RNNs, GNNs, RL, hybrid models) and their roles in signal optimization, congestion forecasting, and incident management, noting their strengths and limitations.

Real-world case studies like Pittsburgh's SURTRAC, Hangzhou's Alibaba City Brain, and Google's Green Light pilot show AI traffic management benefits. These systems reduce delays, stops, emissions, and demonstrate varied strategies (centralized, distributed, automated, decision-support) for cities to consider.

We introduced a layered architecture combining IoT sensing with AI decision-making and feedback learning. It serves as a blueprint for creating modular, scalable, and improving smart traffic systems. The aim is a closed-loop system where real-time data enables immediate actions, and outcomes refine the control policy—forming an ever-learning traffic "brain" for the city.

Looking ahead, challenges include ensuring data privacy and cybersecurity as city infrastructure becomes connected. Managing mixed autonomous and human-driven vehicles needs innovative strategies for connectivity without unfairness. Scalability and reliability of AI remain engineering challenges, requiring hierarchical control and fail-safes. Human oversight, transparency, and alignment with societal goals will influence public acceptance of AI-managed streets.

Intelligent urban traffic management using AI and IoT offers a powerful solution to congestion, adapting to demand in real-time and optimizing objectives. Combining sensing technology, big data, and machine learning can improve traffic flow, reduce pollution, and enhance safety. Cities implementing these systems see benefits like shorter commutes and lower emissions (Gheorghe & Soica, 2025; Michailidis,

Michailidis, Lazaridis, & Kosmatopoulos, 2025). As research advances, AI-driven traffic control will move from pilots to standard smart city infrastructure, increasing efficiency and making cities safer, more responsive, and less congested. Today's intelligent intersections foreshadow the self-managing city of tomorrow.

REFERENCES

Asad, U., & Martin, Y. (2024). "The Role of Edge Computing in Modern Traffic Management and Speed Detection," Dec. 2024. DOI: DOI: 10.13140/RG.2.2.23474.00961

Gheorghe, C., & Soica, A. (2025, February). Revolutionizing Urban Mobility: A Systematic Review of AI, IoT, and Predictive Analytics in Adaptive Traffic Control Systems for Road Networks. *Electronics (Basel)*, *14*(4), 719. DOI: 10.3390/electronics14040719

"Green Light – Using Google AI to Reduce Traffic Emissions," Google Research Project Page (accessed Jul. 2024). https://sites.research.google/gr/greenlight

Karim, A., Raza, M., Alharthi, Y., Abbas, G., Othmen, S., Hossain, M., Nahar, A., & Mercorelli, P. (2024, August). Visual Detection of Traffic Incident through Automatic Monitoring of Vehicle Activities. *World Electric Vehicle Journal*, *15*(9), 382. Advance online publication. DOI: 10.3390/wevj15090382

Kolat, M., Kovari, B., Becsi, T., & Aradi, S. (2023). Multi-agent reinforcement learning for traffic signal control: A cooperative approach. *Sustainability (Basel)*, *15*(4), 3479. DOI: 10.3390/su15043479

Leffer, L. (2024). "Can Google Make Stoplights Smarter? – Google's Project Green Light Uses AI to Take on City Traffic," Scientific American, Aug. 15, 2024. https://www.scientificamerican.com/article/googles-project-green-light-uses-ai-to-take-on-city-traffic/

Li, M., Pan, X., Liu, C., & Li, Z. (2025, November). Federated deep reinforcement learning-based urban traffic signal optimal control. *Scientific Reports*, *15*(1), 11724. Advance online publication. DOI: 10.1038/s41598-025-91966-1 PMID: 40188158

Liu, J., & Xu, S. (2024). "Advancements of Graph Neural Networks in Urban Traffic Prediction," in *Proceedings of the 1st International Conference on Engineering Management, Information Technology and Intelligence (EMITI 2024)*, Jan. 2024, pp. 62–66. SCITEPRESS. DOI: 10.5220/0012902200004508

Maurya, S., Pal, O., & Sarvakar, K. (2024). *"Layered Architecture of IoT," Apr 2024*. Secure and Intelligent IoT-Enabled Smart Cities., DOI: 10.4018/979-8-3693-2373-1.ch009

Michailidis, P., Michailidis, I., Lazaridis, C. R., & Kosmatopoulos, E. (2025, May). Traffic Signal Control via Reinforcement Learning: A Review on Applications and Innovations. *Infrastructures*, *10*(5), 114. Advance online publication. DOI: 10.3390/infrastructures10050114

Mutambik, I. (2025). IoT-Enabled Adaptive Traffic Management: A Multiagent Framework for Urban Mobility Optimisation. *Sensors (Basel)*, *25*(13), 4126. DOI: 10.3390/s25134126 PMID: 40648380

People's Daily. (2019). "City Brain helps boost efficiency in Chinese cities," People's Daily Online, Apr. 24, 2019. https://en.people.cn/n3/2019/0424/c90000-9571706.html

Peyman, M., Copado, P. J., Tordecilla, R. D., Martins, L. C., Xhafa, F., & Juan, A. A. (2021). "Edge Computing and IoT Analytics for Agile Optimization in Intelligent Transportation Systems," Energies, vol. 14, no. 19, Article 6309, Oct. 2021. MDPI https://www.mdpi.com/1996-1073/14/19/6309

Sharma, V., & Shah, A. (2025, January). A Survey on Smart Traffic Monitoring System using AI. *International Journal of Innovative Research in Computer and Communication Engineering*, *13*(1). Advance online publication. DOI: 10.15680/IJIRCCE.2025.1301129

Smith, B., Barlow, G. J., Xie, X., & Rubinstein, M. H. (2013). "SURTRAC: Scalable Urban Traffic Control – Performance Evaluation of a Multi-Agent Adaptive Signal Control System," in Transportation Research Board 92nd Annual Meeting, Paper 13-0315, 2013. https://www.ri.cmu.edu/pub_files/2013/1/13-0315.pdf

Zhang, J., Hua, X., Huang, J., Shen, X., Chen, J., Zhou, Q., Fu, Z., & Zhao, Y. (2019). City Brain: Practice of Large-Scale Artificial Intelligence in the Real World. *IET Smart Cities*, *1*(1), 78–84. DOI: 10.1049/iet-smc.2019.0034

Chapter 8
Smart City Energy Management via IoT, Wind Power, and Electric Vehicles

Siwar Khemakhem
Faculty of Sciences of Gafsa, Tunisia

Mouna Rekik
https://orcid.org/0000-0003-1869-2394
National Engineering School of Sfax, Tunisia

Lotfi Krichen
National Engineering School of Sfax, Tunisia

ABSTRACT

The Internet of Things (IoT) is a key technology for addressing major challenges through its network of interconnected objects with sensors, electronics, and software. It forms a crucial foundation for smart city development by enabling efficient information and communication management. This chapter explores IoT-enabled energy management in smart cities, focusing on the integration of Plug-in Electric Vehicles (PEVs) and Wind Turbines (WTs) as renewable energy sources. It highlights how smart homes and neighborhoods can collaborate for peak shaving and valley filling, aligning energy demand with WT generation and PEV storage capabilities. A case study demonstrates the design of an energy management system that monitors total energy use while smoothing the daily power curve through WT and PEV

DOI: 10.4018/979-8-3373-5727-0.ch008

integration, thus enhancing grid stability. Simulation results validate the proposed control strategy under various load scenarios, showing its effectiveness in flattening load profiles, reducing load variance, and preventing blackouts in smart grids.

INTRODUCTION

With increasing population density, limited availability of conventional energy resources, and growing global concerns over greenhouse gas emissions and climate change, the need for sustainable urban development has become more urgent than ever. Traditional energy systems are no longer sufficient to meet the rising demands for electricity, mobility, and environmental sustainability. As a result, homes and cities are being driven by the emergence of advanced technologies such as the Internet of Things (IoT), Artificial Intelligence (AI), renewable energy systems, and smart grids to evolve into intelligent, adaptive ecosystems. These smart environments are expected to provide efficient, reliable, and environmentally responsible services and infrastructure that enhance the quality of life for inhabitants. Intelligent homes, equipped with automated energy management, real-time monitoring, and integrated renewable sources, are not only responsive to user behavior but also capable of interacting with the broader smart grid. Similarly, smart cities leverage digital technologies to optimize resource usage, reduce carbon footprints, support sustainable mobility, and ensure resilient infrastructure, paving the way toward a greener and more connected future. This necessitates smart keys to address their critical demands such as mobility, energy, and so on (Salah et al., 2019; Nunes et al., 2016; Ashton., 2009; Kabalci & Kabalci, 2019).

Therefore, the utilization of modern technologies serves as a crucial foundation for the development of Smart Homes and Smart Cities, enabling enhanced connectivity, automation, and sustainability across various urban domains. Among these technologies, the integration of intelligent transportation systems particularly through the widespread penetration of Plug-in Electric Vehicles (PEVs) represents a key component in the transition toward cleaner and more efficient cities. PEVs not only offer an environmentally friendly alternative to internal combustion vehicles but also play a dual role in energy systems by acting as mobile energy storage units. When integrated into smart grid infrastructures, PEVs can support Vehicle-to-Grid (V2G) and Vehicle-to-Home (V2H) operations, enabling bi-directional energy flow that contributes to demand response, grid stability, and load balancing. Their deployment within intelligent transportation frameworks, supported by IoT connectivity, real-time data exchange, and predictive control algorithms, enhances traffic efficiency, reduces emissions, and aligns transportation energy consumption with renewable generation availability. As such, PEV penetration is not merely a technological advancement but a strategic lever for realizing the full potential of smart urban ecosystems (Ben Arab et al., 2023; Khemakhem et al., 2019). So, Smart City maintained by the Internet of Things (IoT) technologies presents a good solution to achieve this target (Hu et al., 2019; Dave et al., 2016; Sidid & Gaur, 2017; Guinard., 2011). The current success in the IoT will progress Smart Homes and Smart Cities from hype to reality especially it is powered by information and communication technologies to offer access to the Internet for physical things using modern devices such as sensors as well as to develop a ubiquitous system which ensures all-embracing services for information incorporation, storing, processing and sharing (Zhang et al., 2014; Bekara., 2014; Rath & Pati, 2019). IoT is the main approach to make a lot of traditional communication systems intelligent (Al-Fuqaha et al., 2015). To this end, smart grid, smart city and evenly smart home have become a key challenge notion for electrical power networks and energy management structures to ensure an effective energy efficiency and to enhance security (Irfan et al., 2017; Tuballa & Abundo, 2016; Shirazi et al., 2015), especially with the increasing IoT technologies progress and the innovative two-way communication and information flow (Benzi et al., 2011; Cominola et al., 2015). On the other hand, the more environmentally friendly PEVs are undertaken as a future alternative for reducing greenhouse emissions, fossil fuels consumptions and carbon footprint. Undoubtedly, this must have coordinated PEVs to significantly raise the peak load demand which causes power grid burden, severe voltage fluctuations, harmonics process and blackouts (Fernandez et al., 2010). The combination of IoT technologies based on different number and type of sensors, web servers, IoT paradigm collecting information and scheduling systems; makes the IoT applications widely studied and the PEVs a smart element. In this domain, IoT has been implemented in modern cities for different management models that optimize its application with PEVs and the distributed renewable energy (Jayaraman et al., 2016; Garcia et al., 2016; Rashid et al., 2016). Indeed, Smart Grid concentrates principally on energy saving for cities and homes based on the electrical grid to gather appliances' usage data (Smaoui et al., 2025; Mayouf et al., 2022). In particular, IoT technologies address the algorithms and approaches integration to achieve the equilibrium between the energy from the power plant and the in-home power sources, such as solar panels, PEV, Wind turbine. Hence, this will fulfill a better choice about energy consumption, a flexible IoT models architecture and a smart grid based IoT systems for energy management (Khajenasiri et al., 2017; Rathore et al., 2016). Currently, various endeavors in the research have pointed different aspects of IoT in smart network frameworks. For example, studies in (Botta et al., 2016; Atzori et al., 2010; Gubbi et al., 2013; Boulos & Al-Shorbaji, 2014; Borgia., 2014; Mohanty et al., 2016; Talari et al., 2017) have presented the different technologies, aspects, protocols, advantages, challenges related to describing IoT and investigated the different IoT applications in modern cities. In particular, researchers in (Koundinya et al., 2016) have focused on advanced metering infrastructure (AMI) overview and meter checking advances in order to check the power quality in smart grid context associated with IoT technology. Additionally, authors have addressed how IoT technologies should be linked to electric energy organization and innovative strategies frameworks especially with rechargeable power sources. In this topic, (AlFaris et al., 2017) have explored the main contribution of the internet of things technology with smart home concept integrating the renewable energy to diminish the energy demand and ameliorate its performance by reaching the "low" or "net zero" energy smart home.

The authors in (Kabalci., 2016; Menouar et al., 2017; Xiong et al., 2012) have concentrated on smart city concept and IoT frameworks as emerging concepts. Indeed, they have presented all basics of a smart grid with communication techniques including their upgrades as well as the diverse alternatives to deal with PEVs integration for managing Vehicle-to-Home (V2H) and Home-to-Vehicle (H2V) notions. For the same target, (Zhenga et al., 2019) have displayed a literature review on optimal methodology scheduling about bidirectional power transactions between PEVs and power grid, controlled Centralized/Decentralized methods for PEVs and optimally V2G procedures implementing. A study in (Yao et al., 2014) has investigated a multi-objective collaborative scheduling strategy to address the optimal arrangement problem for PEV charging/discharging systems to join electrical power distribution employing an evolutionary algorithm to find solutions. In (Ioakimidis et al., 2018), an appropriate model is formulated for PEV charging station problem. This study aims to develop four methods to tackle this algorithm, therefore, their performance is estimated taking inti account different practical and artificial cases. (Shirazi et al., 2015) have dealt with a Mixed-Integer Non-Linear optimization approach for optimal introduction and sizing of the fast charging/discharging stations, each station development cost, PEV energy loss as well as smart gird loss. Moreover, this approach takes into consideration the PEV substations and urban roads' location. Authors in (Shi et al., 2019) have proposed an energy management strategy to control injected/absorbed electrical vehicles power in order to smooth the residential loads demands of a smart city. Others in (Rajakaruna et al., 2015) have interested in energy market contribution, they have developed a predictive model approach aiming to minimize both power generation cost and electrical vehicle charging cost all depending on both consumers requirements and PEV power demands. Other authors (Rekik & Krichen, 2021) proposed a PEVs Central Controller "(PEVCC)" model applied for PEVs to enhance their power quality as well as contribute to the electricity market. This study has doted a Particle-Swarm-Optimization (PSO) algorithm which was divided into two layers. The aim of the first optimization layer was to determine the adequate PEV reference power applied to each converter in order to fulfill the utility power demand smoothness, the smart grid frequency regulation and the total cost minimization for each PEV charging/discharging operation. The second one is elaborated to regulate all controllers' parameters of a dynamic PEV model. Collaborative energy supervision among PEV, smart houses and neighbors' interaction for flattening power household profile was investigated in (Khemakhem et al., 2020). This study focuses on a local algorithm according to the specific household power profile and the PEVs' availability as well as a global algorithm according to the power profile of this home, the power demand of neighbors and their PEVs neighbors' availability in order to flatten the demand profile to the greatest extent possible. As has been shown, these works cited above, and others that are not said here have highlighted varied tools and approaches to deal with PEVs contributing purposes for power demand regulating energy loss reducing and electricity cost diminishing. Though, it is important to indicate that these existing methods have not been taken into account the two ways of PEVs services coupled with wind power conversion system (WT) to reduce fluctuations and flatten the total power demand for the smart city context. Given the speedily upgrading capabilities of the IoT, smart City aims to connect users in a city to all ''Smart Technologies'' in order to carry real time information for people with accurate details, moderate its energy problems complexity and population requirements.

To target the same objective, this chapter focuses on the key contribution of the IoT technology in relation to the smart city evolution. In this context, we proposed an innovative framework to minimize the load variance in each household by scheduling the PEV power charging/discharging patterns and the WT power conversion system. Then, it has been proven that load variances will be minimized, loss will be approximately reduced as well as load power factors will be maximized for the considered smart city as new emerging framework. This work can be established for each node of the smart city; hence it will be naturally generalized for the whole smart grid as a new concept.

More precisely, the following are the main contributions of this work to the smart city for energy balance, energy-efficiency achievement and economic growth as highlighted in this chapter:

- Develop an optimal energy management of multiple numbers of houses integrated a flexible number of PEVs and/or wind power system to achieve a flattened smart city power load profile.
- Ensure the energy balance and avoid stress in the whole smart city by minimizing fluctuations and demand peak intensity to shave the peak load and fill the valley as well as minimizing accordingly the load power variance.
- Guarantee control of the WT power generation and an optimal dynamic PEV power scheduling by adjusting their power storing during off peak hours or

their power injecting to the main grid specially when there are surplus in demand.

Control the PEV bidirectional energy flow exchanging to fulfil an optimal V2H and H2V power ways all based on the PEV real-time information to join the smart city concept.

Maintain the PEV battery State of Charge (SOC) in allowable thresholds to prevent the PEV deep discharging and the PEV overcharging aiming to augment the battery's lifetime.

In section **2**, the general architecture of the studied system including the concept of household smart city will be presented. Next, the problem formulation will be introduced in section **3** to reduce the load power variance of each smart home and attain a flattened power curve. After that, analysis and simulation results will be devoted to section **4**. Finally, the conclusion will be developed in section **5**.

STUDIED SYSTEM ARCHITECTURE

A smart city has to scope with the significant incorporation of Distributed Energy Resources, Controllable Loads and PEVs. In fact, the PEVs integration in smart systems exhibits challenges on technical, regulatory and economic issues that must be monitored with new algorithms, architectures and procedures. In this context, wind turbine power generation (WT) and PEVs offer an important opportunity to achieve energy security, reliability, air quality and economic growth.

So far, with the increasing IoT technologies growth and the innovative two-way communication way, smart networks have become a potential challenge for electrical power systems and management structures to boost total energy efficiency and security. Accurately, the "develop smarter" notion includes smart homes as well as smart cities which based on bidirectional communication and information technologies interconnectivity. In this way, our research work is organized with a group of smart homes proprietors coming together to create a smart city which incorporated smart loads, WT energy conversion system and flexible number of PEVs. These PEVs can operate as loads or as distributed sources in notions known as H2V or V2H operating mode, respectively. Figure 1 exposes an overview of a studied smart city structure, in which we can see that clients exchange data with the utility as well as they can feedback the energy generated to the grid in some homes.

Figure 1. Overview of a studied smart city structure.

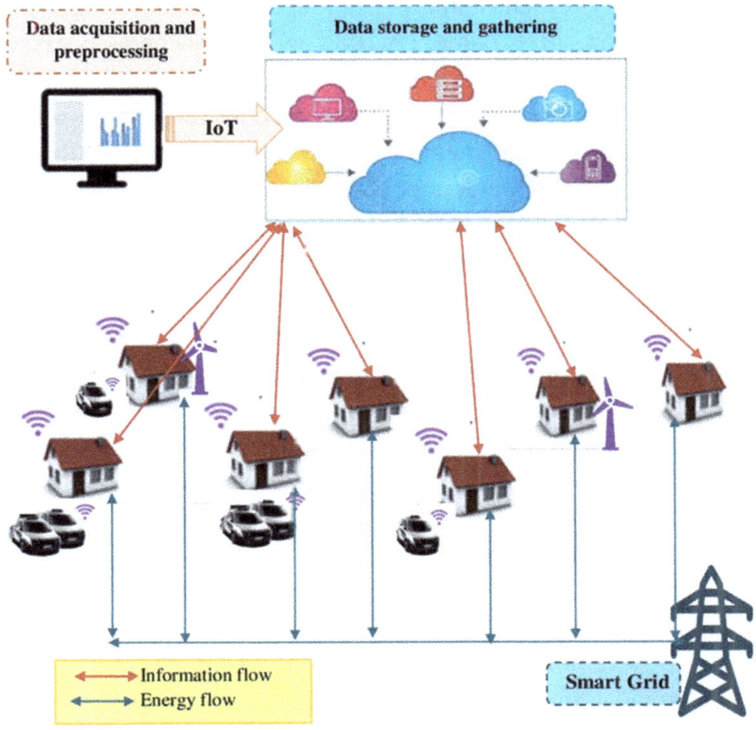

Household Smart City Concept

We consider that most smart homes are equipped with a wide range of AC-powered smart appliances—such as televisions, refrigerators, air conditioners, computers, and washing machines—which require connection to the AC power grid via AC-DC power converters. In addition, smart homes typically include an increasing number of native DC-powered devices, such as LED lighting systems, home security alarms, sensors, and communication equipment, which further necessitate careful energy conversion and management strategies. To enhance energy autonomy and sustainability, many smart homes are also supported by on-site renewable energy sources, particularly small-scale wind turbine distributed generators that provide variable DC or AC outputs. Furthermore, the penetration of Plug-in Electric Vehicles (PEVs) is becoming more prominent, with vehicles powered by lithium-ion (Li-ion) battery packs that can interact with home energy systems through bidirectional chargers. These chargers enable both charging from the grid and discharging energy back to the home or grid under Vehicle-to-Home (V2H) or Vehicle-to-Grid (V2G) configurations. Each smart home is also assumed to be equipped with an intelligent energy management system and a smart meter, capable of monitoring, recording, and optimizing energy flows in real time. This comprehensive energy ecosystem allows for coordinated control among various energy sources and loads, enhances power quality, supports demand response programs, and contributes to greater efficiency and sustainability in the overall smart city infrastructure. Additionally, the property of 'smart' designs the Smart Meter implication and the Energy Management System (EMS). For that, several efforts have been developed to progress wireless sensor network for smart city and residential applications to measure, acquire comprehensive information on electricity usage within each household and send these information measurements to the EMS (Abbas et al., 2020; Anda & Temmen, 2014). This later is concerned with regulating the energy flow within the smart city according to the information acquired by the smart meter. A control algorithm is performed by the EMS to satisfy the power demand curve smoothness, reduce grid reinforcing, and fulfill highly effective operation of the whole smart city system.

IoT Applications in Smart City

Modern cities have become increasingly "smarter" and better organized, largely due to the rapid development of Information and Communications Technology (ICT), which provides innovative solutions to address complex urban challenges such as traffic congestion, energy management, waste handling, and public safety. A smart city can be fundamentally defined as a "settlement powered by digital data development, with the capacity to interact dynamically with residents and utilities." Central to this transformation is the Internet of Things (IoT), which has gained significant attention for its ability to connect a wide range of devices, systems, and users over the Internet. By enabling real-time communication and control, IoT facilitates enhanced monitoring, automation, and data-driven decision-making from virtually anywhere. Its architecture—comprising communication protocols, cloud-based platforms, and intelligent data processing—ensures secure and efficient information storage, analysis, and response. Moreover, the IoT ecosystem incorporates mechanisms for data protection, resilience against third-party interference, and support for scalable integration of new technologies, making it a cornerstone of sustainable smart city development. In general, IoT is oriented by technology while smart city is operator oriented. Each is affecting one other with a principal goal to provide greatest services for modern city (Diaz et al., 2016; Stojkoska & Trivodaliev, 2017; Rath & Tomar, 2021). It is noteworthy that IoT modernization will make the smart grid progressively productive, increasingly canny, gradually secure, and potentially advantageous.

ENERGY MANAGEMENT STRATEGY

PEV and WT Power Generation Model

The successful integration of Plug-in Electric Vehicles (PEVs) and renewable energy sources into smart homes represents one of the key challenges in realizing the vision of a fully functional smart city. This integration must address the dynamic nature of both energy demand and supply, particularly the intermittent generation of renewable sources such as Wind Turbines (WTs). A critical aspect involves the intelligent control of PEV charging and discharging cycles, as well as the optimal utilization of WT-generated power, to perform valley filling during low demand periods and peak shaving during high demand intervals. These strategies aim to align energy consumption patterns with sustainable generation profiles. Within the scope of this study, an advanced energy management strategy has been developed to minimize the power load variance across individual households in a smart city. By coordinating PEV storage capacity and real-time wind energy inputs, this strategy contributes significantly to flattening the overall load profile, reducing short-term power fluctuations, improving the reliability of power delivery, and ultimately alleviating stress on the smart grid infrastructure. Additionally, such management enhances grid resilience, supports demand-response mechanisms, and promotes energy efficiency at the neighborhood and city scale. This is benefit for applying the desired merits of V2H concept and the generated WT power surplus. Firstly, a WT based on a Permanent Magnetic Synchronous Generator (PMSG), an AC/DC converter placed at the PMSG controlling the WT variable speed functioning and the wind speed are well modelled (Rekik & Krichen, 2014; Rekik et al., 2015; Abdelkafi & Krichen, 2014). Secondly, specific attention was given to the PEV structure, considering principally, the battery packs compartment, the battery type, technology and charging/discharging characteristics. The newest technology chosen in this study was the Li- ion battery as the most popular for PEVs applications. In addition, this type of technology was selected thanks to its significant performance parameters, important energy density, high voltage and specific power, long life-time, negligeable waste of charge and lower cost (Sarasketa-Zabala et al., 2016; Jaguemont et al., 2016; Abada et al., 2016). The studied simulation PEV operates as the model presented in Figure 2.

Figure 2. Basic Li-ion battery model for PEV power generation.

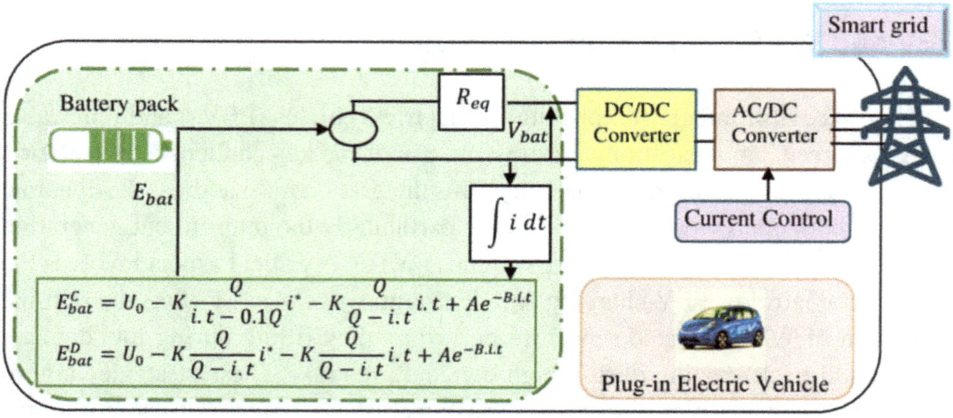

The PEVs are equipped with a pack of batteries which can offer a charge/discharge electrical operation mode. Two converters are assumed to regulate and transmit the bi-directional PEVs energy flows to feedback, if necessary, the required electricity dumped in their batteries pack to the smart home or to store energy for power grids as movable load. In that, the battery cell is established as a series of a voltage source E_{bat} and an internal resistance R_{eq}. It is notable that there an equation E^C_{bat} for the controlled voltage source which defines the discharging process when the battery current was positive and an equation E^D_{bat} for the charging operation indicating a negative battery current (Khemakhem et al., 2021). Thirdly, each PEV involved in joining this new concept should have its full charging/discharging information thoughtfully very well. Consequently, each Customer's PEV should have to create its PEV information connectivity as its number "PEV_i", its model "$PEV_{i,model}$", its battery technology "$PEV_{i,type}$", its battery capacity "PEV_{cap}", its tolerable battery Sate Of Charge lowest limit "SOC_{min}", the tolerable battery SOC high limit "SOC_{max}", its time disconnection "$T_{plug-out}$" and connection at home "$T_{plug-in}$" and the value of the initial SOC when the PEV is connected to home "SOC_{in}". Finally, the daily load power of each home was modeled by its specific profile referring to their appliances power consumption considering the home appliances need priority without affecting their comfortable lifestyle during one day. To facilitate calculation, the day ahead from 00h00 to 24h00 is chosen to be divided into Ti time-intervals, and the duration of each time-interval is assumed by Δt=15 min. Consequently, there are 96 time-intervals through one day and each PEV is expected to begin charging/discharging at the initial some time-interval and finish charging/discharging at the end of another time-interval.

To this end, each PEV involved could exhibit its specific charging/discharging power curve and its battery SOC profile for one day cycle conferring to the data exchanging between each PEV proprietors, smart home and constraints designated by the control strategy. In this way, the V2H and H2V scheduling notions will be performed. For these purposes, The PEVs penetration correlated with the WT power generation could greatly play a potential role in the energy management strategy thanks to their skills to monitor the energy supply and storage, control the energy consumption, alleviate the peak demand and smooth out the daily load demand curve.

The WT parameters are presented as follows.

Nominal power 3 kW
Number of pole pairs 4
Stator resistance 0.82 Ω
Stator inductance 15.1 mH
Inertia 99 10^{-4} $Kg\ m^2$
Friction coefficient 10^{-3} $N\ m\ s\ rad^{-1}$
Permanent Magnetic Flux 0.5 Wb

The PEV characteristics studied are displayed as follows.

U_0 3.366 V
R 0.01 Ω
K 0.0076 Ω
A 0.26422 V
B 26.5487 $(Ah)^{-1}$
η_{PEV} $\frac{1}{6}(KW/Km)$
Battery capacity 14KWh
P_{min} -2 KW
P_{max} 2 KW

Problem Formulation

A smart city operates as a dynamic ecosystem where electrical power distribution is tightly coupled with real-time information transmission, enabling homeowners in each smart home to interconnect, communicate, and make informed decisions regarding the operation and scheduling of their household appliances. This interconnected infrastructure is significantly empowered by the growing penetration of Plug-in Electric Vehicles (PEVs), which introduces new opportunities and challenges for energy management. The increasing number of PEVs not only adds to the overall energy demand but also offers potential energy storage and flexible load options when properly coordinated. In this context, our study focuses on integrating Wind Turbines (WTs) as a form of distributed renewable energy generation and leveraging smart, coordinated PEV charging/discharging operations to achieve a more balanced and manageable energy usage profile across the smart city. The proposed energy management strategy relies on bidirectional information flow between the smart homes and the central control system, which enables optimal decision-making. Through continuous data acquisition—such as real-time power demand, WT generation forecasts, and PEV battery states—the control system determines the optimal amount of WT power to dispatch and schedules the PEVs' charging or discharging accordingly. This results in enhanced load balancing, renewable energy utilization, and peak load reduction. Figure 3 presents a flowchart of the proposed energy management system, detailing the interaction between energy sources, smart meters, user preferences, and control logic that collectively drive intelligent power flow optimization in the smart city.

Figure 3. Flowchart of the proposed energy management strategy.

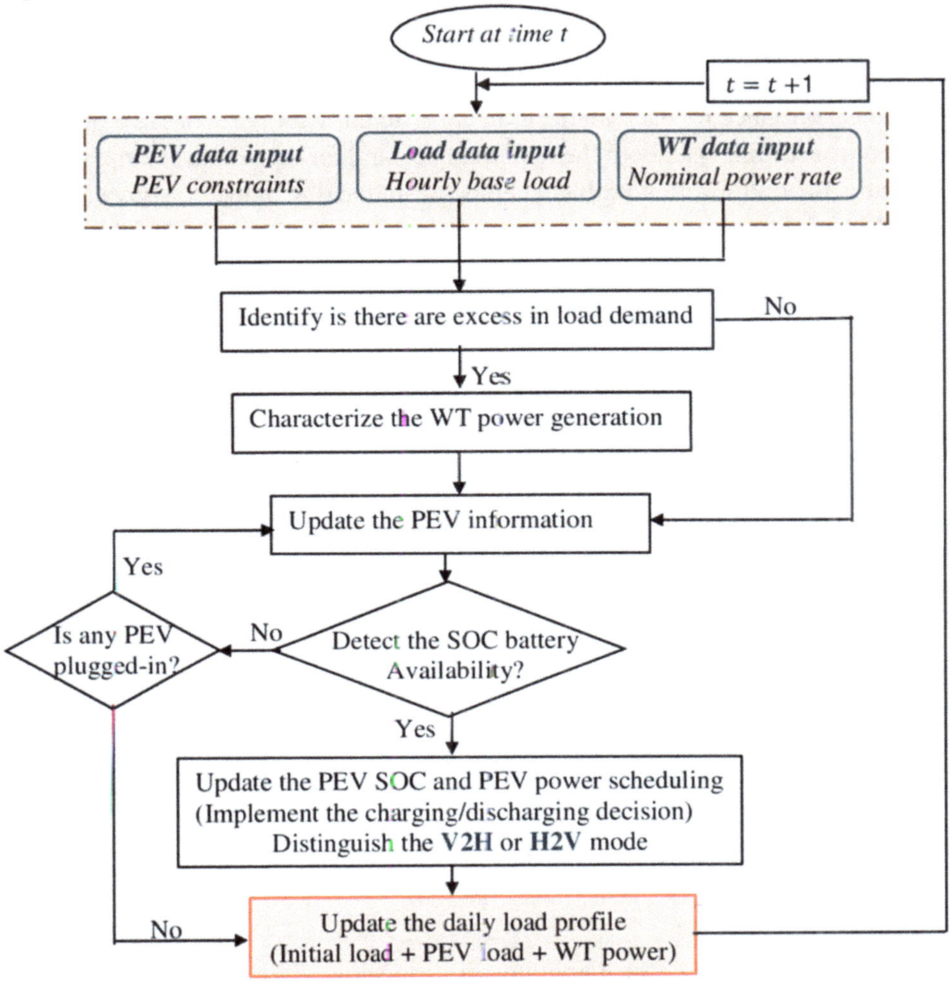

The first step is to present the input needed data such as: the hourly load power profile of each smart home "$P_{SH}(t)$", the total WT power generation profile "$P_{WT}(t)$", the number of PEVs that are being to join the smart city concept, when each PEV is parked, the initial SOC value of each PEV "$SOC_{ini,i}$" as well as the SOC upper "SOC_{min}" and lower "SOC_{max}" boundaries during each time- interval. The second step is to implement a smart meter to deal with PEV data gathering and WT power storage. The role of the smart metering infrastructure is to collect information about PEVs, WT and load consumption, send it to the grid operator, and to establish the required charging/discharging results from the operator to the PEV chargers. Based on this information, the model formulation for this control strategy is applied to make a coordinated charging and discharging operating mode that promises maximum energy moved to the PEV client and minimum power grid losses.

The final step is to apply these developed data to decrease the total power consumption in residential applications, particularly during peak hour, to minimize the load power variance for each home, to make efficiently the V2H or H2V operating mode and accordingly smooth the total load demand for the whole smart city. Consequently, the optimization problem can be formulated with the objective function using Matlab-Simulink as follows:

$$var\ SH_h = min \sum_{t=1}^{Ti} \left[\frac{1}{T} (P_{SH}(t) + P_{WT}(t) + \sum_{i=1}^{n_{PEV}} P_{PEV,i}(t) - P_{SHm})^2 \right] \text{(Eq.1)}$$

with:

$$P_{SH}(t) + P_{WT}(t) + \sum_{i=1}^{n_{PEV}} P_{PEV,i}(t) \leq P_{SHmax}(t)\ t \in [1, Ti]\ ,\ i \in [1, n_{PEV}]\ \text{(Eq.2)}$$

$$P_{SHmean} = \frac{1}{T} \sum_{t=1}^{Ti} (P_{SH}(t) + P_{WT}(t) \sum_{i=1}^{n_{PEV}} P_{PEV,i}(t))\ t \in [1, Ti]\ ,\ i \in [1, n_{PEV}]\ \text{(Eq.3)}$$

$$P_{min,i} \leq P_{PEV,i}(t) \leq P_{max,i}\ t \in [T_{plug-in,i}, T_{plug-out,i}],\ i \in [1, n_{PEV}]\ \text{(Eq.4)}$$

$$P_{PEVtrav,i} = \mu_{PEV,i} . AD_{PEV,i}\ i \in [1, n_{PEV}]\ \text{(Eq.5)}$$

$$SOC_{PEV,i}(t) = \left[SOC_{int,i}(t) - \left(\frac{\int idt}{Q} \right) * 100 \right] (\%)\ t \in [T_{plug-in,i}, T_{plug-out,i}],\ \in [1, n_{PEV}]\ \text{(Eq.6)}$$

$$SOC_{min,i} \leq SOC_{PEV,i}(t) \leq SOC_{max,i}\ t \in [T_{plug-in,i}, T_{plug-out,i}],\ i \in [1, n_{PEV}]\ \text{(Eq.7)}$$

Where:

Equation (2) describes the maximum operating power of the studied smart home at each time-interval, equation (3) defines the average power of the whole smart home in one-day time, equation (4) clarifies the maximum operating power "$P_{max,i}$" which should be injected to the grid during the adequate time-interval, and the minimum operating power "$P_{min,i}$" which should be stored from the smart home and supplied to the available PEV, equation (5) explains the energy required by each PEV needed for their journeys $P_{PEVtrav,i}$ which depends both on the distance running for the trip "$AD_{PEV,i}$" and the PEV efficiency "$\mu_{PEV,i}$ (Khemakhem et al., 2019). Equation (6) explicates the PEV State Of Charge (SOC) which is the existing state that can conceivably be discharging, charging, or lasting in idle state. Obviously, the battery SOC is definite as a percentage of the current capacity according to maximum battery capacity at the present moment integration into the home in order to create the PEV battery capacity evolution during all time-intervals, equation (7) denotes the

SOC minimum and maximum boundaries that maintain the well safeguarding of the Li-ion battery performance with its great operation range of functioning to avoid the state of PEV fully charged or deeply discharged. Therefore, the SOC should be maintained between 20% and 80% and the initial value must be given to the power grid (Khemakhem et al., 2019).

SIMULATION RESULTS AND DISCUSSION

In order to evaluate the proposed approach effectiveness and demonstrate the outcome of the WT power conversion system on peak load shaving as well as the efficiency of the cooperativity of PEVs and WT power generation for flattening the load power profile and load variance minimizing, the simulation of various the households power load curve with their specific constraints was carried out. In this chapter, we consider a smart city associated with 300 smart homes in which some are equipped with a flexible number of PEV, others are beneficed by a WT power conversion system and others which are implemented in the smart grid technology with a renewable distrusted energy generation and PEVs penetration. Namely, we discretize the day-ahead into 96 time-intervals in which data inputs and simulation results are checked at each Δt=15-min.

In fact, the input information which are the specific load power of each smart home under study, the power generated by the WT, each PEV charging information availability (its start/end home parking and its battery energy accessibility to achieve the bi-directional power exchanging) should be achieved to simulate the real conditions to minimize the total load variance and to satisfy a smooth load power curve of the whole smart city. For simplify our work, some suggestions have to be involved such as, for all the PEVs involved, the tolerable maximum and minimum SOC values are made-up to be equal to 80% and 20% respectively. In this chapter, we consider a WT operated power with 3 kW, two types of PEVs with 14kWh and 11kWh battery capacity, Correspondingly, the suitable maximum power to be inserted to the grid and the acceptable minimum power to be deposited to the PEV battery are established to 2 KW and -2 KW, respectively.

Figure 4 shows the input data for the studied smart homes concerning their base load demand curve, the WT energy generation profile in which the WT products its nominal rate power when the wind speed is fixed at 9 m/s, the PEV availability and home parking time for one day cycle.

Figure 4. Specific each household smart city characteristics.

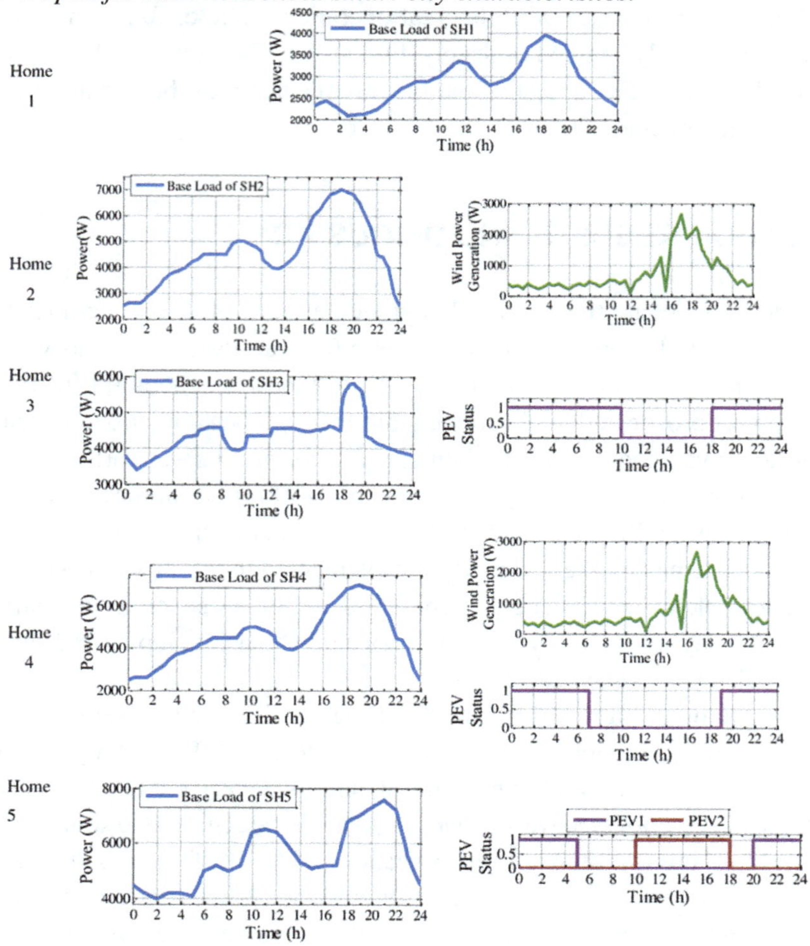

It can be observed that each PEV is involved to plug-in at the specific time-interval with its initial SOC value to guarantee the bidirectional power exchanging and plug-out at the adequate time-interval with its desired value of battery SOC. It is revealed that Smart Home 1 (SH1) is not affiliated neither by a WT generator nor with PEVs. It is remarkable that its base daily load power curve has various peaks creating negative disturbances on the smart city stability and therefore on power grid quality.

The key parameters among each studied home are exhibited as follows.

Home1 Home 2 Home 3 Home 4 Home 5

$P_{SHmax,h}$ (KW) 4.5 7.5 6.5 8 8

$P_{WTnominal,h}$(KW) 0 3 0 3 0
N_{PEV} 0 0 1 1 2
$Tin_{PEVi,h}$ - - 6p.m 7p.m 8p.m; 10a.m
$Tout_{PEVi,h}$ - - 10a.m 7a.m 5a.m; 6p.m
$CAP_{PEVi,h}$ (KWh) - - 14 14 11; 14
$P_{min,i,h}$(KW) - - -2 -2 -2
$P_{max,i,h}$(KW) - - 2 2 2
$SOC_{in,i,h}$ - - 0.2 0.3 0.4; 0.3
$SOC_{out,i,h}$ - - 0.665 0.8 0.8; 0.695
$SOC_{min,i,h}$ - - 0.2 0.2 0.2
$SOC_{max,i,h}$ - - 0.8 0.8 0.8

Figure 5 exposes the characteristics of the PEV parked at home 3. According to the PEV power profile and the chosen convention, it should be revealed that the negative power verifies the charging operation mode where the PEV available to join the V2G concept is able to act as load to satisfy demand and absorb power from Smart Home 3 during the power demand deficit periods: It is H2V mode. Although the positive power confirms the discharging operation mode. This PEV home-parking works to deliver energy to its convenable home during the extreme demand periods, thus aiming to reduce the power load variance and achieve the smart city energy balance: It is V2H mode.

As has been shown, the State of Charge profile of the PEV simulated in Figure 5 (b) demonstrates that during the one-day cycle, the PEV battery operates to be charged or discharged while the initial PEV battery SOC does not reach its maximum and minimum SOC boundaries. If the PEV is parked at home and the battery SOC attains its 80% or 20% of the SOC limits, then there aren't powers exchange between home and PEV. Therefore, the proposed approach not only guarantees energy management, but also improves the PEV battery lifespan since it acts within the desired restrictions.

Figure 5. Characteristics of the PEV parked at home 3. (a) PEV power profile (b) State of Charge profile of the PEV.

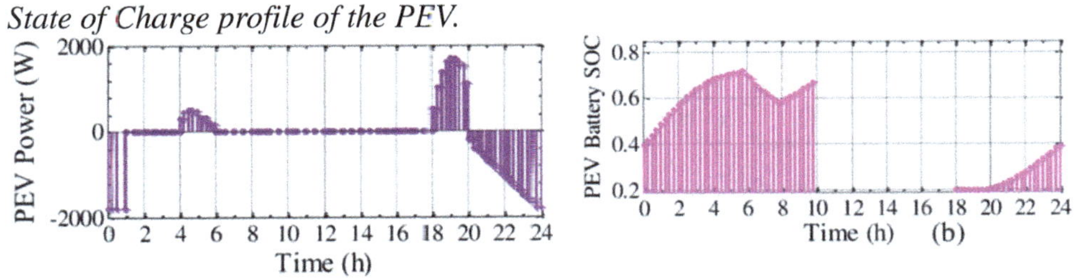

Figure 6 presents the characteristics of the PEV parked at home 4. Indeed, Figure 6 (c) and Figure 6 (d), show the PEV current profile and its dynamic voltage respectively. As has been observed, the PEV voltage is increased in H2V mode, decreased in V2H operation and stayed constant when the PEV attains its SOC value limits. Additionally, it is noteworthy, from the PEV power curve, that the PEV doesn't exist at home 4 during [7a.m, 7p.m]. Thus, it is unavailable to stock or deliver electrical power and consequently to ameliorate its load power during this time interval. In fact, Figure 6 (e) and Figure 6 (f), illustrate the modulated currents "I_{m-bat}"and "I_{m-inv} of the PEV home4-parking which are nearly equal to the DC bus voltage and controlled at its reference value "U_{DC_ref} = 400V" presenting a very little deviation owing to the various operating modes transitions.

Figure 6. Characteristics of the PEV parked at home 4: (a) PEV power profile, (b) State of Charge profile of the PEV, (c) PEV battery voltage, (d) PEV battery current, (e) PEV modulated currents, (f) DC bus voltage.

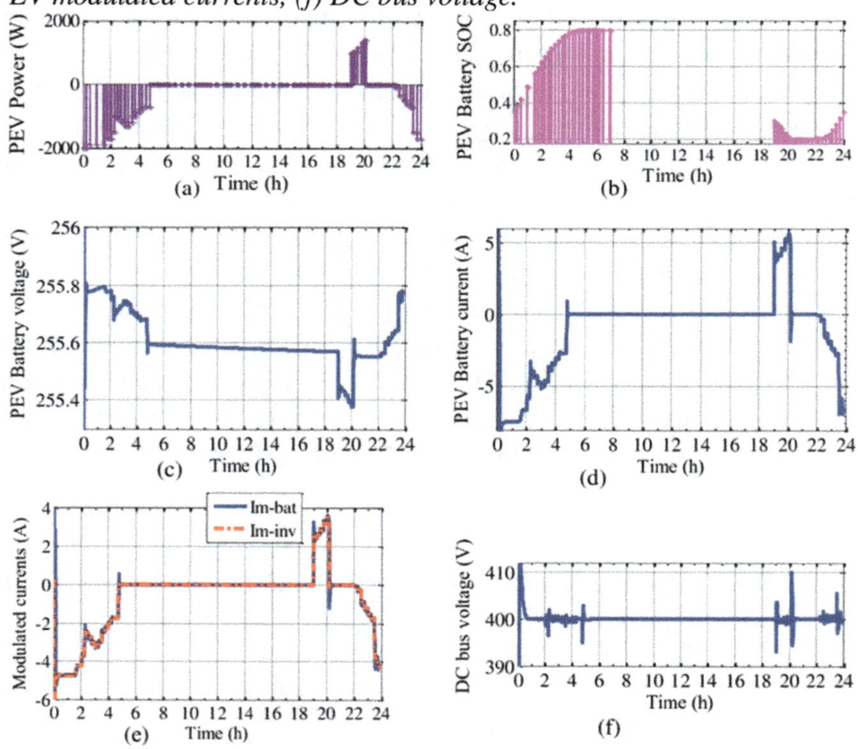

The one-day cycle power curve and battery SOC simulation results of two PEVs connected to homes 5 presented in Figure 7.

Figure 7. Characteristics of the PEV parked at home 5. (a) PEV power profile, (b) State of Charge profile of the PEV.

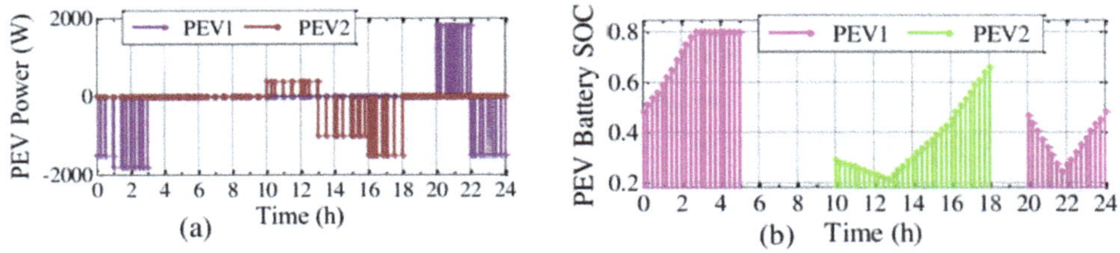

Figure 8 explains the comparison of the load power curves in the studied home. As can be shown, the peak demand in home 2 is significantly reduced thanks to WT power contribution. It can be observed that the valley load filling of the household 3 curve is carried out thanks to PEVs stocking energy when the charging operation is performed. It can be deducted that the power load profile of home 4 can obviously flattened thanks to WT power conversion system implementation and PEV penetration. The same result is detected for the Smart Home 5 when the peak load shaving and the valley filling was achieved thanks to the energy contribution of two PEVs home 5-parking. Whereas the power demand stands unchanged once PEVs are not parked at home or not manageable to stock/supply power from/to the Smart Home. As exposed, home 1 keeps its base power demand profile while there are no WT and PEVs existing at home-parking.

Figure 8. Comparison of total power load curves: (a)Home 2, (b)Home 3, (c)Home 4, (d)Home 5.

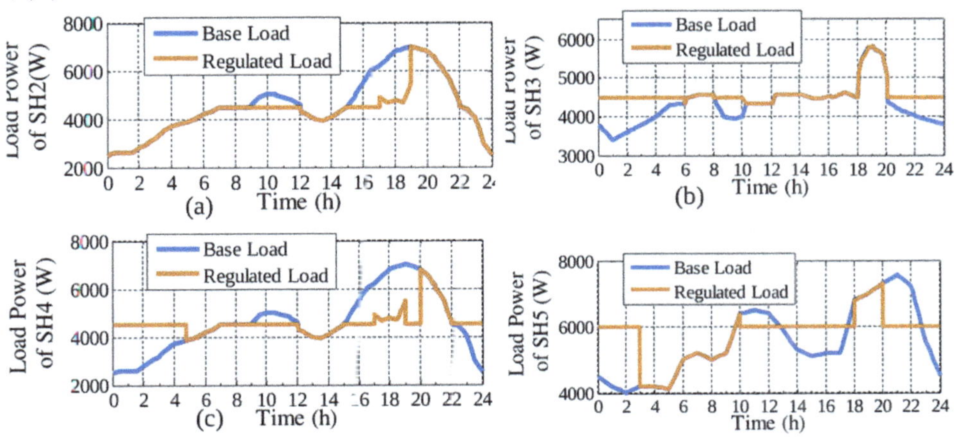

Figure 9 designates the simulation of the total load profile of 300 residential clients. As can be stated, the severe peak was noticeably reduced, compared to the smart city base load profile. Thanks to WT power conversion system managing, PEVs discharging process during rush time-intervals and PEVs charging scheduling to fill valley during off-peak hours, the proposed smart city can considerably fulfill a flattened power profile and subsequently reduce stress on the smart city as well as ensure the smart grid stability.

Figure 9. Smart city total load profile evolution.

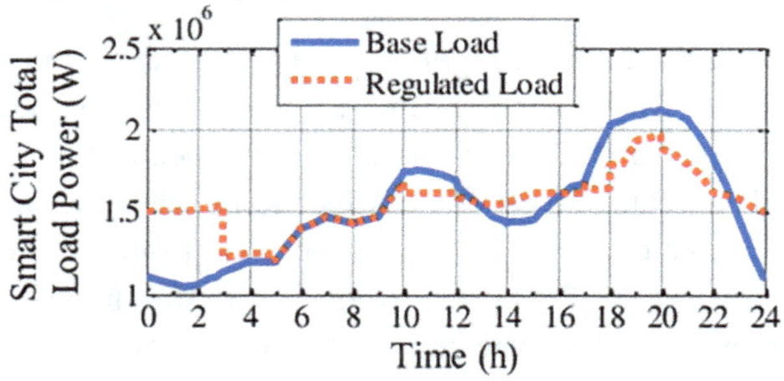

Indeed, a little comparison are explained to illustrate the effectiveness of the proposed approach in the studied smart city. It is clear that the pic demand is considerably decreased, the valley is significantly filled in some domestic application, the load fluctuation is generally reduced, and the load power variance is greatly minimized.

Base Case After Energy Management Approach:

Valley demand filling $1.0521.10^6$ W $1.5.10^6$ W
Peak demand shaving $2.225.10^6$ W $1.892.10^6$ W
Load power variance 0.00547 0.00132

CONCLUSION

The unmanaged WT and PEVs integration into the future smart city can tremendously overload the smart grid disruption. In this topic, a problem formulation for an energy management of WT power generation and PEV power charging/discharging is proposed. This current chapter dealt with a bidirectional power exchanging performing the Vehicle-to-home V2H concept for a multiple number of residential customers within the smart city context aiming to smooth the total load curve and reduce the load power variance. This leads to manage the WT power production and the PEV power under the base consumption of each smart home, the nominal rate power of WT, the PEV accessibility to store/inject power, its start and end time-interval home parking and its initial PEV battery SOC. The power management strategy effectiveness was valued through the load valley filling by storing energy during the lack in demand and the peak shaving by feedbacking power to the considered home during the excess in load demand time-intervals. The gained results accentuate the control strategy performance to reduce principally the extreme peak intensity, reduce the smart city load power profile fluctuation, flatten therefore the considered curve, minimize the total load variance and increase smart city stability as well as diminish regularly the stress on electrical smart grid, accordingly, avoiding the existing power grid blackouts.

REFERENCES

Abada, S., Marlair, G., Lecocq, A., Petit, M., Sauvant-Moynot, V., & Huet, F. (2016). Safety focused modeling of lithium-ion batteries: A review. *Journal of Power Sources*, *306*, 178–192. DOI: 10.1016/j.jpowsour.2015.11.100

Abbas, A. M., Youssef, K. Y., & Zekry, A. (2020). NB-IoT optimization for smart meters networks of smart cities: Case study. *Alexandria Engineering Journal*, *59*(6), 4267–4281. DOI: 10.1016/j.aej.2020.07.030

Abdelkafi, A., & Krichen, L. (2014). Energy management optimization of a hybrid power production unit based renewable energies. *International Journal of Electrical Power & Energy Systems*, *62*, 1–9. DOI: 10.1016/j.ijepes.2014.04.012

Al-Fuqaha, A., Guizani, M., Mohammadi, M., Aledhari, M., & Ayyash, M. (2015). Internet of Things: A survey on enabling technologies, protocols, and applications. *IEEE Communications Surveys and Tutorials*, *4*(4), 2347–2376. DOI: 10.1109/COMST.2015.2444095

AlFaris, F., Juaidi, A., & Manzano-Agugliaro, F. (2017). Intelligent homes' technologies to optimize the energy performance for the net zero energy home. *Energy and Building*, *153*, 262–274. DOI: 10.1016/j.enbuild.2017.07.089

Anda, M., & Temmen, J. (2014). Smart metering for residential efficiency: The use of community based social marketing for behavioural change and smart grid introduction. *Renewable Energy*, *67*, 119–127. DOI: 10.1016/j.renene.2013.11.020

Ashton, K. (2009). That 'Internet of Things' thing. *RFID J*, *22*, 97–114.

Atzori, L., Iera, A., & Morabito, G. (2010). The Internet of Things: A survey. *Computer Networks*, *54*(15), 2778–2805. DOI: 10.1016/j.comnet.2010.05.010

Bekara, C. (2014). Security issues and challenges for the IoT-based smart Network grid. *Procedia Computer Science*, *34*, 532–537. DOI: 10.1016/j.procs.2014.07.064

Ben Arab, M., Rekik, M., & Krichen, L. (2023). A priority-based seven-layer strategy for energy management cooperation in a smart city integrated green technology. *Applied Energy*, *335*, 120767. DOI: 10.1016/j.apenergy.2023.120767

Benzi, F., Anglani, N., Bassi, E., & Frosini, L. (2011). Electricity smart meters interfacing the households. *IEEE Transactions on Industrial Electronics*, *58*(10), 4487–4494. DOI: 10.1109/TIE.2011.2107713

Borgia, E. (2014). The Internet of Things vision: Key features, applications and open issues. *Computer Communications*, *54*, 1–31. DOI: 10.1016/j.comcom.2014.09.008

Botta, A., Donato, W., Persico, V., & Pescape, A. (2016). Integration of Cloud computing and Internet of Things: A survey. *Future Generation Computer Systems*, *56*, 684–700. DOI: 10.1016/j.future.2015.09.021

Boulos, M. N. K., & Al-Shorbaji, N. M. (2014). On the Internet of Things, smart cities and the WHO healthy cities. *International Journal of Health Geographics*, *13*(1), 10. DOI: 10.1186/1476-072X-13-10 PMID: 24669838

Cominola, A., Giuliani, M., Piga, D., Castelletti, A., & Rizzoli, A. E. (2015). Benefits and challenges of using smart meters for advancing residential water demand modeling and management: A review. *Environmental Modelling & Software*, *72*, 198–214. DOI: 10.1016/j.envsoft.2015.07.012

Dave, B., Kubler, S., Framling, K., & Koskela, L. (2016). Opportunities for enhanced lean construction management using Internet of Things standards. *Automation in Construction*, *61*, 86–97. DOI: 10.1016/j.autcon.2015.10.009

Diaz, M., Martin, C., & Rubio, B. (2016). State-of-the-art, challenges, and open issues in the integration of Internet of Things and Cloud computing. *Journal of Network and Computer Applications*, *67*, 99–117. DOI: 10.1016/j.jnca.2016.01.010

Fernandez, L. P., Gomez San Roman, T., Cossent, R., Domingo, C. M., & Frias, P. (2010). Assessment of the impact of plug-in electric vehicles on distribution networks. *IEEE Transactions on Power Systems*, *26*(1), 206–213. DOI: 10.1109/TPWRS.2010.2049133

Garcia, C. G., Meana-Llorian, D., Bustelo, B. C. P. G., Lovelle, J. M. C., & Garcia-Fernandez, N. (2016). Midgar: Detection of people through computer vision in the Internet of Things scenarios to improve the security in smart cities, smart towns, and smart homes. *Future Generation Computer Systems*.

Gubbi, J., Buyya, R., Marusic, S., & Palaniswami, M. (2013). Internet of Things (IoT): A vision, architectural elements, and future directions. *Future Generation Computer Systems*, *29*(7), 1645–1660. DOI: 10.1016/j.future.2013.01.010

Guinard, D. (2011). What the Internet of Things Will Mean for the Smart Grid. URL https://smartgrid.ieee.org/june-2011/95-what-the-internet-of-things-will-mean-for-the-smart-grid

Hu, B., Feng, Y., Sun, J., Gao, Y., & Tan, J. (2019). Driving preference analysis and electricity pricing strategy comparison for electric vehicles in smart city. *Information Sciences*, *504*, 202–220. DOI: 10.1016/j.ins.2019.07.039

Ioakimidis, C. S., Thomas, D., Rycerski, P., & Genikomsakis, K. N. (2018). Peak shaving and valley filling of power consumption profile in nonresidential buildings using an electric vehicle parking lot. *Energy*, *148*, 148–158. DOI: 10.1016/j.energy.2018.01.128

Irfan, M., Iqbal, J., Iqbal, A., Iqbal, Z., Ali Riaz, R., & Mehmood, A. (2017). Opportunities and challenges in control of smart grids – Pakistani perspective. *Renewable & Sustainable Energy Reviews*, *71*, 652–674. DOI: 10.1016/j.rser.2016.12.095

Jaguemont, J., Boulon, L., & Dubé, Y. (2016). A comprehensive review of lithium-ion batteries used in hybrid and electric vehicles at cold temperatures. *Applied Energy*, *164*, 99–114. DOI: 10.1016/j.apenergy.2015.11.034

Jayaraman, P. P., Yavari, A., Geogakopoulos, D., Morshed, A., & Zaslavsky, A. (2016). Internet of Things platform for smart farming: Experiences and lessons learnt. *Sensors (Basel)*, *16*(11), 1884. DOI: 10.3390/s16111884 PMID: 27834862

Kabalci, E., & Kabalci, Y. (2019). Internet of things for smart network_grid applications. In Kabalci, E., & Kabalci, Y. (Eds.), *From Smart Network_grid to Internet of Energy, Academic Press, 2019* (pp. 249–307).

Kabalci, Y. (2016). A survey on smart metering and smart grid communication. *Renewable & Sustainable Energy Reviews*, *57*, 302–318. DOI: 10.1016/j.rser.2015.12.114

Khajenasiri, I., Estebsari, A., Verhelst, M., & Gielen, G. (2017). A review on Internet of Things solutions for intelligent energy control in buildings for smart city applications. *Energy Procedia*, *111*, 770–779. DOI: 10.1016/j.egypro.2017.03.239

Khemakhem, S., Rekik, M., & Krichen, L. (2019). Double layer home energy supervision strategies based on demand response and plug-in electric vehicle control for flattening power load curves in a smart grid. *Energy*, *167*, 312–324. DOI: 10.1016/j.energy.2018.10.187

Khemakhem, S., Rekik, M., & Krichen, L. (2019). Impact of Electric Vehicles integration on residential demand response system to peak load minimizing in smart grid. In: *19th International Conference on Sciences and Techniques of Automatic Control and Computer Engineering, STA 2019*, Sousse, Tunisia, 2019, pp. 572-577. DOI: 10.1109/STA.2019.8717304

Khemakhem, S., Rekik, M., & Krichen, L. (2020). A collaborative energy management among Plug-In Electric Vehicle, smart homes and neighbors' interaction for residential power load profile smoothing. *Journal of Building Engineering*, *27*, 1–11. DOI: 10.1016/j.jobe.2019.100976

Khemakhem, S., Rekik, M., & Krichen, L. (2021). A Smart Home Power Supervision Strategy Based on Optimal Plug-in Electric Vehicles Integration into Smart City Context for Peak Loads Profiles Flattening. *Arabian Journal for Science and Engineering*.

Koundinya, A. K., Sharvani, G. S., & Rao, K. U. (2016). Calibrated security measures for centralized IoT applications of smart network_grids, In: *2016 International Conference on Computation System and Information Technology for Sustainable Solutions (CSITSS)*, Bangalore, 2016, pp. 153-157. DOI: 10.1109/CSITSS.2016.7779414

Mayouf, O., Rekik, M., & Krichen, L. (2022). Designing a hybrid system using solar cells and batteries to power a clinic in Tajarhi desert village in the Libya-Niger border off-grid. In: *2022 IEEE 21st International Conference on Sciences and Techniques of Automatic Control and Computer Engineering*, Sousse, Tunisia, 2022. DOI: 10.1109/STA56120.2022.10019222

Menouar, H., Guvenc, I., Akkaya, K., Uluagac, A. S., Kadri, A., & Tuncer, A. (2017). UAV-enabled intelligent transportation systems for the smart city: Applications and challenges. *IEEE Communications Magazine*, *55*(3), 22–28. DOI: 10.1109/MCOM.2017.1600238CM

Mohanty, S. P., Choppali, U., & Kouglanos, E. (2016). Everything you wanted to know about smart cities: The Internet of Things is the backbone. *IEEE Consumer Electronics Magazine*, *16*(3), 2162–2248. DOI: 10.1109/MCE.2016.2556879

Nunes, P., Figueiredo, R., & Brito, M. C. (2016). The use of parking lots to solar-charge electric vehicles. *Renewable & Sustainable Energy Reviews*, *66*, 679–693. DOI: 10.1016/j.rser.2016.08.015

Rajakaruna, S., Shahnia, F., & Ghosh, A. (2015). Plug. In *Electric Vehicles in Smart Grids: Energy Management*. Springer Singapore.

Rashid, Z., Melia-Segui, J., Pous, R., & Peig, E. (2016). Using augmented reality and Internet of Things to improve accessibility of people with motor disabilities in the context of smart cities. *Future Generation Computer Systems*.

Rath, M., & Pati, B. (2019). Security assertion of IoT devices using cloud of things perception. *International Journal of Interdisciplinary Telecommunications and Networking*, *11*(4), 17–31. DOI: 10.4018/IJITN.2019100102

Rath, M., & Tomar, A. (2021). Chapter 7: Smart grid modernization using Internet of Things technology. In: Advances in Smart Grid Power System, 191-212.

Rathore, M. M., Ahmad, A., Paul, A., & Rho, S. (2016). Urban planning and building smart cities based on the Internet of Things using big data analytics. *Computer Networks, 101*, 63–80. DOI: 10.1016/j.comnet.2015.12.023

Rekik, M., Abdelkafi, A., & Krichen, L. (2015). Synchronization of Wind Farm Power System to Utility Grid under Voltage and Frequency Variations. *International Journal of Renewable Energy Research, 5*, 70–81.

Rekik, M., & Krichen, L. (2014). Active power filter based on wind turbine for electric power system quality improvement. In: 2014 IEEE 11th International Multi-Conference on Systems, Signals and Devices, SSD 2014, Barcelona, Spain. DOI: 10.1109/SSD.2014.6808768

Rekik, M., & Krichen, L. (2021). Double Layer Optimization Approach of Plug-in Electric Vehicle for Participation in Smart Grid Ancillary Services and Energy Markets. *International Journal of Renewable Energy Research, 11*, 1107–1123.

Salah, W. A., Albreem, M. A., Alsayid, B., Zneid, B. A., Alkhasawneh, M., Al-Mofleh, A., Sneineh, A. A., & Abu Al-Aish, A. (2019). Electric vehicle technology impacts on energy. *International Journal of Power Electronics and Drive Systems, 10*(1), 1–19. DOI: 10.11591/ijpeds.v10.i1.pp1-9

Sarasketa-Zabala, E., Martinez-Laserna, E., Berecibar, M., Gandiaga, I., Rodriguez-Martinez, L., & Villarreal, I. (2016). Realistic lifetime prediction approach for Li-ion batteries. *Applied Energy, 162*, 839–852. DOI: 10.1016/j.apenergy.2015.10.115

Shi, Y., Tuan, H. D., Savkin, A. V., Duong, T. Q., & Poor, H. V. (2019). Model Predictive Control for Smart Grids with Multiple Electric-Vehicle Charging Stations. *IEEE Transactions on Smart Grid, 10*(2), 2127–2136. DOI: 10.1109/TSG.2017.2789333

Shirazi, E., Zakariazadeh, A., & Jadid, S. (2015). Optimal joint scheduling of electrical and thermal appliances in a smart home environment. *Energy Conversion and Management, 106*, 181–193. DOI: 10.1016/j.enconman.2015.09.017

Sidid, S., & Gaur, S. (2017). *Smart network_grid building automation based on internet of things. In: 2017 Innovations in Power and Advanced Computing Technologies (i-PACT), 1-4.*

Smaoui, M., Rekik, M., & Krichen, L. (2025). Optimal design of an off-grid electrical system in remote areas with different renewable energy scenarios. *Energy, 318*, 134821. DOI: 10.1016/j.energy.2025.134821

Stojkoska, B. L., & Trivodaliev, K. V. (2017). A review of Internet of Things for smart home: Challenges and solutions. *Journal of Cleaner Production, 140*, 1454–1464. DOI: 10.1016/j.jclepro.2016.10.006

Talari, S., Shafie-khah, M., Siano, P., Loia, V., Tommaetti, A., & Catalao, J. P. S. (2017). A review of smart cities based on the Internet of Things concept. *Energies*, *10*(4), 421. DOI: 10.3390/en10040421

Tuballa, M. L., & Abundo, M. L. (2016). A review of the development of Smart Grid technologies. *Renewable & Sustainable Energy Reviews*, *59*, 710–725. DOI: 10.1016/j.rser.2016.01.011

Xiong, Z., Sheng, H., Rong, W., & Cooper, D. E. (2012). Intelligent transportation systems for smart cities: A progress review. *Science China. Information Sciences*, *55*(12), 2908–2914. DOI: 10.1007/s11432-012-4725-1

Yao, W., Zhao, J., Wen, F., Dong, Z., Xue, Y., Xu, Y., & Meng, K. (2014). A multi-objective collaborative planning strategy for integrated power distribution and electric vehicle charging systems. *IEEE Transactions on Power Systems*, *29*(4), 1811–1821. DOI: 10.1109/TPWRS.2013.2296615

Zhang, Y., Zou, W., Chen, X., Yang, C., & Cao, J. (2014). The security for power internet of things: framework, policies, and countermeasures. In: *2014 International Conference on Cyber-Enabled Distributed Computing and Knowledge Discovery*, Shanghai, 2014, pp. 139-142. DOI: 10.1109/CyberC.2014.32

Zhenga, Y., Niua, S., Shanga, Y., Shaoc, Z., & Jian, L. (2019). Integrating plug-in electric vehicles into power grids: A comprehensive review on power interaction mode, scheduling methodology and mathematical foundation. *Renewable & Sustainable Energy Reviews*, *112*, 424–439. DOI: 10.1016/j.rser.2019.05.059

Chapter 9
Industrial IoT Automation, Predictive Maintenance, and Smart Manufacturing

S. Madhu
https://orcid.org/0009-0000-8816-5798
Jyothy Institute of Commerce and Management, India

Sandeep L. Sarkale
https://orcid.org/0000-0001-7144-6892
School of Management and Research, Dr. D.Y. Patil Dnyan Prasad University, India

Nithin Subba Rao
https://orcid.org/0009-0006-3542-9597
Michigan Technological University, USA

Ganesh Pathak
https://orcid.org/0000-0003-0427-2145
Sri Balaji University, Pune, India

Nilesh Anute
https://orcid.org/0000-0001-6599-813X
Sri Balaji University, Pune, India

P. Selvakumar
https://orcid.org/0000-0002-3650-4548
Department of Science and Humanities, Nehru Institute of Technology, Coimbatore, India

T. C. Manjunath
https://orcid.org/0000-0003-2545-9160
Rajarajeswari College of Engineering, India

ABSTRACT

The (IIoT) transformative force is modern driving an unprecedented wave of digitalization and connectivity across sectors of physical machinery with enabled actuators, advanced data analytics. resulting in monitor, exchange, analyze systems, humans, IIoT empowers industries to achieve greater operational efficiency, pre-

DOI: 10.4018/979-8-3373-5727-0.ch009

dictive maintenance, improved product quality, and enhanced safety. Its impact is particularly visible in manufacturing, energy, transportation, healthcare, and supply chain management—industries that historically relied on manual operations and reactive maintenance are now leveraging IIoT to shift toward data-driven, proactive models. One of the most significant contributions of IIoT is in operational efficiency. Through connected devices and smart sensors, organizations gain granular, real-time visibility into every aspect of their operations. Machines can autonomously report performance metrics, detect anomalies, and suggest optimizations, leading to reduced downtime and better resource utilization.

INTRODUCTION

The (IIoT) transformative force is modern driving an unprecedented wave of digitalization and connectivity across sectors of physical machinery with enabled actuators, advanced data analytics, resulting in monitor, exchange, analyze systems, humans, IIoT empowers industries to achieve greater operational efficiency, predictive maintenance, improved product quality, and enhanced safety. Its impact is particularly visible in manufacturing, energy, transportation, healthcare, and supply chain management—industries that historically relied on manual operations and reactive maintenance are now leveraging IIoT to shift toward data-driven, proactive models. One of the most significant contributions of IIoT is in operational efficiency. Through connected devices and smart sensors, organizations gain granular, real-time visibility into every aspect of their operations. Machines can autonomously report performance metrics, detect anomalies, and suggest optimizations, leading to reduced downtime and better resource utilization. For example, in advanced manufacturing facilities or "smart factories," IIoT enables adaptive production lines that can adjust parameters automatically based on incoming data, ensuring optimal throughput and minimal waste. Deviations are detected and corrected immediately, reducing defects and rework. Additionally, data collected across product lifecycles enables continuous improvement and innovation. In sectors like pharmaceuticals, automotive, and consumer electronics, this data-driven quality assurance is essential to maintaining competitiveness and regulatory compliance. Supply chain management also benefits significantly from IIoT RFID tags, GPS devices, and connected transport fleets provides This visibility enables inventory management, logistics. In a globalized economy where pandemics, other factors have widespread impacts, IIoT-powered supply chains are more resilient, agile, and responsive to change.

Furthermore, IIoT supports ethical and transparent sourcing by providing traceability of materials from origin to final product, addressing increasing consumer and regulatory demands for accountability. In terms of worker safety and ergonomics,

IIoT technologies contribute by monitoring environmental conditions and human-machine interactions, can fatigue levels, exposure to hazardous substances. Smart helmets, vests, or exoskeletons equipped with IIoT sensors can prevent accidents by alerting workers and supervisors in real time. In mining, oil and gas, and construction industries with high-risk environments, these innovations play a critical role in protecting human life while enhancing productivity. However, alongside its benefits, the widespread deployment of IIoT introduces risks. Cybersecurity is a concern, as connected devices expand the potential for malicious actors. Ensuring IIoT requires robust security frameworks, encryption protocols, and continuous monitoring. Interoperability is another challenge, given the diversity of devices, protocols, and standards across different industries and vendors. Without seamless integration, the full potential of IIoT cannot be realized. In conclusion, IIoT is reshaping the industrial landscape by transforming how organizations operate, compete, and innovate. Its impact spans operational efficiency, maintenance, quality, supply chains, safety, sustainability, and beyond. While challenges remain, the continued evolution of IIoT, supported by emerging technologies and robust governance, will drive the next wave of industrial transformation. As industries embrace IIoT, they are not only improving productivity and profitability but also laying the foundation for more intelligent, sustainable, and human-centric industrial ecosystems.

BENEFITS OF INDUSTRIAL IOT: UNLOCKING THE VALUE OF IIOT ADOPTION

The adoption of delivers a wide array of transformative benefits for organizations across diverse sectors. As industries increasingly embrace IIoT, they productivity, safety, sustainability, and innovation. IIoT connects machines, sensors, systems, and people, creating an intelligent industrial ecosystem driven by real-time data and actionable insights. By equipping machinery and production systems with sensors and connectivity, industries gain continuous visibility into processes and equipment performance vibration, and output allows for precise control and rapid adjustments. Automation powered by IIoT reduces manual intervention, lowers error rates, and streamlines workflows (Gadicha & Gadicha, 2025). Adaptive production lines can self-optimize in response to changing conditions or demand fluctuations. As a result, organizations achieve higher throughput, reduced cycle times, and more consistent output, all of which contribute to increased competitiveness. Closely related to efficiency is predictive maintenance, which represents a paradigm shift in asset management. Instead of relying on scheduled or reactive maintenance—which often leads to either unnecessary servicing or unexpected failures—IIoT enables a condition-based approach. Machines equipped with intelligent sensors continuously report health indicators, while advanced analytics predict wear and potential breakdowns. Maintenance teams can intervene only when necessary, extending asset life and avoiding unplanned downtime. This not only reduces maintenance costs but also maximizes equipment availability, an advantage in industries such as manufacturing, energy, transportation, and mining (Endres et al, 2024). Improved product quality is another key benefit of IIoT. Real-time data collection and analysis during production allow for tighter quality control. Sensors monitor critical variables—such as temperature, humidity, torque, and alignment—ensuring adherence to specifications at every stage. When deviations are detected, immediate corrections can be made to prevent defects. Over time, IIoT data enables continuous process improvements through trend analysis and root cause identification. In industries like automotive, aerospace, pharmaceuticals, and consumer electronics—where product quality is paramount—IIoT helps maintain the highest standards while reducing scrap.

IIoT adoption also fosters energy efficiency and sustainability, aligning industrial practices with global environmental goals. Connected energy meters, HVAC systems, lighting, and production equipment enable granular monitoring of energy consumption. This data allows companies to implement energy-saving strategies. By reducing use and carbon emissions, industries contribute to a greener future while benefiting from cost savings and enhanced brand reputation. In addition to operational improvements, IIoT drives business agility and innovation. By unlocking a wealth of data across the organization, IIoT empowers leadership to make informed, data-driven decisions (Despande et al, 2024). Insights from IIoT analytics can reveal new market opportunities, inform product development, and guide strategic planning. For instance, manufacturers can leverage usage data from connected products to design features that better meet customer needs. Service-based business models, such as equipment-as-a-service (EaaS), become viable with IIoT-enabled remote monitoring and performance tracking. This fosters closer customer relationships and new revenue streams. Scalability and flexibility are further advantages of IIoT. Cloud-based IIoT platforms allow organizations to scale deployments as needed, adding new devices, production lines, or facilities without major infrastructure changes. Edge computing enhances flexibility by processing data locally, reducing latency, and ensuring continuity even in disconnected environments. Whether an organization operates a single factory or a global network of sites, IIoT provides the adaptability to evolve with changing business needs. Lastly, IIoT adoption supports workforce empowerment. While some fear that automation may displace workers, in reality, IIoT often augments human capabilities. Maintenance staff can use augmented reality (AR) tools powered by IIoT data for guided repairs. Operators can manage complex systems through intuitive dashboards (Arputharaj et al., 2024). By equipping workers with modern digital tools, organizations foster higher job satisfaction, engagement, and productivity. In conclusion, the benefits of Industrial IoT adoption are extensive and transformative. From operational efficiency and predictive maintenance to product quality, supply chain optimization, energy sustainability, workplace safety, and innovation, IIoT delivers substantial value to organizations. It not only improves competitiveness and profitability but also enables industries to become more agile, customer-centric, and environmentally responsible. As IIoT technologies continue to evolve—integrating AI, 5G, edge benefits only expand, driving future intelligent, connected industries.

INDUSTRIAL AUTOMATION WITH IIOT: UNLOCKING NEW BENEFITS

By seamlessly connecting machines, sensors, software platforms, and human operators, IIoT-enabled automation offers unprecedented levels of intelligence, adaptability, and efficiency across industrial processes. Traditional automation, often based on isolated programmable logic controllers (PLCs) and rigid workflows, is now giving way to dynamic, data-driven systems powered by IIoT. The benefits of this transformation are manifold, ranging from increased operational efficiency and cost savings to enhanced quality, flexibility, safety, and innovation. Embedded sensors and smart devices continuously capture data from all parts of the production environment, including machines, workpieces, environmental conditions, and human interactions (Lins & Givigi, 2021). This data is then transmitted via wired or wireless networks to centralized platforms or processed locally using edge computing. The result is a live digital representation of the factory floor, enabling intelligent monitoring, diagnostics, and control. Automation systems can instantly respond to this data, dynamically adjusting parameters to maintain optimal performance. This closed-loop feedback empowers organizations to achieve higher throughput and more stable production processes. Intelligent automation systems can optimize resource allocation, balance production loads, and reduce energy consumption in real time. Machines and equipment are no longer operated in isolation; instead, they are integrated into coordinated networks that continuously adjust to changing conditions. Automated scheduling and workflow management ensure that equipment operates at peak efficiency with minimal idle time. The result is higher productivity, reduced waste, and lower operating costs (Kumar et al, 2024). Energy optimization is a further benefit of IIoT-enhanced automation. Industrial processes often account for significant energy consumption, which represents both an environmental impact and a major cost driver. IIoT sensors and smart controllers enable granular monitoring of energy usage across machines, lines, and facilities. Automated systems can adjust operations to minimize energy waste, optimize load balancing, and shift consumption to off-peak periods. For industries pursuing sustainability goals and carbon reduction targets, IIoT-driven energy management is an indispensable tool. In terms of workplace safety, IIoT-powered automation plays a vital role. Automated systems can monitor hazardous environments and detect unsafe temperature extremes before they pose risks to personnel. Robotic systems and collaborative robots (co-bots) take on dangerous or repetitive tasks, reducing the potential for human injury. Wearable devices and environmental sensors can provide real-time feedback on worker health and exposure. As a result, IIoT automation helps create safer, healthier work environments, reducing accident rates and enhancing regulatory compliance.

Supply chain integration is another major advantage of IIoT-based automation. Automated processes connected to external suppliers, logistics providers, and customers create more transparent and responsive supply chains. Real-time visibility into inventory levels, work-in-progress, and order status enables more accurate planning and coordination. Automated ordering and replenishment systems reduce stockouts and overstocking often disrupt chains. IIoT-enhanced automation improves agility and resilience. IIoT also fosters far from workers, modern automation systems augment human capabilities by providing actionable insights and advanced tools (Ling et al, 2025). Operators can use augmented reality (AR) displays, voice interfaces, and mobile devices to interact with automated systems. Maintenance technicians can receive step-by-step guidance based on IIoT data. Engineers can remotely monitor and optimize operations from centralized control rooms or even off-site locations. This human-machine collaboration leads to smarter decision-making and more engaged, empowered workforces. Another important benefit is scalability and adaptability. IIoT-based automation systems are designed to be modular and extensible. Finally, IIoT automation supports strategic digital transformation. By integrating operational technology (OT) with information technology (IT), IIoT enables organizations to break down traditional silos and adopt more holistic, data-driven business models. Real-time visibility into operations supports better decision-making at all levels, from the shop floor to executive leadership (Coates et al, 2025). IIoT data also provides the foundation for advanced analytics, artificial intelligence (AI), and digital twin technologies that further enhance automation and innovation. In conclusion, the integration of IIoT with industrial automation delivers transformative benefits that go far beyond simple efficiency gains. It enables smarter, more flexible, and more resilient operations that are better aligned with today's complex market demands, goals IIoT-driven by edge computing, 5G connectivity, and machine learning—the potential benefits of automation will only continue to grow. Organizations that embrace IIoT-driven automation are not only improving productivity and profitability but also building the intelligent, connected factories of the future.

ROBOTICS AND MACHINE-TO-MACHINE COMMUNICATION: SYNERGY IN INDUSTRIAL AUTOMATION

Integration of robotics is one of the defining features of modern industrial automation and a core component of the Industry 4.0 revolution. Robotics has long played a vital role in industrial settings, performing repetitive, hazardous, or precision tasks with speed and accuracy. However, the capabilities of robotics have expanded dramatically through advancements in M2M communication, which allows machines, including robots, sensors, actuators, and control systems. Together, robotics and M2M create a highly responsive, adaptive, and intelligent manufacturing ecosystem (Moor et al, 2021). This synergy offers enormous benefits in terms of efficiency, flexibility, quality, scalability, and innovation, driving the transformation of industries around the world. At its essence, Machine-to-Machine communication involves the automated exchange of data between physical devices without industrial M2M networks, conveyors, automated guided vehicles (AGVs), sensors, and other equipment. These networks use various communication protocols—such as OPC UA, MQTT, PROFINET, EtherCAT, and 5G wireless—to ensure fast, reliable, and secure data transmission. By connecting these devices, M2M enables seamless coordination across complex processes, allowing machines to collaborate in ways that were previously impossible. Industrial robotics encompasses a wide range of technologies, from traditional welding to mobile robots that transport goods autonomously. Historically, robots operated in isolated, pre-programmed environments with little flexibility or real-time adaptability. M2M communication has changed this dynamic (Javed A R., et al, 2021). Today's robots can receive live data from sensors, other robots, and centralized control systems, allowing them to adjust their actions on the fly based on current conditions. This level of responsiveness greatly enhances their versatility and value in modern factories. One of the most significant benefits of combining robotics with M2M communication is enhanced operational efficiency. Through continuous data exchange, robots can synchronize their actions with upstream and downstream processes. This ensures higher accuracy and faster cycle times. Material-handling robots can dynamically adjust their routes based on production flow and inventory status. Such coordination minimizes bottlenecks, reduces idle time, and optimizes resource utilization.

Quality is greatly improved when robotics and M2M communication work together. Robots equipped with vision systems and force sensors can perform in-line inspections and quality checks in real time. M2M data exchange ensures that inspection results are instantly communicated to other machines or supervisory systems. If a defect is detected, robots can automatically sort out faulty products or adjust process parameters to correct deviations. This closed-loop feedback improves first-pass yield rates and ensures consistent product quality, highly industries aerospace, pharmaceuticals, food, and beverage. The synergy between robotics and M2M also enhances safety and relies heavily on M2M communication. They constantly monitor sensor inputs, including force, torque, and proximity data, to ensure safe interactions with human coworkers. M2M networks can alert cobots to the presence of humans or hazardous conditions, allowing them to adjust speed, force, or motion paths in real time (Nisha Angeline et al, 2025). This capability fosters safer, more ergonomic workplaces while enhancing overall productivity. Scalability is another key benefit. M2M communication allows robotic systems to be easily expanded or reconfigured as production needs evolve. New robots or devices can be added to the network with minimal disruption. Centralized orchestration systems can automatically recognize and integrate new assets, ensuring seamless operation. This scalability supports agile manufacturing strategies and helps organizations quickly respond to market demands or business growth. Supply chain integration is also improved through robotics and M2M. Automated material handling systems—including AGVs and AMRs—can communicate with production lines to ensure that materials and components are delivered just in time. Robots can dynamically adjust picking and packing operations based on live order data. This real-time coordination enhances supply chain efficiency, reduces inventory levels, and shortens order-to-delivery cycles. Energy optimization is another area where robotics and M2M communication offer benefits (Gomaa, 2025). Robots can adjust their power consumption based on production demands or energy pricing signals. M2M networks can coordinate energy-intensive operations to avoid peak loads or balance power usage across the factory. By optimizing energy consumption, organizations can reduce operating costs and support sustainability goals. Workforce empowerment is an often-overlooked but highly valuable benefit. M2M communication allows human operators and engineers to interact more effectively with robotic systems. Through intuitive human-machine interfaces (HMIs), mobile apps, and augmented reality (AR) tools, workers can monitor robot status, adjust parameters, and receive real-time feedback. Cost savings from robotics and M2M integration are substantial. Improved efficiency, reduced downtime, lower maintenance costs, enhanced quality, optimized energy usage, and greater flexibility all contribute to a significant return on investment (ROI). Moreover, the ability to produce customized products and respond quickly to market changes creates new revenue opportunities and strengthens customer relationships.

PREDICTIVE MAINTENANCE WITH IIOT: UNLOCKING BENEFITS THROUGH DATA-DRIVEN ASSET MANAGEMENT

Predictive maintenance (PdM), powered by one most transformative advancements in modern industrial operations. It enables organizations to schedule equipment IIoT-connected sensors leveraging advanced analytics, predictive maintenance delivers a wide array of benefits, including reduced downtime, lower maintenance costs, extended asset lifespan, improved safety, higher product quality, and enhanced operational efficiency. As industries increasingly adopt IIoT technologies, predictive maintenance is emerging as a key enabler of operations. Another major benefit is lower maintenance costs. Traditional preventive maintenance relies on fixed schedules or usage-based intervals, often leading to unnecessary maintenance of healthy equipment or late interventions on deteriorating assets (Jha & Tripathi, 2025). Predictive maintenance optimizes resource allocation by targeting maintenance efforts only where and when they are truly needed. This reduces labor hours, spare parts inventory, and maintenance-related disruptions. Over time, organizations realize substantial cost savings while maintaining or even improving asset performance. Predictive maintenance also helps to extend equipment lifespan. By continuously monitoring asset condition and addressing issues early, organizations prevent minor defects from escalating into major failures that cause irreversible damage. This maximizes the return on investment (ROI) for capital assets and reduces the need for costly replacements. In industries with high-value or long-lived equipment—such as aerospace, heavy manufacturing, and utilities—this benefit is particularly significant. Energy efficiency is another area where predictive maintenance provides advantages. Faulty or poorly maintained equipment often consumes more energy than well-maintained assets. For example, a misaligned pump may require more power to achieve the desired flow rate, or a clogged air filter in an HVAC system may increase fan energy consumption (Mon et al, 2025). By ensuring optimal equipment performance, predictive maintenance helps reduce energy waste, lower utility costs, and support sustainability goals. In the context of rising energy prices and increasing environmental regulations, this benefit is increasingly valuable. Predictive maintenance also plays a vital role in optimizing spare parts inventory. By accurately predicting when components will require replacement, organizations can align inventory levels with actual needs. This reduces carrying costs, minimizes obsolescence, and prevents shortages that could delay maintenance activities. The result is a leaner, more efficient supply chain that supports just-in-time maintenance operations.

Another important benefit of predictive maintenance is greater operational transparency and decision support. IIoT platforms aggregate data from across the enterprise, providing a holistic view of asset health and performance. Maintenance managers, engineers, and executives can access dashboards and analytics that enable better planning, budgeting, and risk management. Historical data and predictive insights support informed decision-making about equipment upgrades, capital investments, and lifecycle management. By transforming maintenance from a reactive function into a strategic asset, predictive maintenance aligns with broader business objectives. Scalability and flexibility are inherent in IIoT-powered predictive maintenance solutions (Chhotaray et al, 2023). Cloud-based platforms can accommodate large numbers of assets and facilities, enabling enterprises to scale their maintenance programs across multiple sites. Modular architectures and open standards ensure that predictive maintenance systems can integrate with existing enterprise resource planning (ERP), computerized maintenance management systems, and manufacturing execution systems (MES). As organizations grow or adopt new technologies, predictive maintenance solutions can evolve with them. Moreover, predictive maintenance with IIoT supports workforce empowerment. Technicians and engineers gain access to real-time data, predictive alerts, and diagnostic tools that enhance their effectiveness and efficiency. Finally, predictive maintenance contributes to competitive advantage and innovation. Organizations that excel in predictive maintenance achieve higher operational reliability, lower costs, and better customer satisfaction. They can offer more flexible production schedules, shorter lead times, and higher-quality products (Upadhyaya et al, 2025). Predictive maintenance data also supports broader innovation initiatives, such as digital twin development, machine learning model training, and continuous improvement programs. In a rapidly evolving industrial landscape, the ability to leverage IIoT and predictive maintenance effectively can differentiate market leaders from lagging competitors. In conclusion, predictive maintenance with IIoT unlocks a wide array of benefits that extend across the entire enterprise. By transforming into a proactive capability, organizations can achieve reduced downtime, lower costs, extended asset life, improved safety and quality, and greater energy competitiveness. IIoTdriven by advances in AI, connectivity, the capabilities of predictive maintenance will only grow stronger. For forward-looking organizations, investing in predictive maintenance is a critical step toward building the smart factories of the future.

SMART MANUFACTURING AND IIOT: REVOLUTIONIZING INDUSTRIAL PRODUCTION THROUGH CONNECTIVITY, INTELLIGENCE, AND FLEXIBILITY

Smart manufacturing, powered by represents an approach to production where data, connectivity, automation, and intelligence converge to create highly efficient, agile, and customer-centric operations. Moving beyond traditional manufacturing models, smart manufacturing integrates advanced technologies such as IIoT, digital twins, to predictive capabilities, and unprecedented levels of adaptability. This new paradigm aligns with Industry 4.0 principles, emphasizing flexibility, efficiency, mass customization, sustainability, and continuous improvement (Thirunagalingam & Selvakumar, 2025). At its foundation, smart manufacturing leverages the pervasive connectivity provided by IIoT. Sensors and actuators embedded throughout the factory floor—on machines, production lines, supply chains, and products themselves—continuously collect and transmit parameters, environmental This vast stream of information is processed in real time by advanced analytics platforms, allowing stakeholders to quickly. The seamless flow of data across machines (Machine-to-Machine, or M2M), operators, and enterprise systems creates a digital thread that links every aspect of manufacturing, from design to delivery. Quality management is another area where smart manufacturing excels. IIoT-connected quality sensors, vision systems, and advanced analytics enable in-line quality control and real-time feedback. Defects can be detected and addressed immediately, reducing scrap rates and rework. AI-powered quality systems can also identify root causes of defects by analyzing process data across multiple variables, leading to continuous process improvement (Sahoo & Lo, 2022). This real-time, closed-loop approach to quality ensures consistent product performance and supports compliance with stringent industry standards. Energy and resource efficiency are integral to smart manufacturing strategies. IIoT technologies enable detailed monitoring of energy consumption, water usage, raw materials, and waste streams across the factory. Manufacturers can reduce their environmental footprint, lower operational costs, and align with sustainability goals. In an era of rising energy prices and growing environmental awareness, these capabilities are increasingly critical for both competitiveness and corporate responsibility.

Figure 1. The Role of IoT Solutions in the Manufacturing Industry

Supply chain integration is another hallmark of smart manufacturing. IIoT connects manufacturing operations with suppliers, logistics providers, and customers through cloud platforms and digital supply chain networks. Manufacturers can synchronize production schedules with supplier deliveries, optimize inventory levels, and provide customers with accurate order status updates. This end-to-end integration enables just-in-time manufacturing, faster order fulfillment, and greater supply chain resilience. Digital twins play a pivotal role in smart manufacturing, providing a virtual mirror of physical assets and processes. By integrating real-time IIoT data with simulation models, digital twins enable predictive analytics, scenario planning, and process optimization (Bu et al, 2021). Engineers can test changes virtually, anticipate outcomes, and fine-tune operations without disrupting actual production. Over time, digital twins evolve into comprehensive digital ecosystems that enhance decision-making and drive continuous improvement. The scalability and interoperability of IIoT platforms ensure that smart manufacturing solutions can grow with the business. Cloud-native architectures and open standards allow manufacturers to connect new equipment, integrate emerging technologies, and expand their capabilities without major disruptions. Whether a single factory or a global network of production sites, smart manufacturing systems provide a unified, scalable foundation for digital transformation. They can respond rapidly to market changes, capitalize on new opportunities, and build resilient operations capable of withstanding disruptions. In highly competitive industries, these advantages translate into stronger market positions and long-term success (Wyciślak, 2024). In conclusion, smart manufacturing—powered by a revolutionary leap forward in industrial production. By harnessing connectivity, intelligence, and flexibility, smart factories transcend the limitations of traditional manufacturing. They enable real-time visibility, predictive capabilities, agile production, and continuous improvement. As IIoT technologies continue to evolve, with advances in AI, 5G, and edge computing, the potential will only grow stronger. For forward-thinking organizations, embracing smart manufacturing is not just a path to efficiency—it is a strategic imperative for thriving in the digital age of industry.

REAL-TIME MONITORING AND CONTROL: ENHANCING INDUSTRIAL PERFORMANCE THROUGH CONTINUOUS VISIBILITY AND INSTANTANEOUS RESPONSE

Real-Time Monitoring and Control represents one of the most valuable capabilities unlocked by fundamentally transforming the way operations are managed and optimized. At its core, real-time continuous transmission, operational from equipment, processes, and environmental conditions. Real-time control refers to the ability to act on this data instantaneously—either through automated system responses or human intervention—thereby improving performance, increasing responsiveness, and enhancing decision-making at all levels of an organization. Together, these capabilities drive substantial reliability, industries (Naldi et al, 2025). The foundation of real-time monitoring and control is pervasive connectivity enabled by IIoT. Modern industrial environments are equipped with a vast array of smart sensors, actuators, controllers, and edge computing devices that monitor every aspect of operations. These devices continuously measure key parameters such as temperature, vibration, pressure, flow rates, voltage, current, humidity, noise, gas concentrations, and more. Through wired or wireless networks—including Ethernet, Wi-Fi 6, 5G, LoRaWAN, and industrial fieldbus systems platforms, local control units. The result is a digital nervous system that provides real-time visibility into every component. Traditional industrial systems often relied on periodic inspections or batch data collection, which introduced delays between data acquisition and analysis. This gap made it difficult to catch problems before they escalated into failures. With real-time monitoring, any deviation from normal operating conditions—such as unusual vibrations, temperature spikes, or pressure drops—can be detected immediately (Javed et al, 2021). Automated alert systems notify operators, maintenance teams, or supervisory systems, enabling rapid response. In environments involving hazardous materials, extreme temperatures, or high-pressure systems, process upsets personnel and the environment. Real-time monitoring of critical parameters allows for unsafe gas leaks, overheating, or structural stresses. Automated control systems can initiate emergency shutdown procedures or activate safety systems in milliseconds, preventing accidents and mitigating hazards. In addition, wearables and connected personal protective equipment (PPE) can monitor worker health and exposure in real time, further enhancing workplace safety.

Real-time monitoring and control also contribute to energy efficiency and sustainability. Many industrial processes consume large amounts of energy, with inefficiencies often going unnoticed. By continuously monitoring energy consumption, equipment loads, and environmental conditions, IIoT platforms can identify opportunities for improvement. Control systems can automatically adjust lighting, HVAC, compressed air, and production schedules, costs. Real-time data also supports dynamic load balancing, helping organizations meet sustainability goals and regulatory requirements. Remote monitoring and control capabilities—enabled by IIoT and cloud technologies—further extend the benefits of real-time data (Arputharaj et al, 2024). Operators, engineers, and managers can access live data and control systems from remote locations using mobile devices, tablets, or centralized control rooms. This capability is especially valuable in geographically distributed operations, such as oil and gas pipelines, wind farms, mining operations, and global manufacturing networks. Remote visibility enables faster troubleshooting, reduces travel requirements, and allows organizations to deploy expert resources more efficiently. In times of crisis—such as natural disasters or pandemics—remote capabilities ensure business continuity and operational resilience. Real-time monitoring and control also foster continuous improvement and support methodologies such as Lean, Six Sigma, and Total Productive Maintenance (TPM). By providing immediate feedback on process performance, IIoT systems enable teams to identify inefficiencies, track improvement efforts, and validate the impact of changes accurately, eliminating guesswork at all levels of the organization. Over time, this drives higher productivity, better quality, and more agile operations (Hubballi et al, 2025). Human-machine collaboration is further enhanced through real-time data. Operators equipped with wearable devices, augmented reality (AR) displays, or advanced human-machine interfaces (HMIs) can interact with machines and systems more effectively. They receive instant feedback, guided instructions, and alerts, and safer work. In conclusion, enabled by Industrialare's transformation of industrial operations across the globe. By providing continuous visibility, immediate feedback, and dynamic responsiveness, these capabilities enhance efficiency, quality, reliability, safety, and IIoT advances, with innovations in edge computing, 5G, digital twins, the potential of real-time monitoring and control will only grow. For forward-thinking organizations, investing in these capabilities is a strategic imperative that drives operational excellence and long-term competitiveness.

DATA MANAGEMENT AND ANALYTICS: HARNESSING INDUSTRIAL DATA FOR INSIGHT, INNOVATION, AND OPERATIONAL EXCELLENCE

The era emerged as the most valuable asset in the modern manufacturing landscape. As factories, supply chains, and industrial assets become increasingly connected, they have expanded exponentially. From sensors and machines to enterprise systems and cloud platforms, data flows continuously across every operational layer. Managing this vast and complex data ecosystem—and extracting actionable insights from it just it is Effective data management and analytics lie at the core of smart manufacturing, enabling organizations to optimize performance, reduce downtime, drive innovation, improve quality, and make faster, more informed decisions (Maheswaran et al, 2025). At its core, data management involves the collection, storage, integration, organization, and governance of IIoT systems, including structured data (e.g., temperature readings, cycle counts, and output volumes), unstructured data (e.g., images, video, audio, maintenance logs), and semi-structured data (e.g., JSON from APIs, machine-generated XML files). The challenge lies not only in handling the enormous volume of data but also in ensuring its accuracy, relevance, security, and accessibility. A foundational element of IIoT-enabled data management is data acquisition. Sensors, programmable logic controllers (PLCs), edge devices, and human-machine interfaces (HMIs) generate real-time operational data from every part of the industrial environment. This data must be collected with high fidelity and minimal latency to preserve its value. Edge computing often plays a critical role at this stage, including usage, filtering noise, and ensuring low-latency decision-making. Once acquired, data must be organized and stored efficiently. Industrial data management platforms use time-series databases, data lakes, and hybrid cloud infrastructures to accommodate diverse data types and formats (Gamal, 2025). These platforms must be highly scalable and designed for high availability and redundancy to support mission-critical operations. Metadata tagging, schema design, and indexing ensure that stored data remains searchable, interpretable, and usable across applications. Proper data architecture is essential for integrating operational technology (OT) with information technology (IT), bridging the gap between plant-level systems and enterprise-wide analytics. Effective data management also requires interoperability and integration. Industrial environments typically consist of heterogeneous systems from multiple vendors—machines, ERP platforms, MES software, SCADA systems, CMMS tools, and more. Seamless data exchange between these systems is enabled through APIs, OPC UA protocols, middleware, and industrial IoT gateways. This interoperability is vital for breaking down silos, enabling end-to-end visibility, and creating unified data ecosystems that support enterprise-wide intelligence.

Optimal simulations, constraints, and performance targets, prescriptive models help organizations choose the best course of action in complex scenarios. For example, in a chemical plant, prescriptive analytics might advise adjusting feedstock ratios or temperature setpoints to maximize yield while minimizing energy usage and emissions. Descriptive analytics, while less complex, remains essential. Dashboards, reports, and visualizations summarize key performance indicators (KPIs), productivity metrics, and quality trends. Real-time dashboards powered by IIoT data provide immediate situational awareness, enabling operators to respond quickly to deviations and production managers to make fast, evidence-based decisions. Diagnostic analytics helps uncover root causes when problems occur. By analyzing historical data in context, these tools can trace the origin of defects, process deviations, or unexpected downtime. For instance, if a product batch fails a quality test, diagnostic analytics can pinpoint whether the cause was a sensor miscalibration, material variation, or operator error, facilitating faster resolution and prevention of recurrence (Buja, 2025). Fed by real-time IIoT data, digital twins simulate equipment behavior, process dynamics, and system interactions. Engineers use these models for testing, optimization, and predictive scenario planning, drastically reducing development costs and risk. Analytics-powered digital twins enable continuous innovation by linking design, operation, and performance feedback in a closed-loop system. Quality control and assurance benefit significantly from data analytics. By correlating process variables with product outcomes, analytics tools identify the most influential factors in achieving consistent quality. AI-driven vision systems and sensors detect defects in real time, while machine learning models predict future quality issues before they occur. This enables closed-loop quality control systems that ensure high first-pass yield and minimal waste. Energy management and sustainability are other domains where analytics provide crucial value. By analyzing data from energy meters, equipment loads, environmental sensors, and utility interfaces, organizations can uncover patterns of energy usage, inefficiencies, and waste. These include data silos, integration difficulties, cybersecurity risks, talent shortages, and data overload. Overcoming these requires not just the right technologies, but also a coherent strategy that aligns people, processes, and platforms (Kumar et al, 2024). Organizations must invest in data literacy, cross-functional collaboration, robust IT/OT convergence, and secure data infrastructures. In conclusion, data management and analytics are the twin pillars of industrial transformation in the IIoT era. Effective data management ensures that information is accurate, secure, integrated, and accessible, while analytics unlock the insights that drive action and innovation. Together, they empower organizations to optimize operations, anticipate and prevent problems, accelerate decision-making, intelligent, resilient industrial systems. IIoT, those who master the art and science of data will not only lead in operational excellence, but also shape the future of industry.

CYBERSECURITY CONCERNS: NAVIGATING THE COMPLEX LANDSCAPE OF IIOT SECURITY IN INDUSTRIAL ENVIRONMENTS

As manufacturing operations move into intelligent, data-driven networks, it also significantly expands the digital attack surface. While IIoT brings unprecedented gains in productivity, efficiency, and flexibility, it simultaneously introduces complex cybersecurity risks that traditional industrial control systems were never designed to handle. These vulnerabilities, if exploited, can cause financial losses, data breaches, reputational damage, safety hazards, and even national security threats. Therefore, cybersecurity has become a central concern for all stakeholders involved in IIoT-enabled environments—from manufacturers and operators to policymakers and technology providers. Legacy systems pose one of the greatest threats (Axon et al, 2022). Industrial machinery, SCADA systems, and PLCs (programmable logic controllers) often operate for decades and were built before cybersecurity became a critical concern. They may lack basic authentication protocols, support for encryption, or patching mechanisms. Integrating such systems with modern IIoT platforms introduces inherent risks, especially if these systems are connected directly to public networks or exposed through insecure APIs. Attackers can exploit these vulnerabilities to infiltrate control systems and manipulate physical processes, as seen in the infamous Stuxnet attack that targeted Iran's nuclear program. Another major concern is insufficient network segmentation and access control. In many industrial networks, once an attacker gains entry—often through compromised endpoints or phishing attacks—they can move laterally across the system due to poor segmentation between critical and non-critical assets (Rojas et al, 2024). This lateral movement enables attackers to escalate privileges, access sensitive data, or interfere with industrial processes. Supply chain vulnerabilities further exacerbate cybersecurity risks in IIoT systems. Modern industrial operations rely on a vast network of hardware, firmware. Any compromise within this supply chain—whether through counterfeit components, infected software updates, or backdoors—can introduce hidden threats into the network. The 2020 SolarWinds cyberattack underscored how even trusted software platforms can become vectors for widespread compromise. In IIoT environments, where devices and systems are deeply integrated and continuously exchanging data is both difficult and essential.

Data security and privacy are also critical concerns. IIoT platforms collect massive amounts of sensitive operational and business data, including production metrics, equipment status, performance indicators, location data, user credentials, and proprietary processes. If not adequately protected, this data can be intercepted, stolen, or manipulated, leading to industrial espionage, intellectual property theft, or competitive sabotage. Moreover, the integration of IIoT with customer-facing systems (e.g., in energy utilities or smart manufacturing) raises concerns about the protection of such as other data privacy laws. Cloud and edge computing architectures, while essential for scalable IIoT deployments, introduce additional cybersecurity complexities (Chen et al, 2024). Cloud platforms must be secured against unauthorized access, misconfigurations, data leakage, and denial-of-service (DoS) attacks. Edge computing devices, which process data closer to the source, are often deployed in physically accessible and vulnerable environments, increasing the risk of tampering and physical compromise. Ensuring end-to-end encryption, secure boot processes, firmware integrity, and regular updates is vital for both cloud and edge infrastructure. Human error and insider threats remain persistent vulnerabilities. Despite sophisticated technologies, many breaches occur due to misconfigured devices, weak passwords, outdated firmware, or untrained personnel. Employees may unknowingly connect unsecured devices, click on malicious links, or neglect software updates. Additionally, disgruntled insiders or negligent contractors can abuse their access privileges to compromise systems. A robust cybersecurity strategy must therefore include continuous employee training, security awareness programs, and monitoring of user behavior. Organizations must also develop robust incident response and recovery of a breach; having clear protocols for containment, eradication, recovery, and communication is vital to minimize impact (Kotte et al, 2025). Cybersecurity is not a one-time initiative but an ongoing effort that requires continuous assessment, adaptation, and investment. In conclusion, cybersecurity in the IIoT era is a complex, high-stakes challenge that demands comprehensive and adaptive strategies. As industrial environments become more interconnected, the risks of cyberattacks multiply, requiring organizations to rethink traditional approaches and embrace a proactive, multilayered defense model. Cybersecurity must be built into the DNA of IIoT architectures from the outset, rather than added as an afterthought. With the right mix of technologies, policies, training, and culture, organizations can secure their IIoT ecosystems and confidently harness the transformative power of the digital industry while safeguarding their assets, data, operations, and people.

EMERGING TRENDS IN IIOT: CHARTING THE FUTURE OF INDUSTRIAL CONNECTIVITY, INTELLIGENCE, AND INNOVATION

IIoThas already catalyzed fundamental transformation way industries operate, data collection, and automation, enhanced across sectors. However, the journey of IIoT is far from complete. As technology evolves rapidly and industries adapt to changing demands, several emerging trends are shaping the next generation of IIoT solutions. These trends are not only technological but also strategic and cultural, encompassing shifts in connectivity, intelligence, business models, sustainability, and workforce dynamics. Understanding and adapting to these trends, future-proof and maintain competitiveness in the age of Industry 4.0 and beyond (Sakthivel et al, 2025). Edge Computing and Distributed Intelligence: One of the most significant trends in IIoT is that Traditional IIoT architectures often relied on transmitting vast amounts of raw data to centralized cloud platforms for processing. However, as the volume and velocity of data continue to grow—along with the demand for ultra-low latency and real-time decision-making—processing data at the edge, closer to the source, has become essential. Edge by keeping data local, and supports autonomous decision-making in time-sensitive industrial environments. Modern edge devices are becoming more powerful, capable of running sophisticated analytics, AI models, and control functions without cloud dependency. This shift enables distributed intelligence, where decisions are made locally by machines or microcontrollers based on real-time conditions. Edge computing is especially predictive maintenance, inspection, and autonomous robotics, where milliseconds can make a significant difference (Anusha et al, 2024). Digital Twins and Virtual Commissioning: Digital twins—virtual replicas of physical assets, processes, or systems—are increasingly being used in conjunction with IIoT to simulate, monitor, and optimize performance in real Powered AI, simulation models, digital twins enable continuous visibility into the health, behavior, and future performance of machines, lines, or entire factories. Emerging use cases include predictive simulations, what-if analysis, and virtual commissioning, where new systems are tested and validated in a digital environment before actual deployment. This reduces downtime, accelerates time to market, and minimizes the risk of costly errors. As computational power grows and modeling tools improve, digital twins are becoming more granular, interactive, and integrated with enterprise systems.

Cybersecurity by Design: As IIoT systems become more connected and complex, cybersecurity continues to be a top priority—and an evolving trend. Organizations are moving beyond reactive security models to embrace "security by design"—integrating robust cybersecurity measures into the core architecture of IIoT devices and platforms. Emerging security trends include: Zero Trust Architecture (ZTA) for IIoT networks, Hardware-based security modules, Blockchain for secure device identity and data integrity, AI-driven threat detection and response, Remote attestation and secure boot processes. With growing threats to critical infrastructure and supply chains, regulatory bodies are also stepping in to mandate stricter IIoT security standards, further shaping industry practices (Wu et al, 2022). Sustainability and Green IIoT: Sustainability is no longer a corporate buzzword—it is becoming a regulatory, ethical, and business imperative. IIoT is playing a pivotal role in driving green manufacturing by enabling real-time monitoring of energy use, emissions, water consumption, and waste. Through smart sensors, energy analytics, and intelligent controls, companies can optimize resource usage, reduce carbon footprints, and meet environmental targets. Autonomous Operations and Industrial Metaverse: Looking even further ahead, IIoT is paving the way for autonomous operations, where systems self-monitor, self-optimize, and self-heal with minimal human intervention. Combined with AI, robotics, and digital twins, this autonomy could transform how factories, refineries, and infrastructure systems operate. The rise of the industrial metaverse—a convergence of physical and digital worlds—offers immersive environments for training, simulation, design, and collaboration (Jamil et al, 2022). Powered by IIoT data, AR/VR technologies, and cloud computing, the industrial metaverse is expected to redefine industrial productivity, innovation, and customer engagement. The future of IIoT is being shaped by a dynamic convergence of connectivity, intelligence, sustainability, and human-centric design. From edge AI and 5G to digital twins and green manufacturing, these emerging trends signal a shift toward more autonomous, adaptive, and responsive industrial systems. Organizations that recognize and strategically embrace these developments build resilience, innovation, and lead in a new era of industrial transformation.

The journey ahead will require agility, openness to change, and a commitment to integrating people, processes, and technology in a unified vision for the connected enterprise of tomorrow.

CONCLUSION

(IIoT) is reshaping the fabric of modern industry, creating a powerful fusion of connectivity, computation, and cognitive capability. While IIoT facilitates the acquisition vast array of connected sensors, systems, AI provides the analytical horsepower to interpret that data, generate actionable insights, and enable automated, intelligent decision-making. Together, AI and IIoT form the backbone of smart manufacturing, industrial autonomy. As the face of growing complexity, competitive pressures, and demands for sustainability, AI's role in IIoT is becoming not just valuable but indispensable. Another critical application is process optimization. In complex industrial operations—such as chemical processing, energy production, or semiconductor fabrication—dozens or even hundreds of variables must be controlled in tandem to achieve optimal output. AI algorithms, particularly those based on reinforcement learning or deep learning, can analyze historical process data to determine ideal parameter settings and continuously adjust them in real-time to maximize yield, minimize waste, or maintain quality within strict tolerances. This kind of adaptive optimization leads to self-tuning systems that operate at peak efficiency with minimal human oversight. Natural Language Processing (NLP) and conversational AI are also making inroads into IIoT environments. Voice assistants, chatbots, and intelligent interfaces are being used to streamline human-machine interaction. Field technicians can ask an AI-powered assistant for a machine's diagnostic history or receive guided instructions during a maintenance procedure, without having to sift through manuals or databases. The evolution of AI at the edge is further expanding IIoT capabilities. This allows for immediate, localized decision-making, which is critical in latency-sensitive environments like robotics, process control, and autonomous vehicles. Edge AI reduces network bandwidth usage, enhances privacy and data security, and enables operations in remote or connectivity-challenged locations. Combined with TinyML (machine learning on microcontrollers), edge AI is enabling smarter and more autonomous industrial devices than ever before. In the realm of supply chain and logistics, AI-enabled IIoT systems are improving forecasting, asset tracking, and inventory management. Predictive analytics helps forecast demand and optimize procurement cycles, while computer vision and RFID-integrated systems powered by AI provide real-time visibility of goods in transit. AI algorithms also evaluate supplier performance, identify bottlenecks, and model supply chain disruptions to improve resilience and responsiveness.

From a cybersecurity perspective, AI enhances IIoT defense by detecting and responding to threats more effectively. With traditional rule-based systems struggling to keep up with sophisticated cyberattacks, AI-powered security solutions use behavioral analytics, anomaly detection, and pattern recognition to identify malicious activities in real-time. These systems can automatically quarantine affected devices, isolate network segments, or initiate alerts, minimizing the impact of potential breaches and protecting sensitive operational data. However, the vast deployment of IIoT includes: Data quality and availability: AI systems are only as good as the data they are trained on. Inconsistent, incomplete, or noisy data from industrial sensors can impair model accuracy. Model explainability and transparency: In mission-critical environments, AI decisions must be interpretable and auditable. Black-box models can pose compliance and safety risks. Integration with legacy systems: Many industrial environments rely on outdated equipment and infrastructure that are not conducive to AI deployment. Skill gaps: Developing, deploying, and maintaining AI systems domain knowledge, engineering, and resources that are often scarce in traditional manufacturing settings. To address these challenges, organizations are adopting best practices such as: Implementing data governance frameworks to ensure data integrity, Using hybrid AI approaches, combining rules-based logic with machine learning, Leveraging AutoML platforms to simplify model development, Fostering collaboration between IT, OT, and data science teams, Investing in training and upskilling the workforce. Artificial Intelligence is no longer a futuristic concept in the context of IIoT—it is a driving force behind its most transformative capabilities. From predictive maintenance and intelligent automation to anomaly detection and digital twins, AI empowers organizations to convert vast quantities of sensor data into strategic decisions. IIoT enables factories, supply chains, and industrial systems to become more adaptive, autonomous, and resilient. Mature and integrate deeply into industrial workflows, and they will only become more central. Organizations that embrace AI-enabled IIoT today are not merely optimizing operations—they are the foundation enterprises of tomorrow.

REFERENCES

Anusha, S., Ezhilvandan, M., Gayathri, S., Sivaraman, V., Anish, T. P., & Subramanian, R. S. (2024). Emerging Trends in Edge AI for Industry 4.0 and 5.0: Technologies, Applications, and Challenges. In *AI Innovations in Service and Tourism Marketing* (pp. 211–228). IGI Global., DOI: 10.4018/979-8-3693-7909-7.ch010

Arputharaj, J. V., William, B. N. J., Haruna, A. A., & Prasad, D. D. (2024). Exploring the Synergy of IIoT, AI, and Data Analytics in Industry 6.0. In *Industry 6.0: Transforming Manufacturing Through Smart Technologies* (pp. 1–36). CRC Press., DOI: 10.1201/9781003517993-1

Axon, L., Fletcher, K., Scott, A. S., Stolz, M., Hannigan, R., El Kaafarani, A., Goldsmith, M., & Creese, S. (2022). Emerging Cybersecurity Capability Gaps in the Industrial Internet of Things: Overview and Research Agenda. *Digital Threats : Research and Practice*, *3*(4), 34. Advance online publication. DOI: 10.1145/3503920

Bu, L., Zhang, Y., Liu, H., Yuan, X., Guo, J., & Han, S. (2021). An IIoT-Driven and AI-Enabled Framework for Smart Manufacturing System Based on Three-Terminal Collaborative Platform. *Advanced Engineering Informatics*, *50*, 101370. Advance online publication. DOI: 10.1016/j.aei.2021.101370

Buja, A. (2025). *Cybersecurity of Industrial Internet of Things (IIoT)* (1st ed.). CRC Press., DOI: 10.1201/9781003383253

Chen, J., He, J., Chen, F., Lv, Z., Tang, J., Li, W., Liu, Z., Yang, H. H., & Han, G. (2024). "Towards General Industrial Intelligence: A Survey on IIoT-Enhanced Continual Large Models." arXiv preprint, arXiv:2409.01207. DOI: /arXiv.2409.01207DOI: 10.48550

Chhotaray, P., Behera, B. C., Moharana, B. R., Muduli, K., & Sephyrin, F. T. R. (2023). Enhancement of Manufacturing Sector Performance with the Application of Industrial Internet of Things (IIoT). In *Smart Technologies for Improved Performance of Manufacturing Systems and Services* (pp. 1–19). CRC Press., DOI: 10.1201/9781003346623-1

Coates, J. M., Santhanalakshmi, K., Dinesh, N., Gandhimathi, S., & Chandel, P. S. Mohit & Selvakumar P, (2025). "Assessment and Evaluation in Transnational Staff Development." In G Morris & J Ye (Eds.), Training and Development in Transnational Higher Education (pp. 387–406). IGI Global Scientific Publishing. DOI: 10.4018/979-8-3373-1897-4.ch019

Despande, P., Chandar, S. K., & Elangovan, N. (2024). "Enhancing Small and Medium OEMs' Adoption of IIoT Technologies." In *Proceedings of the 8th International Conference on Inventive Systems and Control (ICISC)*, IEEE, pp. 656–661. DOI: DOI: 10.1109/ICISC62624.2024.00115

Endres, H., Indulska, M., & Ghosh, A. (2024). Unlocking the Potential of Industrial Internet of Things (IIOT) in the Age of the Industrial Metaverse: Business Models and Challenges. *Industrial Marketing Management*, *119*, 90–107. DOI: 10.1016/j.indmarman.2024.03.006

Gadicha, A. B., & Gadicha, V. B. (2025). Unveiling the Potential of IoT and IIoT Industrial Technologies in Cybersecurity: Trends, Applications, and Future Prospective. In *Advancing Cybersecurity in Smart Factories Through Autonomous Robotic Defenses* (pp. 377–396). IGI Global., DOI: 10.4018/979-8-3373-0583-7.ch016

Gamal, H. (2025). Technological Integration and Innovative Strategies Harnessing Artificial Intelligence for Operational Excellence. In *Building Business Knowledge for Complex Modern Business Environments* (pp. 237–270). IGI Global., DOI: 10.4018/979-8-3693-6447-5.ch009

Gomaa, A. H. (2025). Optimizing Asset Integrity for Critical Manufacturing Systems Using Advanced Proactive Maintenance Strategies. *International Journal of Emerging Science and Engineering*, *13*(4), 21–33. DOI: 10.35940/ijese.B2026.13040325

Hubballi, R. B., Selvakumar, P., Seenivasan, R., Basava Aradhya, S. G., Dinesh, N., & Seelam, P. K. (2025). Overview of Current AI Technologies in Education. In Kyei-Blankson, L., & Ntuli, E. (Eds.), *Transformative AI Practices for Personalized Learning Strategies* (pp. 1–26). IGI Global Scientific Publishing., DOI: 10.4018/979-8-3693-8744-3.ch001

Jamil, S., Rahman, M. U., & Fawad, . (2022). A Comprehensive Survey of Digital Twins and Federated Learning for Industrial Internet of Things (IIoT), Internet of Vehicles (IoV) and Internet of Drones (IoD). *Applied System Innovation*, *5*(3), 56. Advance online publication. DOI: 10.3390/asi5030056

Javed, A. R., Sarwar, M. U., Beg, M. O., Asim, M., Baker, T., & Tawfik, H. (2021). Cloud-based intelligent predictive maintenance system for industrial equipment using machine learning. *Journal of Ambient Intelligence and Humanized Computing*, *12*, 4413–4429. DOI: 10.1007/s12652-020-02669-4

Jha, V., & Tripathi, P. (2025). An Energy-Efficient Knowledge Discovery in Cognitive IoT. *OPSEARCH*, *62*, 1–18. DOI: 10.1007/s12597-025-00977-z

Kotte, K. R., Attaluri, V., & Selvakumar, P. (2025). Revolutionising Retail Banking with AI and Virtual Technologies AI-Driven Financial Training. In Nair, A. J., Manohar, S., Limbu, Y. B., & Huhmann, B. A. (Eds.), *Intersecting Natural Language Processing and FinTech Innovations in Service Marketing* (pp. 421–440). IGI Global Scientific Publishing., DOI: 10.4018/979-8-3693-9944-6.ch020

Kumar, S., Kameswari, Y. L., Rao, K. R., Moram, V., & Shital, S. (2024). Risk Management in IIoT. In *Industrial Internet of Things Security* (pp. 43–60). CRC Press., DOI: 10.1201/9781003466284-3

Kumar, S. S., Kumar, S. R., & Ramesh, G. (2024). From Industry 4.0 to 5.0: Enriching Manufacturing Excellence Through Human–Robot Interaction and Technological Empowerment. In *Industrial Internet of Things for Smart Manufacturing* (pp. 28–55). Taylor & Francis., DOI: 10.1201/9781032642789-2

Ling, S., Wang, B., Rong, Y., & Huang, G. Q. (2025). Human–Machine Synchronization for Smart Manufacturing Towards Industry 5.0. In *Human-Centric Smart Manufacturing Towards Industry 5.0* (pp. 341–357). Springer., DOI: 10.1007/978-3-031-82170-7_15

Lins, R. G., & Givigi, S. N. (2021). Cooperative Robotics and Machine Learning for Smart Manufacturing: Platform Design and Trends Within the Context of Industrial Internet of Things. *IEEE Access : Practical Innovations, Open Solutions*, *9*, 95444–95455. DOI: 10.1109/ACCESS.2021.3094374

Maheswaran, K., Selvakumar, P., Shaikh, M., Shashank, G., Gupta, S., & Manjunath, T. C. (2025). The Evolution of Smart Urban Mobility With Emerging Technologies. In Ponnusamy, S., Assaf, M., Youssef, B., Jeon, G., & Kalyanaraman, S. (Eds.), *Leveraging VANETs and Blockchain Technology for Urban Mobility* (pp. 1–28). IGI Global Scientific Publishing., DOI: 10.4018/979-8-3373-0265-2.ch001

Mon, B. F., Hayajneh, M., Ali, N. A., Ullah, F., Ullah, H., & Alkobaisi, S. (2025). Digital Twins in the IIoT: Current Practices and Future Directions Toward Industry 5.0. *Computers, Materials & Continua*, *83*(3), 3675–3712. DOI: 10.32604/cmc.2025.061411

Moor, M., Vaher, K., Riives, J., Kangru, T., & Otto, T. (2021). "Modern Robot Integrated Manufacturing Cell According to the Needs of Industry 4.0." IOP Conference Series: Materials Science and Engineering, Vol. 1140, Article 012012. DOI: DOI: 10.1088/1757-899X/1140/1/012034

Nisha Angeline, M., Selvakumar, P., Gupta, M. K., Nazar, N., Elumalai, D., & Bhattacharya, S. (2025). Emerging Materials for Energy Storage and Environmental Applications. In Muduli, K., Moharana, B., Ales, S., & Biswal, D. (Eds.), *Using Computational Intelligence for Sustainable Manufacturing of Advanced Materials* (pp. 53–72). IGI Global Scientific Publishing., DOI: 10.4018/979-8-3693-7974-5.ch003

Rojas, E., Carrascal, D., Lopez-Pajares, D., Alvarez-Horcajo, J., Carral, J. A., Arco, J. M., & Martinez-Yelmo, I. (2024). A Survey on AI-Empowered Softwarized Industrial IoT Networks. *Electronics (Basel)*, *13*(10), 1979. Advance online publication. DOI: 10.3390/electronics13101979

Sahoo, S., & Lo, C. Y. (2022). Smart Manufacturing Powered by Recent Technological Advancements: A Review. *Journal of Manufacturing Systems*, *64*, 236–250. DOI: 10.1016/j.jmsy.2022.06.008

Sakthivel, M., Jaganathan, A., Mohanraj, M., & Selvakumar, P. Mohit & Manjunath T, (2025). "Brand Loyalty through AI and Personalisation AI-Generated Content for Financial Marketing." In A J Nair, S Manohar, Y B Limbu, & B A Huhmann (Eds.), Intersecting Natural Language Processing and FinTech Innovations in Service Marketing (pp. 77–96). IGI Global Scientific Publishing. DOI: 10.4018/979-8-3693-9944-6.ch006

Thirunagalingam, A., & Selvakumar, P. (2025). Application of Automation and Robotics in the Oil and Gas Industry. In Abdullayev, V., & Khang, A. (Eds.), *Revolutionizing AI and Robotics in the Oil and Gas Industry* (pp. 229–244). IGI Global Scientific Publishing., DOI: 10.4018/979-8-3693-8156-4.ch011

Upadhyaya, A. N., Debnath, P., Rani, P., Selvakumar, P., Manjunath, T. C., & Bhattacharya, S. (2025). AI-Based Disaster Prediction and Early Warning Systems. In Mobo, F. (Ed.), *AI and Emerging Technologies for Emergency Response and Smart Cities* (pp. 55–74). IGI Global Scientific Publishing., DOI: 10.4018/979-8-3693-9770-1.ch003

Wu, Y., Dai, H. N., Wang, H., Xiong, Z., & Guo, S. (2022). A Survey of Intelligent Network Slicing Management for Industrial IoT: Integrated Approaches for Smart Transportation, Smart Energy, and Smart Factory. *IEEE Communications Surveys and Tutorials*, *24*(2), 1175–1211. DOI: 10.1109/COMST.2022.3158270

Chapter 10
AI-Driven IoT Surveillance Framework for Signal-Resilient Environmental Monitoring in Extreme Condition

Dhruv Dhayal
https://orcid.org/0009-0000-8346-7834
Institute of Information Technology and Management, GGSIP University, New Delhi, India

Pratham Aggarwal
https://orcid.org/0009-0007-8093-4047
Institute of Information Technology and Management, GGSIP University, New Delhi, India

Uvesh Husain
Mazoon College, Muscat, Oman

Manzoor Ansari
https://orcid.org/0000-0002-3429-5615
Institute of Information Technology and Management, GGSIP University, New Delhi, India

Masood Alam
Center for Preparatory Studies, Sultan Qaboos University, Muscat, Oman

DOI: 10.4018/979-8-3373-5727-0.ch010

ABSTRACT

This article introduces a signal-resilient surveillance platform for monitoring extreme environments. Its tele-operated robotic vehicle operates in perilous locations inaccessible to humans, featuring a critical radar-enabled system for real-time obstacle detection and directional mapping. This data is displayed on a simple red/green interface, enabling rapid operator interpretation under stress. A closed IP connection ensures secure, reliable data transfer, maintaining constant communication despite structural or electromagnetic interference. Key use cases include rescue missions in confined spaces like borewells and hazard detection in underground mines. Experiments confirmed the system maintains signal integrity and real-time data feedback in harsh conditions. By prioritizing signal resilience and operational simplicity, this platform overcomes critical limitations of existing technologies, paving the way for future autonomous systems in disaster management and environmental hazard detection.

1 INTRODUCTION

Surveillance systems in extreme environments such as borewells, coal mines, and disaster zones face numerous challenges, foremost in communication signal weakness, structural barriers, and electromagnetic interference. Such conditions pose a hindrance to real-time data transmission, accurate navigation, and continuous operator control, which are serious detractors for mission success.(Jeon et al., 2024).Often conventional monitoring platforms lack the capability to withstand such conditions reliably (Li et al., 2024). To bridge this gap, we suggest a conceptual design for a signal-resilient surveillance platform specifically intended for extreme environmental monitoring. This system consists of a teleoperated robotic vehicle with multimodal sensors and a radar for recognizing obstacles and giving angular and directional feedback (Pirmagomedov et al., 2018; Fascista, 2022). The simple red-green parameter guided interface ensures better comprehension by operators, while the protected private IP communication channel confirms a strong and uninterrupted data flow (Sood, 2015). The design considerations focus on real-time object detection, signal integrity, and the ease of deployment in confined or dangerous sites. Areas of application would include confined space rescue missions and underground hazard detection, where human entry would be limited or unsafe. This framework is representative of the next generation of remote surveillance technologies through operational simplicity coupled with technical resilience, fostering improved situational awareness in mission-critical environments.

To address the growing need for autonomous monitoring in inaccessible or hazardous environments, this study proposes a secure, signal-resilient, and intelligent surveillance framework based on a layered architectural design. The framework integrates advanced IoT sensor fusion, edge computing, and encrypted communication protocols to ensure uninterrupted data acquisition and decision-making in real-time. Unlike conventional systems that fail under electromagnetic interference, structural barriers, or confined settings, the proposed model incorporates adaptive hardware, dual-mode communication channels, and AI-driven analytics to overcome these limitations. This multi-layered solution is designed to support critical applications such as borewell rescues, underground hazard detection, military operations, and smart city infrastructure maintenance, thereby establishing a new benchmark for autonomous remote surveillance platforms in extreme operational conditions.

a. Background and Related Work

Many remote surveillance systems exist, but most lack signal resilience, obstacle detection, and adaptability in extreme environments. Table 1 presents a comprehensive overview of the smart IoT-Cloud Framework, connecting it with other relevant studies through various classifications, including Cloud Computing (CC), Internet of Things (IoT) In addition, the chapter assesses these techniques, privacy and security (PS), AI-based technique (AI), and decision-making model (DMM). Table 1 provides a comprehensive summary of the ascertainable status of these parameters as derived from prior academic investigations indicates inclusion, while an X denotes exclusion (Jeon et al., 2024; Li et al., 2024; Fascista, 2022).

Table 1. Comparative Analysis for IoT-Cloud Enabled

Parameters							
S.No	**Ref.**	**Key-Contributions**	**IoT**	**CC**	**AI**	**PS**	**DMM**
1.	Khan et al.	Implemented AI-enabled IoT-cloud platform for smart agriculture and anomaly detection	✓	✓	✓		✓
2.	Verma et al.	Developed a privacy-preserving smart city model integrating cloud services and IoT	✓	✓		✓	✓
3.	Kaginalkar et al.	Designed a climate modeling framework using IoT-cloud integration for real-time weather analysis	✓	✓	✓		✓
4.	Majhi et al.	Developed a neural network model with optimization for earthquake magnitude prediction			✓		✓

continued on following page

Table 1. Continued

Parameters							
S.No	**Ref.**	**Key-Contributions**	**IoT**	**CC**	**AI**	**PS**	**DMM**
5.	Pirmagomedov et al.	Applied animal behavior sensing via IoT for early seismic detection	✓		✓		
6.	Sood et al.	Introduced a social cloud-IoT model for analyzing and forecasting flood conditions	✓	✓		✓	
7.	Kaur et al.	Presented an energy-efficient IoT-cloud solution for real-time flood management	✓	✓	✓		✓
8.	Pérez et al.	Designed a fog-cloud hybrid framework to process mobility data for traffic forecasting	✓	✓	✓		

Many studies have established frameworks for improving the surveillance and monitoring systems in hazardous and compromised signal environments. Khan et al. have concentrated on the development of wireless sensor networks whose data acquisition is more robust against disasters (Jeon et al., 2024), whereas Verma et al. established robotic rescue systems for confined places like borewells(Fascista, 2022; Verma et al., 2015). Kaginalkar et al. focused on developing electromagnetic-resistant communication modules that contribute to the stability of signals underground operations as in coal mines (Pirmagomedov et al., 2018; Kaginalkar et al., 2021). Majhi et al. have further addressed challenges of structural interference by incorporating adaptive signal-routing strategies in complicated industrial sites (Li et al., 2024; Majhi et al., 2023). Pirmagomedov et al. have studied radar-based sensing systems designed for real-time object detection with precision in angels, a technique that is becoming very important in high-risk terrains with low visibility (Pirmagomedov et al., 2018). According to Sood et al., intuitive human- remote operability and decision-making process in systems such as emergency response systems (Panneer Selvam & Al-Humairi, 2024; Sood, 2015). Kaur et al. also examined the aspects of multi-sensor fusion regarding better obstacle perception in robotic vehicles (Panneer Selvam & Al-Humairi, 2024); Pérez et al., of strong communication frameworks that could carry on ensuring data integrity across distributed surveillance nodes in harsh environments (Sood, 2015).

These studies advance their case separately, but none is comprehensive enough to fuse signal resilience, adaptive sensing, and operator-friendly designs for the extreme environmental applications. This points toward creating a conceptual frame to plug these holes wherein the mentioned features can be scanned and integrated-a radar sensor, secure Private IP-based communication, and modular control layers and then consolidated into a cohesive, layered architecture. Such an approach would effectively address the real-world challenges such as borewell rescue operations, hazard detection in mines, environmental monitoring in disaster zones, being a scalable solution mission-ready, and pushing beyond current systems while resolving age-old field problems. Several studies have been conducted by researchers to address the challenges mentioned above that have proposed such innovative techniques to gain reliability, adaptability, and safety in surveillance platforms across extreme environments. For example, Khan et al. define a new form of wireless sensor network that is actually able to support the signal transmission in disaster marks without deterioration for robust transmission of data (Jeon et al., 2024). Contained rescue operations such as borewell rescues have been tackled by Virtually Operated Robotic Systems designed by Verma et al. in terms of demonstrating success in reduced human risk while application reachability (Verma et al., 2015). Kaginalkar et al. investigated antenna configurations to establish optimal signal propagation through dense geological layers to enable reliable communication conditions when underground, wherein coal mines would specifically find application (Kaginalkar et al., 2021). Majhi et al. implemented a noise-proof design and signal-filtering mechanisms to have robust circuitry that ensures clearer data transmission in a cluttered industrial environment against structural and electromagnetic interference (Li et al., 2024; Majhi et al., 2023). "Meanwhile," Pirmagomedov added, "all These paradigms do not negate the fact that advanced solutions are still quite unattainable in practice and sparse in literature (Pirmagomedov et al., 2018)."water level and soil condition monitoring in agricultural applications (Khan et al., 2022; Majhi et al., 2023). The system can be commissioned into the military and defense for surveillance and navigation missions in inhospitable lands where conventional means fall short . Thus, metro maintenance should contribute to urban infrastructure and smart cities by improving underground cable inspection, sewer monitoring, and other similar efforts to achieve high efficiency as well as maintaining human safety in the cities (Kaginalkar et al., 2021; Pérez-Chacón et al., 2010; Fascista, 2022). This is an IoT-based monitoring and navigation system that has extensive applications in various important areas. It ensures worker safety in mining and tunneling, protecting from toxic gases and structural instabilities with live communication possibilities in severely signal-degraded locations machine interfaces (HMIs) should be used to aid the user's.

Figure 1. Operational Workflow

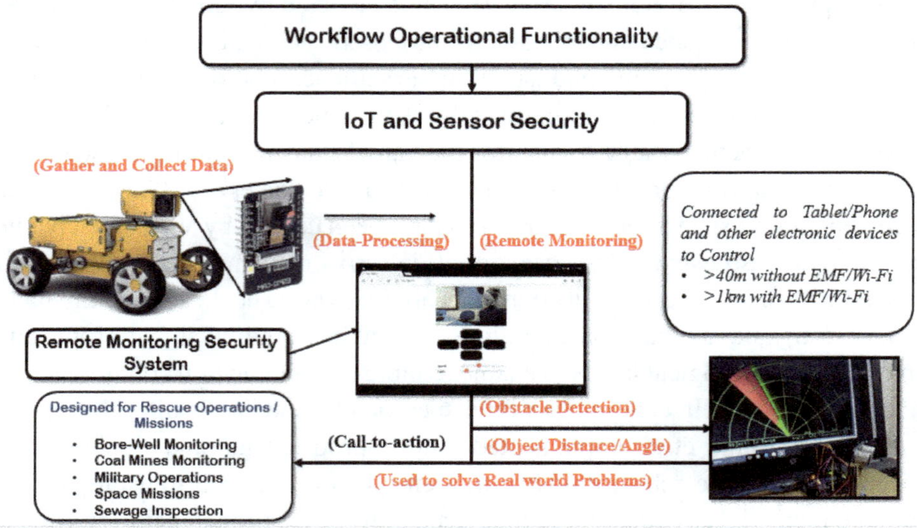

2 PROBLEM DEFINITION

Despite the advancements in surveillance and monitoring technologies, existing systems continue to face substantial challenges when deployed in extreme, hazardous, or signal-compromised environments such as borewells, coal mines, disaster-affected zones, and subterranean infrastructures (Khan et al., 2022; Verma et al., 2015). One of the most critical issues is the degradation or complete loss of communication signals due to structural obstacles, electromagnetic interference, and limited network availability, which severely impacts the reliability of real-time data transmission and remote control operations (Kaur et al., 2021). Moreover, conventional systems often lack autonomous functionality and depend heavily on continuous human supervision, making them unsuitable for scenarios where human access is either restricted or dangerous. The inability to adapt dynamically to environmental variations, coupled with a rigid system architecture, further limits their effectiveness across diverse use cases like rescue missions, underground hazard detection, military surveillance, and smart city infrastructure monitoring. In addition to these technical limitations, most existing platforms do not incorporate adequate data privacy and security mechanisms, leaving sensitive information vulnerable to interception, tampering, or unauthorized access. Furthermore, the absence of intuitive human-machine interfaces leads to delayed decision-making during critical moments, increasing the risk to both human lives and infrastructure. These multifaceted shortcomings highlight the urgent need for a next-generation surveillance system that integrates signal-resilient communication protocols, modular and scalable design, autonomous processing capabilities, and end-to-end data security. Such a system must be capable of operating reliably in harsh conditions while providing real-time, intelligent insights and ensuring minimal dependence on human operators, thereby addressing both operational and ethical requirements in modern surveillance and rescue frameworks.

3 OPERATIONAL MODEL AND WORKFLOW APPROACH

The IoT-and sensor Security based on Remote Monitoring System is data workflow-based integration of hardware and software modules to address real-time applications. The entire process is initiated by a mobile robotic platform acquiring multiple sensors (cameras, infrared, ultrasonic, or gas) and actuators to record measurement environmental data in real time. The data is directed to an onboard microcontroller or processing unit (Arduino, Raspberry Pi, etc.) where it undergoes primary processing (Khan et al., 2022; Kaginalkar et al., 2021; Majhi et al., 2023). The sensors measure temperature, obstacle presence, gas levels, structural anomalies, and the raw data is filtered, analyzed, and made meaningful which is subsequently filtered, analyzed, and transformed into actionable insights (Khan et al., 2022; Pirmagomedov et al., 2018). usually a tablet or smartphone, via wireless communication modules to allow an operator to observe the environment and control the robot and also make decisions based on real-time visual and sensor feedback (Kaur et al., 2021; Pérez-Chacón et al., 2010). Short-range communication (up to 40 m away) and long-distance communication (up to 1 km away) depending on whether there is EMF or Wi-Fi available, are used in this system. The GUI shows real-time video, distance readings, and object detection status for the user to send a call-to-action command, including movement control, object avoidance, or emergency alerts (Jeon et al., 2024). It helps to reduce the risk of human death or solves problems by monitoring confined or narrow spaces where non-human intervention is not possible.

With the incorporation of all these components, it guarantees a closed-loop system in which data acquisition, processing, transmission, and decision-making are highly integrated (Panneer Selvam & Al-Humairi, 2024). This methodology addresses real-world issues in high-risk situations where human presence is not needed. For example, when a well is to be saved, the robot will go down a narrow shaft and feed images preventing accidents; similar actions can be done in coal mines to detect earlier gas leakages and collapse; incremental surveillance and mapping are carried out in inaccessible locations for military and space operations (Pirmagomedov et al.,

2018); and in the case of sewage inspection, it can detect blockages or poisonous environments. Thus, it will enhance safety, efficiency, and responsiveness for rescue missions and industrial operations, thus reducing the chance for human casualties. Having understood the operational workflow and integration of hardware and software, we now discuss the Conceptual Framework. It provides a high-level view of the system's core components—sensors, processors, communication modules, and user interfaces—along with their data flow and interaction. This framework helps explain system behavior, scalability, and customization for different applications (Khan et al., 2022; Kaginalkar et al., 2021). It highlights how physical devices connect with digital processing, clarifying the flow from sensing to action. Figure 2 illustrates the complete structure, showing how data and control move through the system for effective decision-making in real-time scenarios. Which is shown in the following complete conceptual diagram given below.

Figure 2. Operational Workflow

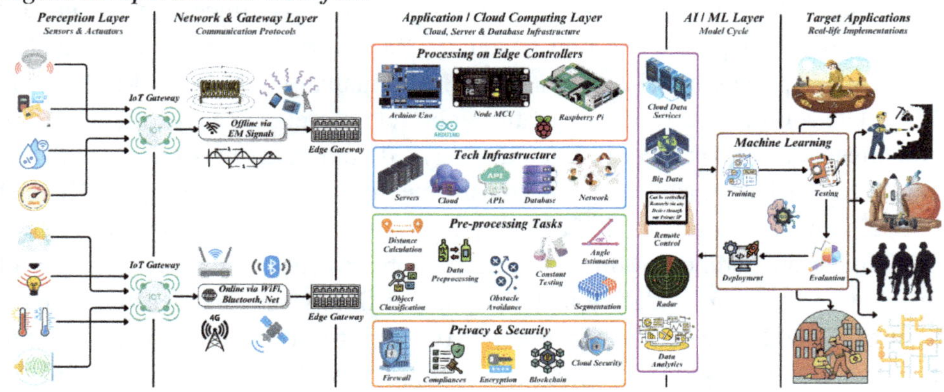

The operational model of the proposed surveillance platform adopts a structured and modular workflow aimed at ensuring real-time, secure, and resilient monitoring in extreme environments. The system initiates operation with autonomous bootstrapping of all onboard sensors, including radar, gas, ultrasonic, temperature, and obstacle detection modules. These sensors collaborate through sensor fusion techniques, which combine inputs from multiple modalities to improve environmental awareness, object recognition, and obstacle mapping accuracy, especially in confined or signal-compromised spaces. Once data is collected, it is processed at the edge using microcontrollers like Arduino, Raspberry Pi, or NodeMCU. This edge-level processing helps in noise reduction, normalization, obstacle classification, angular estimation, and the early detection of hazards such as gas leaks or structural instabilities. By performing critical computations locally, the system reduces latency and bandwidth consumption, which is essential in mission-critical scenarios where internet access may be unstable or unavailable.

The platform features a dual-mode encrypted communication system, ensuring uninterrupted data transmission. In scenarios where network availability is limited, the system uses low-frequency electromagnetic signals for offline transmission, while in connected environments, it employs WiFi, 4G, or satellite communication for real-time streaming. All transmitted data is secured using encryption protocols such as TLS or RSA to prevent interception and ensure data integrity. After transmission, the data reaches cloud servers where it undergoes further analysis using AI algorithms for pattern recognition, risk prediction, and event detection. Simultaneously, a graphical user interface (GUI), usually accessed via tablets or smartphones, presents live feeds, telemetry data, and control options. The GUI is optimized for high-stress operations, using intuitive controls, visual markers (such as red-green status indicators), and preloaded commands that simplify decision-making and reduce operator fatigue. A feedback loop is established wherein either the operator or the AI logic triggers appropriate actions. These commands are relayed back to the robotic unit, allowing real-time movement, obstacle avoidance, camera panning, signaling, or task-specific actions like object retrieval. This bi-directional loop ensures both autonomy and operator supervision, depending on the situational demands.

Furthermore, the system includes self-diagnostic and fail-safe features that constantly monitor hardware status, communication health, battery levels, and sensor calibration. In the event of a detected failure, the system activates a predefined fail-safe routine that either alerts the operator, triggers an emergency shutdown, or initiates return-to-base navigation. The platform is designed to be modular and customizable, allowing for adaptation to various domains including underground maintenance, disaster response, and military reconnaissance. With rugged casing and energy-efficient operation, it supports prolonged missions in hazardous, remote, or structurally complex locations, significantly enhancing both the scope and safety of environmental monitoring.

4 SECURE DATA TRANSACTIONS CONCEPTUAL FRAMEWORK FOR SYSTEM DESIGN LOGIC

The conceptual framework is structured around six functional layers: Perception, Network & Gateway, Application & Cloud Computing, Privacy & Security, AI/ML Intelligence, and Target Application. Each layer plays a distinct role in establishing a closed-loop system that ensures data collection, processing, secure transmission, and intelligent response. This is an impressive architecture diagram for your remote racking and monitoring system. Each layer is analyzed in detail:

a. Perception Layer (Sensors & Actuators)

This layer acts as the system's interface with the physical environment. It comprises a variety of sensors (e.g., temperature, humidity, gas, obstacle detection, pressure, and light) and actuators to capture environmental parameters and facilitate robotic actions. This foundational layer connects the physical world to your digital system. Sensors continuously monitor physical conditions and send analog or digital signals to the onboard controller. Actuators respond to system commands to enable movement, adjustment, or mitigation actions (e.g., object avoidance, shaft descent). It includes:

- Various sensors (temperature, pressure, gas, moisture, light sensors) collecting environmental data
- Actuators enabling the system to physically interact with the environment
- These components serve as the "eyes and ears" of your surveillance system in confined spaces

b. Network & Gateway Layer (Communication Protocols)

This layer handles the secure and resilient transmission of data between the perception layer and the cloud or control interfaces. Sensor data is routed through IoT gateways that filter and pre-process it. Communication takes place over dual channels Offline- Electromagnetic signalling is used where internet connectivity is unavailable (e.g., underground). Online-WiFi, Bluetooth, 4G/5G, and satellite links enable long-distance transmission. Edge gateways ensure consistent connectivity between local devices and the cloud infrastructure, even during disruptions.

This layer handles data transmission through:

- IoT Gateways that process and filter sensor data before forwarding
- Dual communication paths:
  - o Offline via electromagnetic signals (crucial for underground operations)
  - o Online via WiFi, Bluetooth, 4G, and satellite connectivity
- Edge Gateways that manage the connection between field devices and cloud infrastructure

c. Application/Cloud Computing Layer

This layer provides a scalable environment for real-time data processing and decision-making. Data is then forwarded to cloud servers for deeper analytics, storage, and system integration via APIs. Processed outputs are displayed on operator interfaces, allowing live monitoring and control. This layer handles data processing through three key components:

1. **Edge Controllers**:
   - o Arduino Uno boards for basic control functions
   - o Node MCUs (ESP8266/ESP32) for WiFi-enabled processing
   - o Raspberry Pi devices for more complex computing tasks
   - o These perform initial data processing to reduce latency and bandwidth requirements
2. **Tech Infrastructure**:
   - o Servers for data storage and processing
   - o Cloud platforms for scalable computing
   - o APIs for system integration
   - o Databases for structured data storage
   - o Network management for reliable communication
3. **Pre-processing Tasks**:
   - o Distance calculation for spatial awareness
   - o Object classification to identify items of interest
   - o Data preprocessing to clean and format incoming data
   - o Obstacle avoidance for autonomous movement
   - o Constant testing to ensure reliable operation
   - o Angle estimation for orientation
   - o Segmentation for breaking down complex data
4. **Privacy & Security**:
   - o Firewalls to prevent unauthorized access
   - o Compliance measures to meet regulatory requirements
   - o Encryption to protect sensitive data
   - o Blockchain technology for tamper-proof logs

d. AI/ML Layer (Model Cycle)

Security is embedded within the architecture to protect the system against unauthorized access, data breaches, and manipulation. This intelligence layer includes:

- Cloud data services for distributed processing
- Big data analytics for pattern recognition
- Machine learning workflow (training, testing, deployment, evaluation)
- Remote control capabilities for operator intervention[
- Radar systems for ehanced spatial awareness
- Data analytics for extracting actionable insights

e. Target Applications (Real-life Implementations)

This layer adds computational intelligence, allowing the system to make autonomous or assisted decisions based on environmental inputs. Your system is designed for high-risk environments:

- Bore-well rescue operations
- Military surveillance missions
- Space exploration
- Coal mine monitoring
- Confined space inspection (sewage systems, pipelines)
- Any environment too dangerous or inaccessible for human entry
- The harsh conditions of underground or confined spaces require the integration of energy-efficient solutions to ensure prolonged operation.

This comprehensive architecture creates a system that can autonomously monitor hazardous environments while providing real-time data to operators, potentially saving lives by replacing human presence in dangerous situations. The multi-layered approach ensures redundancy, reliability, and flexibility across various deployment scenarios (Khan et al., 2022; Kaginalkar et al., 2021).

1. **Inadequate Human-Machine Interface (HMI)**
    - Complex interfaces greatly delay response times during critical operations. There is a need for more intuitive systems that can reduce decision-making time during emergencies
    - These are not able to present any intuitive alerts that would allow operators to make fast decisions with little training under normal circumstances, and none during emergencies
2. **Lack of System Adaptability**
    - Different scenarios for surveillance will range massively, from disaster strikes to environmental monitoring to industrialists. The system, because of a rigid design and non-modular structure, does not support task-specific adaptations
    - The system, because of a rigid design and non-modular structure, does not support task-specific adaptations

Change is urgently needed in these difficult domains by way of interdisciplinary methods, integrating strong system design, advanced sensing technologies, secured communications (Pérez-Chacón et al., 2010), and human-centric control systems, as the primary elements underpinning our conceptual framework (Li et al., 2024;

Fascista, 2022). The evaluation of this framework considers human factors and organizational requirements, which bear with them broader implications on its applicability and the interdisciplinary nature (Kaginalkar et al., 2021; Sood, 2015).

5 SYSTEM SCALABILITY AND MODULARITY

Scalability and modularity are essential characteristics of any robust surveillance and monitoring platform intended for deployment across dynamic and unpredictable environments.(Verma et al., 2015) The proposed framework is designed with a highly modular architecture, allowing seamless integration or replacement of components such as sensors, processors, communication modules, and power units based on specific use-case requirements. This design flexibility ensures that the core functionality of the system remains intact while enabling customized configurations for various terrains, environmental challenges, and operational domains. For example, in high-temperature zones, heat-resistant sensors and cooling units can be easily added, while in aquatic conditions, waterproof casings and sonar modules can be integrated without overhauling the entire system. Additionally, the platform supports vertical and horizontal scalability. (Sood, 2015)

Vertical scalability is achieved by upgrading internal computing power—from microcontrollers like Arduino to more powerful units like Raspberry Pi or Jetson Nano—thus enabling the processing of larger datasets and more complex AI algorithms on the edge. Horizontal scalability allows multiple units to operate in a coordinated manner, sharing data over a distributed network for large-area coverage, such as during flood monitoring or disaster relief operations. Each module communicates via standardized interfaces and protocols, making it interoperable across different vendors and hardware systems. (Kaur et al., 2021; Pérez-Chacón et al., 2010)This approach reduces maintenance costs, shortens deployment time, and future-proofs the architecture by allowing technological upgrades without full system replacement. The scalability and modularity of the system also encourage cross-agency collaboration, where different departments (e.g., military, disaster management, environmental protection) can deploy tailored versions of the platform to meet their specific operational needs.(Jeon et al., 2024) Ultimately, this flexibility enhances system resilience, improves adaptability, and ensures that the solution remains effective and relevant in diverse real-world scenarios.A key strength of the proposed surveillance platform lies in its emphasis on system scalability and modularity, enabling it to adapt across a wide range of operational environments and use-case scenarios without requiring a complete redesign. The modular nature of the system allows individual hardware or software components—such as sensors, communication modules, computing units, and power sources—to be independently upgraded, replaced, or reconfigured based on specific mission requirements. This significantly reduces dependency on proprietary hardware, increases compatibility with third-party modules, and simplifies maintenance and customization. For instance, in a borewell rescue application, gas and thermal sensors might be prioritized, whereas in mining or sewer inspections, structural vibration sensors or sonar devices may be more appropriate. These specialized modules can be hot-swapped or embedded into the base system through standardized interfaces.(Kaur et al., 2021; Pérez-Chacón et al., 2010)

The system is also vertically scalable, meaning it can support upgrades in computational capabilities to match increased data complexity or processing demands. A basic implementation might use an Arduino or NodeMCU for lightweight sensing tasks, but the same architecture can be expanded to support Raspberry Pi, NVIDIA Jetson Nano, or other edge AI boards for running machine learning models, real-time analytics, and video processing. This vertical scaling enables the system to grow in intelligence without modifying its core framework, making it highly future-proof. In terms of horizontal scalability, the platform supports the deployment of multiple units across large geographic regions. Each unit operates semi-independently but communicates via mesh networking, cloud integration, or local gateways to share telemetry, threat detection status, and geospatial data. This allows coordinated, distributed surveillance, useful in scenarios such as flood monitoring, post-disaster search-and-rescue, or wide-area border surveillance.(Sood, 2015)

Furthermore, modularity ensures that software updates, security patches, and AI model retraining can be performed remotely without affecting other functional modules. This is essential for continuous improvement and rapid response to evolving threat landscapes or environmental conditions (Kaginalkar et al., 2021; Majhi et al., 2023). Integration with containerized deployment tools like Docker or edge orchestration platforms like Kubernetes further enhances modular deployment and resource management. From a deployment perspective, this modular and scalable structure reduces both cost and complexity, allowing agencies or organizations to start small and scale as needed. It promotes interdisciplinary collaboration—environmental agencies may focus on pollution sensors, while military use cases may require advanced radar and encryption tools—each adapting the same core platform to fit their strategic needs. Thus, scalability and modularity not only offer technical flexibility but also create a sustainable ecosystem for innovation, reusability, and broad multi-sectoral applicability.

6 ETHICAL, LEGAL AND SECURITY CONSIDERATIONS

The integration of AI, IoT, and remote surveillance technologies in high-risk and sensitive environments introduces a range of ethical, legal, and security challenges that must be systematically addressed to ensure responsible deployment. From an ethical standpoint, the primary concern lies in balancing operational efficiency with respect for human dignity and privacy. Surveillance systems, especially those operating autonomously, must avoid intrusive data collection practices and ensure that individuals' identities and activities are not monitored or stored without informed consent, particularly in civilian or urban settings. Furthermore, ethical deployment demands inclusivity and transparency—stakeholders, including operators and affected communities, must be made aware of the system's capabilities, limitations, and intended usage through clear documentation and training.

From a legal perspective, the system must comply with national and international data protection laws such as the General Data Protection Regulation (GDPR), the Personal Data Protection Bill (India), or sector-specific regulations like IT Act 2000 for cybercrime prevention. Legal compliance requires secure handling of personal and environmental data, proper licensing of software components, and adherence to jurisdiction-specific norms regarding surveillance and data retention. Any misuse or unregulated deployment of such technologies can lead to legal liabilities, including lawsuits, sanctions, or operational shutdowns.

On the security front, the system must be hardened against potential cyber threats including unauthorized access, data breaches, malware injections, and denial-of-service attacks. The use of end-to-end encryption, blockchain-based tamper-proof logs, multi-factor authentication, and firewalls are critical components to establish a secure communication and data management pipeline. Furthermore, the deployment of AI-driven anomaly detection mechanisms can help monitor system integrity and flag any malicious activities or irregular patterns in real-time. Operational systems in defense, emergency response, or industrial inspection must also include fallback mechanisms, remote shutdown capabilities, and regular audits to ensure continual risk assessment. Taken together, ethical, legal, and security considerations are not optional add-ons but foundational pillars for the development and deployment of intelligent surveillance frameworks. A failure to address any of these dimensions can result in not only technical failure but also social distrust, legal repercussions, and national security risks. Therefore, integrating these considerations from the early design phase ensures responsible innovation, greater public acceptance, and sustainable implementation across sectors.

7 CONCLUSION AND FUTURE DIRECTIONS

In the near future, The conceptual framework presents a layered architecture for an extreme environment monitoring platform, integrating radar-based detection, secure communications, and remote operability (Pirmagomedov et al., 2018; Kaur et al., 2021). This design achieves signal resilience through multi-spectrum redundancies and adaptive processing, maintains operational autonomy via modular design, provides environmental adaptability with specialized hardware protection, and ensures secure data management using encryption and distributed storage. Future development priorities include field testing a physical prototype, integrating advanced AI capabilities for autonomous operation during communication disruptions (Jeon et al., 2024; Fascista, 2022), enhancing edge computing architecture for improved efficiency (Sood, 2015), evolving user interfaces for better situational awareness (Kaginalkar et al., 2021), and developing interoperability with existing infrastructure (Pérez-Chacón et al., 2010). This framework aims to create an autonomous intelligent monitoring platform providing reliable data in challenging environments (Kaur et al., 2021; Li et al., 2024).

Building upon the current architecture, future advancements could focus on integrating quantum-secure communication protocols to further fortify data security in hostile environments. Expansion of the framework toward underwater or extraterrestrial applications can also be explored, utilizing adaptive pressure and temperature-resistant modules. There is potential for deploying bio-inspired sensors to detect chemical, biological, or radiological threats, and coupling them with swarm robotics for large-scale autonomous monitoring. Additionally, collaboration with national defense and disaster management agencies can steer the development toward policy-driven deployment models and cross-border interoperability frameworks. Integration with 5G/6G networks and real-time satellite communication will further ensure seamless data transmission in remote or mobile deployments.

REFERENCES

Fascista, A. (2022). Toward integrated large-scale environmental monitoring using WSN/UAV/Crowdsensing: A review of applications, signal processing, and future perspectives. Sensors, 22(5), 1824.

Jeon, H., Kim, H., Kim, D., & Kim, J. (2024). PASS-CCTV: Proactive Anomaly Surveillance System for CCTV Footage Analysis in Adverse Environmental Conditions. *Expert Systems with Applications, 185*, 124391. DOI: 10.1016/j.eswa.2024.124391

Kaginalkar, A., Kumar, S., Gargava, P., & Niyogi, D. (2021). Review of urban computing in air quality management as smart city service: An integrated IoT, AI, and cloud technology perspective. *Urban Climate, 39*, 100972. DOI: 10.1016/j.uclim.2021.100972

Kaur, M., Kaur, P. D., & Sood, S. K. (2021). Energy efficient IoT-based cloud framework for early flood prediction. *Natural Hazards, 109*(3), 2053–2076. DOI: 10.1007/s11069-021-04910-7

Khan, A. I., Alsolami, F., Alqurashi, F., Abushark, Y. B., Sarker, I. H., & Ali, A. M. (2022). Novel energy management scheme in IoT enabled smart irrigation system using optimized intelligence methods. *Sensors (Basel), 22*(18), 6970. PMID: 36146319

Li, Z., Yan, W., & Wang, L. (2024). Measuring Mobility Resilience with Network-Based Simulations of Flow Dynamics Under Extreme Events. *Transportation Research Part D, Transport and Environment, 104*, 104362. DOI: 10.1016/j.trd.2024.104362

Majhi, R., Jena, P. R., Majhi, B., & Kalli, R. (2023). Prediction of crop yield using climate variables in the south-western province of India: A functional artificial neural network modeling (FLANN) approach. *Environment, Development and Sustainability, 25*(10), 11033–11056. DOI: 10.1007/s10668-022-02517-x

Panneer Selvam, A., & Al-Humairi, S. N. S. (2024, December). Environmental Impact Evaluation Using Smart Real-Time Weather Monitoring Systems: A Systematic Review [Online]. *Innovative Infrastructure Solutions, 10*(13).

Pérez-Chacón, J., Luna-Romera, J. M., Troncoso, A., & Martínez-Álvarez, F. (2010). Big data analytics for discovering electricity consumption patterns in smart cities. *Expert Systems with Applications, 37*(12), 8333–8342.

Pirmagomedov, R., Bobrikova, E., Blinnikov, M., Amelyanovich, A., Glushakov, R., Loskutov, S., Koucheryavy, A., & Kirichek, R. (2018). IoT Based Earthquake Prediction Technology. In. Lecture Notes in Computer Science: Vol. 11118. *Internet of Things, Smart Spaces, and Next Generation Networks and Systems* (pp. 535–546). Springer. [Online], Available https://link.springer.com/chapter/10.1007/978-3-030-01168-0_48 DOI: 10.1007/978-3-030-01168-0_48

Sood, S. K. (2015). "Fog aided IoT-enabled patient health monitoring in smart homes," *Proceedings of the International Conference on Intelligent Computing and Internet of Things (ICIT)*, Harbin, China, pp. 41–44, doi: DOI: 10.1109/ICIT.2015.7293596

Verma, R., Govindaraj, J., Chhabra, S., & Gupta, G. (2015). "DF 2.0: An automated, privacy preserving, and efficient digital forensic framework that leverages machine learning for evidence prediction and privacy evaluation," in *Proceedings of the 2015 Conference on Systematic Approaches to Digital Forensic Engineering (SADFE)*, pp. 1–10.

Compilation of References

Abada, S., Marlair, G., Lecocq, A., Petit, M., Sauvant-Moynot, V., & Huet, F. (2016). Safety focused modeling of lithium-ion batteries: A review. *Journal of Power Sources, 306*, 178–192. DOI: 10.1016/j.jpowsour.2015.11.100

Abbas, A. M., Youssef, K. Y., & Zekry, A. (2020). NB-IoT optimization for smart meters networks of smart cities: Case study. *Alexandria Engineering Journal, 59*(6), 4267–4281. DOI: 10.1016/j.aej.2020.07.030

Abd-Alzhra, A. S., & Al- Tamimi, M. S. H. (2022). Image Compression Using Deep Learning: Methods and Techniques. *Iraqi Journal of Science*, 1299–1312. DOI: 10.24996/ijs.2022.63.3.34

Abdelkafi, A., & Krichen, L. (2014). Energy management optimization of a hybrid power production unit based renewable energies. *International Journal of Electrical Power & Energy Systems, 62*, 1–9. DOI: 10.1016/j.ijepes.2014.04.012

Abdusalomov, A., Mirzakhalilov, S., Umirzakova, S., Buriboev, A. S., Meliboev, A., Muminov, B., & Jeon, H. S. (2025). Accessible AI Diagnostics and Lightweight Brain Tumor Detection on Medical Edge Devices. *Bioengineering (Basel, Switzerland), 12*(1), 62. DOI: 10.3390/bioengineering12010062 PMID: 39851336

Abhiram, M. S. D., Kuppili, J., & Manga, N. A. (2020). Smart farming system using IoT for efficient crop growth. 2020 IEEE International Students' *Conference on Electrical, Electronics and Computer Science (SCEECS)*, 1–4. DOI: 10.1109/SCEECS48394.2020.147

Agarwal, M., Rani, G., Kumar, A., & Kumar, P. (2024). Deep Learning for Enhanced Brain Tumor Detection and Classification. *Results in Engineering, 22*, 102117. DOI: 10.1016/j.rineng.2024.102117

Agrawal, S., & Das, M. L. (2011). Internet of Things—A paradigm shift of future Internet applications. *2011 Nirma University International Conference on Engineering*, 1–7. DOI: 10.1109/NUiConE.2011.6153246

Ahmmed, A., Paul, M., Murshed, M., & Taubman, D. (2021, September). Human-machine collaborative video coding through cuboidal partitioning. In *Proceedings of the IEEE International Conference on Image Processing (ICIP)* DOI: 10.1109/ICIP42928.2021.9506150

Akbari, M., Liang, J., & Han, J. (2019). DSSLIC: Deep semantic segmentation-based layered image compression. In *Proceedings of the IEEE International Conference on Acoustics, Speech and Signal Processing (ICASSP)* (pp. 2042–2046) DOI: 10.1109/ICASSP.2019.8683541

Akter, A., Nosheen, N., Ahmed, S., Hossain, M., Abu Yousuf, M., Mohammad, A. A. A., Hasan, K. F., & Moni, M. A. (2024). Robust Clinical Applicable CNN and U-Net Based Algorithm for MRI Classification and Segmentation for Brain Tumor. *Expert Systems with Applications*, *238*, 122347. DOI: 10.1016/j.eswa.2023.122347

Alaa, M., Zaidan, A. A., Zaidan, B. B., Talal, M., & Kiah, M. L. M. (2017). A review of smart home applications based on Internet of Things. *Journal of Network and Computer Applications*, *97*, 48–65. DOI: 10.1016/j.jnca.2017.08.017

Alam, M. R., Reaz, M. B. I., & Ali, M. A. M. (2012). A Review of Smart Homes—Past, Present, and Future. IEEE Transactions on Systems, Man, and Cybernetics. *IEEE Transactions on Systems, Man, and Cybernetics. Part C, Applications and Reviews*, *42*(6), 1190–1203. DOI: 10.1109/TSMCC.2012.2189204

Alanazi, M. F., Ali, M. U., Hussain, S. J., Zafar, A., Mohatram, M., Irfan, M., AlRuwaili, R., Alruwaili, M., Ali, N. H., & Albarrak, A. M. (2022). Brain Tumor/Mass Classification Framework Using Magnetic-Resonance-Imaging-Based Isolated and Developed Transfer Deep-Learning Model. *Sensors (Basel)*, *22*(1), 372. DOI: 10.3390/s22010372 PMID: 35009911

AlFaris, F., Juaidi, A., & Manzano-Agugliaro, F. (2017). Intelligent homes' technologies to optimize the energy performance for the net zero energy home. *Energy and Building*, *153*, 262–274. DOI: 10.1016/j.enbuild.2017.07.089

Al-Fuqaha, A., Guizani, M., Mohammadi, M., Aledhari, M., & Ayyash, M. (2015). Internet of Things: A survey on enabling technologies, protocols, and applications. *IEEE Communications Surveys and Tutorials*, *4*(4), 2347–2376. DOI: 10.1109/COMST.2015.2444095

Ali, S., & Agrawal, J. (2024). Automated Segmentation of Brain Tumour Images Using Deep Learning-Based Model VGG19 and ResNet 101. *Multimedia Tools and Applications*, *83*(11), 33351–33370. DOI: 10.1007/s11042-023-16828-2

Ali, T. E., Ali, F. I., Dakić, P., & Zoltan, A. D. (2025). Trends, prospects, challenges, and security in the healthcare internet of things. *Computing*, *107*(1), 28. DOI: 10.1007/s00607-024-01352-4

Alshammari, A. (2022). Construction of VGG16 Convolution Neural Network (VGG16_CNN) Classifier with NestNet-Based Segmentation Paradigm for Brain Metastasis Classification. *Sensors (Basel)*, *22*(20), 8076. DOI: 10.3390/s22208076 PMID: 36298426

Amaraweera, S. P., & Halgamuge, M. N. (2019). Internet of things in the healthcare sector: overview of security and privacy issues. Security, privacy and trust in the IoT environment, 153-179.

Amin, Javaria, Muhammad Sharif, Anandakumar Haldorai, Mussarat Yasmin, and Ramesh Sundar Nayak. 2022. "Brain Tumor Detection and Classification Using Machine Learning: A Comprehensive Survey." *Complex \& intelligent systems* 8(4): 3161–83.

Anda, M., & Temmen, J. (2014). Smart metering for residential efficiency: The use of community based social marketing for behavioural change and smart grid introduction. *Renewable Energy*, *67*, 119–127. DOI: 10.1016/j.renene.2013.11.020

Anirudh Reddy, R., & Laasya, G. (2017) "Design of a weed detection system for cotton field" Mat Journal of Image Processing and Artificial Intelligence, Volume 3 Issue 3 in 2017.

Anusha, S., Ezhilvandan, M., Gayathri, S., Sivaraman, V., Anish, T. P., & Subramanian, R. S. (2024). Emerging Trends in Edge AI for Industry 4.0 and 5.0: Technologies, Applications, and Challenges. In *AI Innovations in Service and Tourism Marketing* (pp. 211–228). IGI Global., DOI: 10.4018/979-8-3693-7909-7.ch010

Arias, M., Notarnicola, C., Campo-Bescós, M. Á., Arregui, L. M., & Álvarez-Mozos, J. (2023). Evaluation of soil moisture estimation techniques based on Sentinel-1 observations over wheat fields. *Agricultural Water Management*, *278*(Aug), 113372. DOI: 10.1016/j.agwat.2023.108422

Arputharaj, J. V., William, B. N. J., Haruna, A. A., & Prasad, D. D. (2024). Exploring the Synergy of IIoT, AI, and Data Analytics in Industry 6.0. In *Industry 6.0: Transforming Manufacturing Through Smart Technologies* (pp. 1–36). CRC Press., DOI: 10.1201/9781003517993-1

Arumugam, M., Thiyagarajan, A., Adhi, L., & Alagar, S. (2024). Crossover Smell Agent Optimized Multilayer Perceptron for Precise Brain Tumor Classification on MRI Images. *Expert Systems with Applications*, *238*, 121453. DOI: 10.1016/j.eswa.2023.121453

Asad, U., & Martin, Y. (2024). "The Role of Edge Computing in Modern Traffic Management and Speed Detection," Dec. 2024. DOI: DOI: 10.13140/RG.2.2.23474.00961

Ashton, K. (2009). That 'Internet of Things' thing. *RFID J*, *22*, 97–114.

Asif, S., Yi, W., Ain, Q. U., Hou, J., Yi, T., & Si, J. (2022). Improving Effectiveness of Different Deep Transfer Learning-Based Models for Detecting Brain Tumors from MR Images. *IEEE Access : Practical Innovations, Open Solutions*, *10*, 34716–34730. DOI: 10.1109/ACCESS.2022.3153306

Atzori, L., Iera, A., & Morabito, G. (2010). The Internet of Things: A survey. *Computer Networks*, *54*(15), 2778–2805. DOI: 10.1016/j.comnet.2010.05.010

Aurna, N. F., Abu Yousuf, M., & Abu Taher, K. (2022). A Classification of MRI Brain Tumor Based on Two Stage Feature Level Ensemble of Deep CNN Models. *Computers in Biology and Medicine*, *146*, 105539. DOI: 10.1016/j.compbiomed.2022.105539 PMID: 35483227

Axon, L., Fletcher, K., Scott, A. S., Stolz, M., Hannigan, R., El Kaafarani, A., Goldsmith, M., & Creese, S. (2022). Emerging Cybersecurity Capability Gaps in the Industrial Internet of Things: Overview and Research Agenda. *Digital Threats : Research and Practice*, *3*(4), 34. Advance online publication. DOI: 10.1145/3503920

Ayadi, W., Elhamzi, W., Charfi, I., & Atri, M. (2021). Deep CNN for Brain Tumor Classification. *Neural Processing Letters*, *53*(1), 671–700. DOI: 10.1007/s11063-020-10398-2

Bai, S., Kolter, J. Z., & Koltun, V. (2018). "An Empirical Evaluation of Generic Convolutional and Recurrent Networks for Sequence Modeling." *arXiv Preprint arXiv:1803.01271*.

Baker, S. B., Xiang, W., & Atkinson, I. (2017). Internet of Things for Smart Healthcare: Technologies, Challenges, and Opportunities. *IEEE Access : Practical Innovations, Open Solutions*, *5*, 26521–26544. DOI: 10.1109/ACCESS.2017.2775180

Ballé, J., Minnen, D., Singh, S., Hwang, S. J., & Johnston, N. (2018). Variational image compression with a scale hyperprior. In *Proceedings of the 6th International Conference on Learning Representations (ICLR)*

Baltrušaitis, T., Ahuja, C., & Morency, L.-P. (2018). Multimodal Machine Learning: A Survey and Taxonomy. *IEEE Transactions on Pattern Analysis and Machine Intelligence*, *41*(2), 423–443. DOI: 10.1109/TPAMI.2018.2798607

Barocas, S., M. Hardt, and A. Narayanan. (2016). *Fairness and Machine Learning*. fairmlbook.org.

Batool, A., & Byun, Y.-C. (2024). Brain Tumor Detection with Integrating Traditional and Computational Intelligence Approaches across Diverse Imaging Modalities-Challenges and Future Directions. *Computers in Biology and Medicine*, *175*, 108412. DOI: 10.1016/j.compbiomed.2024.108412 PMID: 38691914

Beam, A. L., & Kohane, I. S. (2018). Big Data and Machine Learning in Health Care. *Journal of the American Medical Association*, *319*(13), 1317–1318. DOI: 10.1001/jama.2017.18391

Bedi, G., Venayagamoorthy, G. K., & Singh, R. (2016). Internet of Things (IoT) sensors for smart home electric energy usage management. *2016 IEEE International Conference on Information and Automation for Sustainability (ICIAfS)*, 1–6. DOI: 10.1109/ICIAFS.2016.7946568

Bekara, C. (2014). Security issues and challenges for the IoT-based smart Network grid. *Procedia Computer Science*, *34*, 532–537. DOI: 10.1016/j.procs.2014.07.064

Ben Arab, M., Rekik, M., & Krichen, L. (2023). A priority-based seven-layer strategy for energy management cooperation in a smart city integrated green technology. *Applied Energy*, *335*, 120767. DOI: 10.1016/j.apenergy.2023.120767

Benjamens, S., Dhunnoo, P., & Mesko, B. (2020). The State of Artificial Intelligence-Based FDA-Approved Medical Devices and Algorithms: An Online Database. *NPJ Digital Medicine*, *3*(1), 118. DOI: 10.1038/s41746-020-00324-0

Benzi, F., Anglani, N., Bassi, E., & Frosini, L. (2011). Electricity smart meters interfacing the households. *IEEE Transactions on Industrial Electronics*, *58*(10), 4487–4494. DOI: 10.1109/TIE.2011.2107713

Bhuiyan, M. N., Rahman, M. M., Billah, M. M., & Saha, D. (2021). Internet of things (IoT): A review of its enabling technologies in healthcare applications, standards protocols, security, and market opportunities. *IEEE Internet of Things Journal*, *8*(13), 10474–10498. DOI: 10.1109/JIOT.2021.3062630

Bing, K., Fu, L., Zhuo, Y., & Yanlei, L. (2011). Design of an Internet of Things-based smart home system. 2011 2nd International Conference on Intelligent Control and Information Processing, 921–924. DOI: 10.1109/ICICIP.2011.6008384

Bonawitz, K., Ivanov, V., Ben Kreuter, A. M., McMahan, H. B., Patel, S., Ramage, D., Segal, A., & Seth, K. (2017). "Practical Secure Aggregation for Privacy-Preserving Machine Learning." In *Proceedings of the 2017 ACM SIGSAC Conference on Computer and Communications Security*, 1175–91. CCS '17. New York, NY, USA: Association for Computing Machinery. DOI: 10.1145/3133956.3133982

Borgia, E. (2014). The Internet of Things vision: Key features, applications and open issues. *Computer Communications*, *54*, 1–31. DOI: 10.1016/j.comcom.2014.09.008

Botta, A., Donato, W., Persico, V., & Pescape, A. (2016). Integration of Cloud computing and Internet of Things: A survey. *Future Generation Computer Systems*, *56*, 684–700. DOI: 10.1016/j.future.2015.09.021

Boulos, M. N. K., & Al-Shorbaji, N. M. (2014). On the Internet of Things, smart cities and the WHO healthy cities. *International Journal of Health Geographics*, *13*(1), 10. DOI: 10.1186/1476-072X-13-10 PMID: 24669838

Bruton, S. V., Medlin, M., Brown, M., & Sacco, D. F. (2020). Personal Motivations and Systemic Incentives: Scientists on Questionable Research Practices. *Science and Engineering Ethics*, *26*(3), 1531–1547. DOI: 10.1007/s11948-020-00182-9 PMID: 31981051

Buja, A. (2025). *Cybersecurity of Industrial Internet of Things (IIoT)* (1st ed.). CRC Press., DOI: 10.1201/9781003383253

Bu, L., Zhang, Y., Liu, H., Yuan, X., Guo, J., & Han, S. (2021). An IIoT-Driven and AI-Enabled Framework for Smart Manufacturing System Based on Three-Terminal Collaborative Platform. *Advanced Engineering Informatics*, *50*, 101370. Advance online publication. DOI: 10.1016/j.aei.2021.101370

Caruana, R., Lou, Y., & Gehrke, J.. (2015). "Intelligible Models for Healthcare: Predicting Pneumonia Risk and Hospital 30-Day Readmission." In *Proceedings of the 21th ACM SIGKDD International Conference on Knowledge Discovery and Data Mining*, 1721–30. DOI: 10.1145/2783258.2788613

Chan, M., Estève, D., Escriba, C., & Campo, E. (2008). A review of smart homes—Present state and future challenges. *Computer Methods and Programs in Biomedicine*, *91*(1), 55–81. DOI: 10.1016/j.cmpb.2008.02.001 PMID: 18367286

Chen, J., He, J., Chen, F., Lv, Z., Tang, J., Li, W., Liu, Z., Yang, H. H., & Han, G. (2024). "Towards General Industrial Intelligence: A Survey on IIoT-Enhanced Continual Large Models." arXiv preprint, arXiv:2409.01207. DOI: /arXiv.2409. 01207DOI: 10.48550

Cheng, Z., Sun, H., Takeuchi, M., & Katto, J. (2020). Learned image compression with discretized Gaussian mixture likelihoods and attention modules. In *Proceedings of the IEEE/CVF Conference on Computer Vision and Pattern Recognition (CVPR)* (pp. 7939–7948). DOI: 10.1109/CVPR42600.2020.00796

Chen, I. Y., Pierson, E., Rose, S., Joshi, S., Ferryman, K., & Ghassemi, M. (2021). Ethical Machine Learning in Healthcare. *Annual Review of Biomedical Data Science*, *4*(May), 123–144. DOI: 10.1146/annurev-biodatasci-092820-114757

Chen, Q., Raghu, M., Yedidia, B., Luo, Y., & Tan, W. (2022). "Transformer-Based Patient Triage from Conversational Text." In *Proceedings of the Conference on Empirical Methods in Natural Language Processing*.

Chen, T., Kornblith, S., Norouzi, M., & Hinton, G. (2020). "A Simple Framework for Contrastive Learning of Visual Representations." In *International Conference on Machine Learning*, 1597–607.

Chen, T., Liu, H., Ma, Z., Shen, Q., Cao, X., & Wang, Y. (2021). End-to-end learnt image compression via non-local attention optimization and improved context modeling. *IEEE Transactions on Image Processing*, *30*, 3179–3191. DOI: 10.1109/TIP.2021.3058615 PMID: 33606630

Chhotaray, P., Behera, B. C., Moharana, B. R., Muduli, K., & Sephyrin, F. T. R. (2023). Enhancement of Manufacturing Sector Performance with the Application of Industrial Internet of Things (IIoT). In *Smart Technologies for Improved Performance of Manufacturing Systems and Services* (pp. 1–19). CRC Press., DOI: 10.1201/9781003346623-1

Chi, Q., Yan, H., Zhang, C., Pang, Z., & Xu, L. D. (2014). A Reconfigurable Smart Sensor Interface for Industrial WSN in IoT Environment. *IEEE Transactions on Industrial Informatics*, *10*(2), 1417–1425. DOI: 10.1109/TII.2014.2306798

Cho, K., van Merriënboer, B., Gulcehre, C., Bahdanau, D., Bougares, F., Schwenk, H., & Bengio, Y. (2014). "Learning Phrase Representations Using RNN Encoder-Decoder for Statistical Machine Translation." In *Proceedings of EMNLP*, 1724–34. DOI: 10.3115/v1/D14-1179

Chopade, S. S., Gupta, H. P., & Dutta, T. (2023). Survey on sensors and smart devices for IoT enabled intelligent healthcare system. *Wireless Personal Communications*, *131*(3), 1957–1995. DOI: 10.1007/s11277-023-10528-8 PMID: 37360143

Ciabattoni, L., Cimini, G., Grisostomi, M., Ippoliti, G., & Longhi, S. (2014). An interoperable framework for home automation design, testing and control. *22nd Mediterranean Conference on Control and Automation*, 1049–1054. DOI: 10.1109/MED.2014.6961513

Coates, J. M., Santhanalakshmi, K., Dinesh, N., Gandhimathi, S., & Chandel, P. S. Mohit & Selvakumar P, (2025). "Assessment and Evaluation in Transnational Staff Development." In G Morris & J Ye (Eds.), Training and Development in Transnational Higher Education (pp. 387–406). IGI Global Scientific Publishing. DOI: 10.4018/979-8-3373-1897-4.ch019

Cominola, A., Giuliani, M., Piga, D., Castelletti, A., & Rizzoli, A. E. (2015). Benefits and challenges of using smart meters for advancing residential water demand modeling and management: A review. *Environmental Modelling & Software*, *72*, 198–214. DOI: 10.1016/j.envsoft.2015.07.012

Consortium, International Warfarin Pharmacogenetics. (2013). "Estimation of the Warfarin Dose with Clinical and Pharmacogenetic Data." *New England Journal of Medicine* 360 (8): 753–64. .DOI: 10.1056/NEJMoa0809329

Daliya, V. K., & Ramesh, T. K. (2017). A survey on enhancing the interoperability aspect of IoT based systems. *2017 International Conference on Smart Technologies for Smart Nation (SmartTechCon)*, 581–586. DOI: 10.1109/SmartTechCon.2017.8358438

Dang, V. A., Vu Khanh, Q., Nguyen, V. H., Nguyen, T., & Nguyen, D. C. (2023). Intelligent healthcare: Integration of emerging technologies and Internet of Things for humanity. *Sensors (Basel)*, *23*(9), 4200. DOI: 10.3390/s23094200 PMID: 37177402

Dave, B., Kubler, S., Framling, K., & Koskela, L. (2016). Opportunities for enhanced lean construction management using Internet of Things standards. *Automation in Construction*, *61*, 86–97. DOI: 10.1016/j.autcon.2015.10.009

Davis, S. L., & Dukes, M. D. (2010, December). Irrigation scheduling performance by evapotranspiration-based controllers. *Agricultural Water Management*, *98*(1), 19–28. DOI: 10.1016/j.agwat.2010.07.006

Dax, G., Nagarajan, S., Li, H., & Werner, M. (2023). Compression supports spatial deep learning. *IEEE Journal of Selected Topics in Applied Earth Observations and Remote Sensing*, *16*(5), 702–713. DOI: 10.1109/JSTARS.2022.3226563

de Dombal, F. T., Leaper, D. J., Staniland, J. R., McCann, A. P., & Horrocks, J. C. (1972). Computer-Aided Diagnosis of Acute Abdominal Pain. *BMJ (Clinical Research Ed.)*, *2*(5804), 9–13. DOI: 10.1136/bmj.2.5804.9

Despande, P., Chandar, S. K., & Elangovan, N. (2024). "Enhancing Small and Medium OEMs' Adoption of IIoT Technologies." In *Proceedings of the 8th International Conference on Inventive Systems and Control (ICISC)*, IEEE, pp. 656–661. DOI: DOI: 10.1109/ICISC62624.2024.00115

Dewan, R., Swathi, R., & Sivakumar, V. (2025). A deep dive into detecting and investigating fileless malware. International Journal of Sensors, Wireless Communications and Control, 15(Issue unspecified), e22103279377106. DOI: 10.2174/01 22103279377106250507052410

Dey, A., Bose, S., Bisht, R., Yadav, M., & Mohapatra, H. (2025). Solutions for energy management and environmental conservation in IoT-integrated smart home. *Advances in Computational Intelligence and Robotics*, •••, 389–408. DOI: 10.4018/979-8-3693-9770-1.ch018

Dhanaraju, M., Chenniappan, P., Ramalingam, K., Pazhanivelan, S., & Kaliaperumal, R. (2022). Smart farming: Internet of Things (IoT)-based sustainable agriculture. *Agriculture*, *12*(10), 1745. DOI: 10.3390/agriculture12101745

Dhanvijay, M. M., & Patil, S. C. (2019). Internet of Things: A survey of enabling technologies in healthcare and its applications. *Computer Networks*, *153*, 113–131. DOI: 10.1016/j.comnet.2019.03.006

Diaz, M., Martin, C., & Rubio, B. (2016). State-of-the-art, challenges, and open issues in the integration of Internet of Things and Cloud computing. *Journal of Network and Computer Applications*, *67*, 99–117. DOI: 10.1016/j.jnca.2016.01.010

Ding, D., Cooper, R. A., Pasquina, P. F., & Fici-Pasquina, L. (2011). Sensor technology for smart homes. *Maturitas*, *69*(2), 131–136. DOI: 10.1016/j.maturitas.2011.03.016 PMID: 21531517

Diro, A. A., & Chilamkurti, N. (2018). Distributed attack detection scheme using deep learning approach for Internet of Things. *Future Generation Computer Systems*, *82*, 761–768. DOI: 10.1016/j.future.2017.08.043

Dosovitskiy, A., Beyer, L., Kolesnikov, A., Weissenborn, D., Zhai, X., Unterthiner, T., & Dehghani, M.. (2021). "An Image Is Worth 16x16 Words: Transformers for Image Recognition at Scale." In *International Conference on Learning Representations (ICLR)*. https://arxiv.org/abs/2010.11929

Dwork, C., & Roth, A. (2014). The Algorithmic Foundations of Differential Privacy. *Foundations and Trends in Theoretical Computer Science*, *9*(3–4), 211–407. DOI: 10.1561/0400000042

Eapen, B. R., Sartipi, K., & Archer, N. (2020). Serverless on FHIR: Deploying Machine-Learning Models for Healthcare on the Cloud. *Journal of Digital Health, 6*(1), 1–11. https://arxiv.org/abs/2006.04748

El-Wahab, A. (2023). Btc-Fcnn: Fast Convolution Neural Network for Multi-Class Brain Tumor Classification. *Health Information Science and Systems, 11*(1), 3. DOI: 10.1007/s13755-022-00203-w PMID: 36606077

Endres, H., Indulska, M., & Ghosh, A. (2024). Unlocking the Potential of Industrial Internet of Things (IIOT) in the Age of the Industrial Metaverse: Business Models and Challenges. *Industrial Marketing Management, 119*, 90–107. DOI: 10.1016/j.indmarman.2024.03.006

Epstein, R. S., Moyer, T. P., Aubert, R. E., Teagarden, J. R., Chen, Y., Symm, B., Zhou, Z., Larrabee, P., Razouki, Z., & Sack, J. S. (2010). Warfarin Genotyping Reduces Hospitalization Rates: Results from the Medco–Mayo Warfarin Effectiveness Study. *Journal of the American College of Cardiology, 55*(25), 2804–2812. DOI: 10.1016/j.jacc.2010.03.009

Erdem, Y., Arin, L., Erdem, T., Polat, S., Deveci, M., Okursoy, H., & Gültas, H. T. (2010, December). Crop water stress index for assessing irrigation scheduling of drip irrigated broccoli (Brassica oleracea L. var. italica). *Agricultural Water Management, 98*(1), 148–156. DOI: 10.1016/j.agwat.2010.08.013

Esteva, A., Kuprel, B., Novoa, R. A., Ko, J., Swetter, S. M., Blau, H. M., & Thrun, S. (2017). Dermatologist-Level Classification of Skin Cancer with Deep Neural Networks. *Nature, 542*(7639), 115–118. DOI: 10.1038/nature21056

Fang, X., Wu, X., Li, F., Duan, Y., & Tao, X. (2025). Group image compression for dual use of machine and human vision. *IEEE Transactions on Circuits and Systems for Video Technology, 35*(3), 2820–2831. DOI: 10.1109/TCSVT.2024.3486558

Faust, O., Shenfield, A., Kareem, M., Brechtl, J., & Ciaccio, E. J. (2020). Anomaly Detection for Short Single-Lead ECG Using Autoencoders. *Computers in Biology and Medicine, 124*, 103922. DOI: 10.1016/j.compbiomed.2020.103922

Feng, X., Bi, S., Li, H., Qi, Y., Chen, S., & Shao, L. (2022). Soil moisture forecasting for precision irrigation management using real-time electricity consumption records. *Agricultural Water Management, 254*(Jul), 107014. DOI: 10.1016/j.agwat.2023.108656

Fernandez, L. P., Gomez San Roman, T., Cossent, R., Domingo, C. M., & Frias, P. (2010). Assessment of the impact of plug-in electric vehicles on distribution networks. *IEEE Transactions on Power Systems*, *26*(1), 206–213. DOI: 10.1109/TPWRS.2010.2049133

Finlayson, S. G., Bowers, J. D., Ito, J., Zittrain, J. L., Beam, A. L., & Kohane, I. S. (2019). Adversarial Attacks on Medical Machine Learning. *Science*, *363*(6433), 1287–1289. DOI: 10.1126/science.aaw4399

Fischer, K., Fleckenstein, F., Herglotz, C., & Kaup, A. (2021, June). Saliency-driven versatile video coding for neural object detection. In *Proceedings of the IEEE International Conference on Acoustics, Speech and Signal Processing (ICASSP)*. DOI: 10.1109/ICASSP39728.2021.9415048

Food and Drug Administration. (2022, September 28). Clinical Decision Support Software: Guidance for Industry and Food and Drug Administration Staff. U.S. Department of Health and Human Services.

Franchini, M., Mengoli, C., Cruciani, M., Bonfanti, C., Pizzolo, G., Veneri, D., Lippi, G., & Favaloro, E. J. (2014). Effects on Bleeding Complications of Pharmacogenetic Testing for Initial Dosing of Vitamin k Antagonists: A Systematic Review and Meta-analysis. *Journal of Thrombosis and Haemostasis*, *12*(9), 1480–1490. DOI: 10.1111/jth.12647

Frid-Adar, M., Klang, E., Amitai, M., Goldberger, J., & Greenspan, H. (2018). "Synthetic Data Augmentation Using GAN for Improved Liver Lesion Classification." In *2018 IEEE 15th International Symposium on Biomedical Imaging (ISBI 2018)*, 289–93. DOI: 10.1109/ISBI.2018.8363576

Gadicha, A. B., & Gadicha, V. B. (2025). Unveiling the Potential of IoT and IIoT Industrial Technologies in Cybersecurity: Trends, Applications, and Future Prospective. In *Advancing Cybersecurity in Smart Factories Through Autonomous Robotic Defenses* (pp. 377–396). IGI Global., DOI: 10.4018/979-8-3373-0583-7.ch016

Gaikwad, P. P., Gabhane, J. P., & Golait, S. S. (2015). A survey based on Smart Homes system using Internet-of-Things. 2015 International Conference on Computation of Power, Energy, Information and Communication (ICCPEIC), 0330–0335. DOI: 10.1109/ICCPEIC.2015.7259486

Gamal, H. (2025). Technological Integration and Innovative Strategies Harnessing Artificial Intelligence for Operational Excellence. In *Building Business Knowledge for Complex Modern Business Environments* (pp. 237–270). IGI Global., DOI: 10.4018/979-8-3693-6447-5.ch009

Garcia, C. G., Meana-Llorian, D., Bustelo, B. C. P. G., Lovelle, J. M. C., & Garcia-Fernandez, N. (2016). Midgar: Detection of people through computer vision in the Internet of Things scenarios to improve the security in smart cities, smart towns, and smart homes. *Future Generation Computer Systems*.

Gaur, L., Bhandari, M., Razdan, T., Mallik, S., & Zhao, Z. (2022). Explanation-Driven Deep Learning Model for Prediction of Brain Tumour Status Using MRI Image Data. *Frontiers in Genetics*, *13*, 822666. DOI: 10.3389/fgene.2022.822666 PMID: 35360838

Gazis, A., & Katsiri, E. (2021). Smart Home IoT Sensors: Principles and Applications – A Review of Low-Cost and Low-Power Solutions. *International Journal on Engineering Technologies and Informatics*, *2*(1), 19–23. DOI: 10.51626/ijeti.2021.02.00007

Ghassemi, M., Naumann, T., Schulam, P., Beam, A. L., Chen, I. Y., & Ranganath, R. (2020). A Review of Challenges and Opportunities in Machine Learning for Health. [American Medical Informatics Association.]. *Proceedings of the AMIA Joint Summits on Translational Science*, *2020*, 191–200.

Gheorghe, C., & Soica, A. (2025, February). Revolutionizing Urban Mobility: A Systematic Review of AI, IoT, and Predictive Analytics in Adaptive Traffic Control Systems for Road Networks. *Electronics (Basel)*, *14*(4), 719. DOI: 10.3390/electronics14040719

Gomaa, A. H. (2025). Optimizing Asset Integrity for Critical Manufacturing Systems Using Advanced Proactive Maintenance Strategies. *International Journal of Emerging Science and Engineering*, *13*(4), 21–33. DOI: 10.35940/ijese.B2026.13040325

Gondaliya, T. (2016). A Survey on an Efficient IoT Based Smart Home. *International Journal of Review in Electronics and Communication Engineering*, *4*(1).

Gonzalez, R. C., & Woods, R. E. (2018). *Digital Image Processing*. Pearson.

Goodfellow, I., Pouget-Abadie, J., Mirza, M., Xu, B., Warde-Farley, D., Ozair, S., Courville, A., & Bengio, Y. (2014). "Generative Adversarial Networks." In *Advances in Neural Information Processing Systems (NIPS)*, 27:2672–80.

Greco, L., Percannella, G., Ritrovato, P., Tortorella, F., & Vento, M. (2020). Trends in IoT based solutions for health care: Moving AI to the edge. *Pattern Recognition Letters*, *135*, 346–353. DOI: 10.1016/j.patrec.2020.05.016 PMID: 32406416

Greenwood, D., & Stopczynski, A. (2019). Privacy and Human Behavior in the Age of Information. *Science*, *347*(6221), 509–514.

Gregor, E. G., Smith, A. L., Jones, R. D., Brown, J. M., & Nguyen, L. T. (2015, April). Real-time data integration for optimized irrigation scheduling based on weather forecasts. *Journal of Irrigation and Drainage Engineering*, *141*(4), 04015005. DOI: 10.1061/(ASCE)IR.1943-4774.0000891

Gubbi, J., Buyya, R., Marusic, S., & Palaniswami, M. (2013). Internet of Things (IoT): A vision, architectural elements, and future directions. *Future Generation Computer Systems*, *29*(7), 1645–1660. DOI: 10.1016/j.future.2013.01.010

Guinard, D. (2011). What the Internet of Things Will Mean for the Smart Grid. URL https://smartgrid.ieee.org/june-2011/95-what-the-internet-of-things-will-mean -for-the-smart-grid

Gunge, V., & Yalagi, P. (2016). Smart Home Automation: A Literature Review. International Journal of Computer Applications (0975 – 8887) RTDM 2016.

Gunning, D., & Aha, D. W. (2019). DARPA's Explainable Artificial Intelligence (XAI) Program. *AI Magazine*, *40*(2), 44–58. DOI: 10.1609/aimag.v40i2.2850

Hardt, M., Price, E., & Srebro, N. (2016). Advances in Neural Information Processing Systems: Vol. 3315–23. *Equality of Opportunity in Supervised Learning*.

Hatamizadeh, A., Nath, V., Tang, Y., Yang, D., Roth, H., & Xu, D. (2022). "Swin UNETR: Swin Transformers for Semantic Segmentation of Brain Tumors in MRI Images." *arXiv Preprint arXiv:2201.01266*. https://arxiv.org/abs/2201.01266 DOI: 10.1007/978-3-031-08999-2_22

He, D., Zheng, Y., Sun, B., Wang, Y., & Qin, H. (2021). Checkerboard context model for efficient learned image compression. In *Proceedings of the IEEE/CVF Conference on Computer Vision and Pattern Recognition (CVPR)* (pp. 14771–14780). DOI: 10.1109/CVPR46437.2021.01453

He, K., Fan, H., Wu, Y., Xie, S., & Girshick, R. (2020). "Momentum Contrast for Unsupervised Visual Representation Learning." In *Proceedings of the IEEE/CVF Conference on Computer Vision and Pattern Recognition*, 9729–38. DOI: 10.1109/CVPR42600.2020.00975

He, K., Gkioxari, G., Dollar, P., & Girshick, R. (2017). "Mask r-CNN." In *Proceedings of the IEEE International Conference on Computer Vision (ICCV)*, 2961–69. DOI: 10.1109/ICCV.2017.322

He, K., Zhang, X., Ren, S., & Sun, J. (2016). "Deep Residual Learning for Image Recognition." In *Proceedings of the IEEE Conference on Computer Vision and Pattern Recognition (CVPR)*, 770–78. DOI: 10.1109/CVPR.2016.90

Hochreiter, S., & Schmidhuber, J. (1997). Long Short-Term Memory. *Neural Computation, 9*(8), 1735–1780. DOI: 10.1162/neco.1997.9.8.1735

Hripcsak, G., & Albers, D. J. (2011). Correlating Electronic Health Record Concepts with Healthcare Process Events. *Journal of the American Medical Informatics Association : JAMIA, 18*(4), 509–513. DOI: 10.1136/amiajnl-2011-000153

Huang, G., Liu, Z., van der Maaten, L., & Weinberger, K. Q. (2017). "Densely Connected Convolutional Networks." In *Proceedings of the IEEE Conference on Computer Vision and Pattern Recognition (CVPR)*, 4700–4708. DOI: 10.1109/CVPR.2017.243

Hu, B., Feng, Y., Sun, J., Gao, Y., & Tan, J. (2019). Driving preference analysis and electricity pricing strategy comparison for electric vehicles in smart city. *Information Sciences, 504*, 202–220. DOI: 10.1016/j.ins.2019.07.039

Hubballi, R. B., Selvakumar, P., Seenivasan, R., Basava Aradhya, S. G., Dinesh, N., & Seelam, P. K. (2025). Overview of Current AI Technologies in Education. In Kyei-Blankson, L., & Ntuli, E. (Eds.), *Transformative AI Practices for Personalized Learning Strategies* (pp. 1–26). IGI Global Scientific Publishing., DOI: 10.4018/979-8-3693-8744-3.ch001

Idso, S. B., Jackson, R. D., Pinter, P. J.Jr, Reginato, R. J., & Hatfield, J. L. (1981). Normalizing the stressdegree-day parameter for environmental variability. *Agricultural Meteorology, 24*(Jan), 45–55. DOI: 10.1016/0002-1571(81)90032-7

Ioakimidis, C. S., Thomas, D., Rycerski, P., & Genikomsakis, K. N. (2018). Peak shaving and valley filling of power consumption profile in nonresidential buildings using an electric vehicle parking lot. *Energy, 148*, 148–158. DOI: 10.1016/j.energy.2018.01.128

Irfan, M., Iqbal, J., Iqbal, A., Iqbal, Z., Ali Riaz, R., & Mehmood, A. (2017). Opportunities and challenges in control of smart grids – Pakistani perspective. *Renewable & Sustainable Energy Reviews, 71*, 652–674. DOI: 10.1016/j.rser.2016.12.095

Irmak, E. (2021). Multi-Classification of Brain Tumor MRI Images Using Deep Convolutional Neural Network with Fully Optimized Framework. *Iranian Journal of Science and Technology. Transaction of Electrical Engineering, 45*(3), 1015–1036. DOI: 10.1007/s40998-021-00426-9

Irvin, J., Rajpurkar, P., Ko, M., Yu, Y., Ciurea-Ilcus, S., Chute, C., Marklund, H., Haghgoo, B., Ball, R., Shpanskaya, K., Seekins, J., Mong, D. A., Halabi, S. S., Sandberg, J. K., Jones, R., Larson, D. B., Langlotz, C. P., Patel, B. N., Lungren, M. P., & Ng, A. Y. (2019). CheXpert: A Large Chest Radiograph Dataset with Uncertainty Labels and Expert Comparison. *Proceedings of the AAAI Conference on Artificial Intelligence*, *33*(1), 590–597. DOI: 10.1609/aaai.v33i01.3301590

Islam, M., Reza, M. T., Kaosar, M., & Parvez, M. Z. (2023). Effectiveness of Federated Learning and CNN Ensemble Architectures for Identifying Brain Tumors Using MRI Images. *Neural Processing Letters*, *55*(4), 3779–3809. DOI: 10.1007/s11063-022-11014-1 PMID: 36062060

Jaguemont, J., Boulon, L., & Dubé, Y. (2016). A comprehensive review of lithium-ion batteries used in hybrid and electric vehicles at cold temperatures. *Applied Energy*, *164*, 99–114. DOI: 10.1016/j.apenergy.2015.11.034

Jamil, S., Rahman, M. U., & Fawad, . (2022). A Comprehensive Survey of Digital Twins and Federated Learning for Industrial Internet of Things (IIoT), Internet of Vehicles (IoV) and Internet of Drones (IoD). *Applied System Innovation*, *5*(3), 56. Advance online publication. DOI: 10.3390/asi5030056

Javed, A. R., Sarwar, M. U., Beg, M. O., Asim, M., Baker, T., & Tawfik, H. (2021). Cloud-based intelligent predictive maintenance system for industrial equipment using machine learning. *Journal of Ambient Intelligence and Humanized Computing*, *12*, 4413–4429. DOI: 10.1007/s12652-020-02669-4

Jayaraman, P. P., Yavari, A., Geogakopoulos, D., Morshed, A., & Zaslavsky, A. (2016). Internet of Things platform for smart farming: Experiences and lessons learnt. *Sensors (Basel)*, *16*(11), 1884. DOI: 10.3390/s16111884 PMID: 27834862

Jeon, H., Kim, H., Kim, D., & Kim, J. (2024). PASS-CCTV: Proactive Anomaly Surveillance System for CCTV Footage Analysis in Adverse Environmental Conditions. *Expert Systems with Applications*, *185*, 124391. DOI: 10.1016/j.eswa.2024.124391

Jha, V., & Tripathi, P. (2025). An Energy-Efficient Knowledge Discovery in Cognitive IoT. *OPSEARCH*, *62*, 1–18. DOI: 10.1007/s12597-025-00977-z

Jhawar, V., Rangnekar, Y., & Mohapatra, H. (2025). Survey of fog computing applications in vehicular communication for smart transportation systems. Driving Innovation at the Intersection of Renewable Energy and the Internet of Vehicles, 345-382. .DOI: 10.4018/979-8-3373-3321-2.ch010

Jing, L., & Tian, Y. (2020). Self-Supervised Visual Feature Learning with Deep Neural Networks: A Survey. *IEEE Transactions on Pattern Analysis and Machine Intelligence, 43*(11), 4037–4058. DOI: 10.1109/TPAMI.2020.2992393

Johnson, A. E. W., Bulgarelli, L., Pollard, T. J., Horng, S., Celi, L. A., & Mark, R. G. (2023). MIMIC-IV, a Freely Accessible Electronic Health Record Dataset. *Scientific Data, 10*(1), 23. DOI: 10.1038/s41597-023-02136-9

Johnston, N. (2018). Improved lossy image compression with priming and spatially adaptive bit rates for recurrent networks. In *Proceedings of the IEEE/CVF Conference on Computer Vision and Pattern Recognition (CVPR)* (pp. 4385–4393) DOI: 10.1109/CVPR.2018.00461

Johri, S., Jain, D., Gupta, A., & Sivakumar, V. (2024). LangBridge: Computational AI solutions for global learning. In *2024 International Conference on Emerging Research in Computational Science (ICERCS)* (pp. 1–8). IEEE. DOI: 10.1109/ICERCS63125.2024.10895947

Jung, S., Jin, S., & Kim, H. (2024). A novel side-channel archive framework using deep learning-based leakage compression. *IEEE Access : Practical Innovations, Open Solutions, 12*, 105326–105336. DOI: 10.1109/ACCESS.2024.3416199

Jury, W. A., & Vaux, H. J.Jr. (2007). The emerging global water crisis: Managing scarcity and conflict between water users. *Advances in Agronomy, 95*(Sep), 1–76. DOI: 10.1016/S0065-2113(07)95001-4

Kabalci, E., & Kabalci, Y. (2019). Internet of things for smart network_grid applications. In Kabalci, E., & Kabalci, Y. (Eds.), *From Smart Network_grid to Internet of Energy, Academic Press, 2019* (pp. 249–307).

Kabalci, Y. (2016). A survey on smart metering and smart grid communication. *Renewable & Sustainable Energy Reviews, 57*, 302–318. DOI: 10.1016/j.rser.2015.12.114

Kaginalkar, A., Kumar, S., Gargava, P., & Niyogi, D. (2021). Review of urban computing in air quality management as smart city service: An integrated IoT, AI, and cloud technology perspective. *Urban Climate, 39*, 100972. DOI: 10.1016/j.uclim.2021.100972

Karim, A., Raza, M., Alharthi, Y., Abbas, G., Othmen, S., Hossain, M., Nahar, A., & Mercorelli, P. (2024, August). Visual Detection of Traffic Incident through Automatic Monitoring of Vehicle Activities. *World Electric Vehicle Journal, 15*(9), 382. Advance online publication. DOI: 10.3390/wevj15090382

Kaur, M., Kaur, P. D., & Sood, S. K. (2021). Energy efficient IoT-based cloud framework for early flood prediction. *Natural Hazards*, *109*(3), 2053–2076. DOI: 10.1007/s11069-021-04910-7

Khajenasiri, I., Estebsari, A., Verhelst, M., & Gielen, G. (2017). A review on Internet of Things solutions for intelligent energy control in buildings for smart city applications. *Energy Procedia*, *111*, 770–779. DOI: 10.1016/j.egypro.2017.03.239

Khalaf, R., Mohammed, A., Essa, E., & Ali, H. (2019). Controlling Smart Home Activities Using IoT. *2019 International Conference on Computing and Information Science and Technology and Their Applications (ICCISTA)*, 1–6. DOI: 10.1109/ICCISTA.2019.8830664

Khan, A. I., Alsolami, F., Alqurashi, F., Abushark, Y. B., Sarker, I. H., & Ali, A. M. (2022). Novel energy management scheme in IoT enabled smart irrigation system using optimized intelligence methods. *Sensors (Basel)*, *22*(18), 6970. PMID: 36146319

Khan, M. A., Khan, A., Alhaisoni, M., Alqahtani, A., Alsubai, S., Alharbi, M., Malik, N. A., & Damaševičius, R. (2023). Multimodal Brain Tumor Detection and Classification Using Deep Saliency Map and Improved Dragonfly Optimization Algorithm. *International Journal of Imaging Systems and Technology*, *33*(2), 572–587. DOI: 10.1002/ima.22831

Khan, M., & Tripathi, A. (2024, December). Digital Patient Eye. *IEEE Transactions on Circuits and Systems for Video Technology*, *34*(12).

Khemakhem, S., Rekik, M., & Krichen, L. (2019). Double layer home energy supervision strategies based on demand response and plug-in electric vehicle control for flattening power load curves in a smart grid. *Energy*, *167*, 312–324. DOI: 10.1016/j.energy.2018.10.187

Khemakhem, S., Rekik, M., & Krichen. L. (2019). Impact of Electric Vehicles integration on residential demand response system to peak load minimizing in smart grid. In: *19th International Conference on Sciences and Techniques of Automatic Control and Computer Engineering, STA 2019*, Sousse, Tunisia, 2019, pp. 572-577. DOI: 10.1109/STA.2019.8717304

Khemakhem, S., Rekik, M., & Krichen, L. (2020). A collaborative energy management among Plug-In Electric Vehicle, smart homes and neighbors' interaction for residential power load profile smoothing. *Journal of Building Engineering*, *27*, 1–11. DOI: 10.1016/j.jobe.2019.100976

Khemakhem, S., Rekik, M., & Krichen, L. (2021). A Smart Home Power Supervision Strategy Based on Optimal Plug-in Electric Vehicles Integration into Smart City Context for Peak Loads Profiles Flattening. *Arabian Journal for Science and Engineering*.

Kingma, D. P., & Welling, M. 2013. "Auto-Encoding Variational Bayes." In *International Conference on Learning Representations (ICLR), arXiv Preprint*.

Kodeswaran, P., Kokku, R., Mallick, M., & Sen, S. (2016). Demultiplexing activities of daily living in IoT enabled smarthomes. *IEEE INFOCOM, 2016*, 1–9. DOI: 10.1109/INFOCOM.2016.7524359

Kolat, M., Kovari, B., Becsi, T., & Aradi, S. (2023). Multi-agent reinforcement learning for traffic signal control: A cooperative approach. *Sustainability (Basel), 15*(4), 3479. DOI: 10.3390/su15043479

Kollem, S., Reddy, K. R., Prasad, C. R., Chakraborty, A., Ajayan, J., Sreejith, S., Bhattacharya, S., Joseph, L. M. I. L., & Janapati, R. (2023). AlexNet-NDTL: Classification of MRI Brain Tumor Images Using Modified AlexNet with Deep Transfer Learning and Lipschitz-Based Data Augmentation. *International Journal of Imaging Systems and Technology, 33*(4), 1306–1322. DOI: 10.1002/ima.22870

Komorowski, M., Celi, L. A., Badawi, O., Gordon, A. C., & Faisal, A. A. (2018). The Artificial Intelligence Clinician Learns Optimal Treatment Strategies for Sepsis in Intensive Care. *Nature Medicine, 24*(11), 1716–1720. DOI: 10.1038/s41591-018-0213-5

Kotte, K. R., Attaluri, V., & Selvakumar, P. (2025). Revolutionising Retail Banking with AI and Virtual Technologies AI-Driven Financial Training. In Nair, A. J., Manohar, S., Limbu, Y. B., & Huhmann, B. A. (Eds.), *Intersecting Natural Language Processing and FinTech Innovations in Service Marketing* (pp. 421–440). IGI Global Scientific Publishing., DOI: 10.4018/979-8-3693-9944-6.ch020

Koundinya, A. K., Sharvani, G. S., & Rao, K. U. (2016). Calibrated security measures for centralized IoT applications of smart network_grids, In: *2016 International Conference on Computation System and Information Technology for Sustainable Solutions (CSITSS)*, Bangalore, 2016, pp. 153-157. DOI: 10.1109/CSITSS.2016.7779414

Kramp, T., van Kranenburg, R., & Lange, S. (2013). Introduction to the Internet of Things. In Enabling Things to Talk (pp. 1–10). Springer. DOI: 10.1007/978-3-642-40403-0_1

Kumar, H. S., Moiz, A., Kalshetti, K. V., Jain, V., & Kumar, K. (2025, April–May). *Image compression and enhancement using deep learning algorithm*. In *Proceedings of the 2025 International Conference on Emerging Smart Computing and Informatics (ESCI 2025)*. https://doi.org/DOI: 10.1109/ESCI63694.2025.10988240

Kumar, S. S., Kumar, S. R., & Ramesh, G. (2024). From Industry 4.0 to 5.0: Enriching Manufacturing Excellence Through Human–Robot Interaction and Technological Empowerment. In *Industrial Internet of Things for Smart Manufacturing* (pp. 28–55). Taylor & Francis., DOI: 10.1201/9781032642789-2

Kumar, S., Kameswari, Y. L., Rao, K. R., Moram, V., & Shital, S. (2024). Risk Management in IIoT. In *Industrial Internet of Things Security* (pp. 43–60). CRC Press., DOI: 10.1201/9781003466284-3

Ledley, R. S., & Lusted, L. B. (1959). Reasoning Foundations of Medical Diagnosis. *Science*, *130*(3366), 9–21. DOI: 10.1126/science.130.3366.9

Lee, J., Yoon, W., Kim, S., Kim, D., Kim, S., So, C. H., & Kang, J. (2020). BioBERT: A Pre-Trained Biomedical Language Representation Model for Biomedical Text Mining. *Bioinformatics (Oxford, England)*, *36*(4), 1234–1240. DOI: 10.1093/bioinformatics/btz682

Leffer, L. (2024). "Can Google Make Stoplights Smarter? – Google's Project Green Light Uses AI to Take on City Traffic," Scientific American, Aug. 15, 2024. https://www.scientificamerican.com/article/googles-project-green-light-uses-ai-to-take-on-city-traffic/

Libbrecht, M. W., & Noble, W. S. (2015). Machine Learning Applications in Genetics and Genomics. *Nature Reviews. Genetics*, *16*(6), 321–332. DOI: 10.1038/nrg3920

Li, M., Pan, X., Liu, C., & Li, Z. (2025, November). Federated deep reinforcement learning-based urban traffic signal optimal control. *Scientific Reports*, *15*(1), 11724. Advance online publication. DOI: 10.1038/s41598-025-91966-1 PMID: 40188158

Ling, S., Wang, B., Rong, Y., & Huang, G. Q. (2025). Human–Machine Synchronization for Smart Manufacturing Towards Industry 5.0. In *Human-Centric Smart Manufacturing Towards Industry 5.0* (pp. 341–357). Springer., DOI: 10.1007/978-3-031-82170-7_15

Lins, R. G., & Givigi, S. N. (2021). Cooperative Robotics and Machine Learning for Smart Manufacturing: Platform Design and Trends Within the Context of Industrial Internet of Things. *IEEE Access : Practical Innovations, Open Solutions*, *9*, 95444–95455. DOI: 10.1109/ACCESS.2021.3094374

Litjens, G., Kooi, T., Bejnordi, B. E., Setio, A. A. A., Ciompi, F., Ghafoorian, M., van der Laak, J. A. W. M., van Ginneken, B., & Sánchez, C. I. (2017). A Survey on Deep Learning in Medical Image Analysis. *Medical Image Analysis*, *42*, 60–88. DOI: 10.1016/j.media.2017.07.005

Liu, . (2021). "A Weed Detection System for Crops Based on a Deep Learning Network". November (2021). *Remote Sensing*, *13*, 4486.

Liu, J., & Xu, S. (2024). "Advancements of Graph Neural Networks in Urban Traffic Prediction," in *Proceedings of the 1st International Conference on Engineering Management, Information Technology and Intelligence (EMITI 2024)*, Jan. 2024, pp. 62–66. SCITEPRESS. DOI: 10.5220/0012902200004508

Li, Z., Yan, W., & Wang, L. (2024). Measuring Mobility Resilience with Network-Based Simulations of Flow Dynamics Under Extreme Events. *Transportation Research Part D, Transport and Environment*, *104*, 104362. DOI: 10.1016/j.trd.2024.104362

Lokhande, D. G., & Mohsin, S. A. (2017). Internet of things for ubiquitous smart home system. 2017 1st International Conference on Intelligent Systems and Information Management (ICISIM), 314–320. DOI: 10.1109/ICISIM.2017.8122191

Lucas, P. J. F., van der Gaag, L. C., & Abu-Hanna, A. (2004). Bayesian Networks in Biomedicine and Health-Care. *Artificial Intelligence in Medicine*, *30*(3), 201–214. https://doi.org/https://doi.org/10.1016/j.artmed.2003.11.001. DOI: 10.1016/j.artmed.2003.11.001

Lu, G., Ge, X., Zhong, T., Hu, Q., & Geng, J. (2024). Preprocessing enhanced image compression for machine vision. *IEEE Transactions on Circuits and Systems for Video Technology*, *34*(12), 13556–13568. DOI: 10.1109/TCSVT.2024.3441049

Lundberg, S. M., & Lee, S.-I. (2017). "A Unified Approach to Interpreting Model Predictions." In *Proceedings of the 31st International Conference on Neural Information Processing Systems*, 4768–77. NIPS'17. Red Hook, NY, USA: Curran Associates Inc.

Luo, R., Sun, L., Xia, Y., Qin, T., Zhang, S., Poon, H., & Liu, T.-Y. (2022). BioGPT: Generative Pre-Trained Transformer for Biomedical Text Generation and Mining. *Briefings in Bioinformatics*, *23*(6), bbac409. DOI: 10.1093/bib/bbac409

Mahesh, T. R., & Muskan Gupta, T. A. (2024). An XAI-Enhanced EfficientNetB0 Framework for Precision Brain Tumor Detection in MRI Imaging. *Journal of Neuroscience Methods*, *410*, 110227. DOI: 10.1016/j.jneumeth.2024.110227 PMID: 39038716

Maheswaran, K., Selvakumar, P., Shaikh, M., Shashank, G., Gupta, S., & Manjunath, T. C. (2025). The Evolution of Smart Urban Mobility With Emerging Technologies. In Ponnusamy, S., Assaf, M., Youssef, B., Jeon, G., & Kalyanaraman, S. (Eds.), *Leveraging VANETs and Blockchain Technology for Urban Mobility* (pp. 1–28). IGI Global Scientific Publishing., DOI: 10.4018/979-8-3373-0265-2.ch001

Mahmud, M. I., Mamun, M., & Abdelgawad, A. (2023). A Deep Analysis of Brain Tumor Detection from Mr Images Using Deep Learning Networks. *Algorithms*, *16*(4), 176. DOI: 10.3390/a16040176

Majeed, A. F., Salehpour, P., Farzinvash, L., & Pashazadeh, S. (2024). Multi-Class Brain Lesion Classification Using Deep Transfer Learning with Mobilenetv3. *IEEE Access : Practical Innovations, Open Solutions*, *12*, 155295–155308. DOI: 10.1109/ACCESS.2024.3413008

Majhi, R., Jena, P. R., Majhi, B., & Kalli, R. (2023). Prediction of crop yield using climate variables in the south-western province of India: A functional artificial neural network modeling (FLANN) approach. *Environment, Development and Sustainability*, *25*(10), 11033–11056. DOI: 10.1007/s10668-022-02517-x

Maurya, S., Pal, O., & Sarvakar, K. (2024). *"Layered Architecture of IoT," Apr 2024*. Secure and Intelligent IoT-Enabled Smart Cities., DOI: 10.4018/979-8-3693-2373-1.ch009

Mayouf, O., Rekik, M., & Krichen, L. (2022). Designing a hybrid system using solar cells and batteries to power a clinic in Tajarhi desert village in the Libya-Niger border off-grid. In: *2022 IEEE 21st International Conference on Sciences and Techniques of Automatic Control and Computer Engineering*, Sousse, Tunisia, 2022. DOI: 10.1109/STA56120.2022.10019222

Menouar, H., Guvenc, I., Akkaya, K., Uluagac, A. S., Kadri, A., & Tuncer, A. (2017). UAV-enabled intelligent transportation systems for the smart city: Applications and challenges. *IEEE Communications Magazine*, *55*(3), 22–28. DOI: 10.1109/MCOM.2017.1600238CM

Michailidis, P., Michailidis, I., Lazaridis, C. R., & Kosmatopoulos, E. (2025, May). Traffic Signal Control via Reinforcement Learning: A Review on Applications and Innovations. *Infrastructures*, *10*(5), 114. Advance online publication. DOI: 10.3390/infrastructures10050114

Migliaccio, K. W., Schaffer, B., Crane, J. H., & Davies, F. S. (2010, October). Plant response to evapotranspiration and soil water sensor irrigation scheduling methods for papaya production in south Florida. *Agricultural Water Management*, *97*(10), 1452–1460. DOI: 10.1016/j.agwat.2010.04.012

Miller, R. A.Jr, Pople, H. E.Jr, & Myers, J. D. (1982). INTERNIST-1, an Experimental Computer-Based Diagnostic Consultant for General Internal Medicine. *The New England Journal of Medicine, 307*(8), 468–476. DOI: 10.1056/NEJM198208193070803

Minnen, D., Ballé, J., & Toderici, G. D. (2018). Joint autoregressive and hierarchical priors for learned image compression. In *Advances in Neural Information Processing Systems (NeurIPS)* (pp. 10771–10780).

Miori, V., & Russo, D. (2014). Domotic Evolution towards the IoT. 2014 28th International Conference on Advanced Information Networking and Applications Workshops, 809–814. DOI: 10.1109/WAINA.2014.128

Mittelstadt, B. D., Allo, P., Taddeo, M., Wachter, S., & Floridi, L. (2016). The Ethics of Algorithms: Mapping the Debate. *Big Data & Society, 3*(2), 2053951716679679. DOI: 10.1177/2053951716679679

Mocrii, D., Chen, Y., & Musilek, P. (2018). IoT-based smart homes: A review of system architecture, software, communications, privacy and security. *Internet of Things : Engineering Cyber Physical Human Systems, 1–2*, 81–98. DOI: 10.1016/j.iot.2018.08.009

Mohanty, S. P., Choppali, U., & Kouglanos, E. (2016). Everything you wanted to know about smart cities: The Internet of Things is the backbone. *IEEE Consumer Electronics Magazine, 16*(3), 2162–2248. DOI: 10.1109/MCE.2016.2556879

Mohapatra, H. (2021). Socio-technical Challenges in the Implementation of Smart City. 2021 International Conference on Innovation and Intelligence for Informatics, Computing, and Technologies (3ICT), 57–62. DOI: 10.1109/3ICT53449.2021.9581905

Mohapatra, H. (2025). The fine-tuning journey of ChatGPT in industry applications. *Advances in Computational Intelligence and Robotics*, •••, 263–284. DOI: 10.4018/979-8-3693-8332-2.ch011

Mohite, R. B., Chinchawade, A. J., Waghmode, U. P., Thorat, Z. V., Mane, V. B., & Pawar, V. E. (2023). Smart farm protection system using IoT. 2023 4th International Conference on Signal Processing and Communication (ICSPC), 158–162. DOI: 10.1109/ICSPC57692.2023.10125709

Mon, B. F., Hayajneh, M., Ali, N. A., Ullah, F., Ullah, H., & Alkobaisi, S. (2025). Digital Twins in the IIoT: Current Practices and Future Directions Toward Industry 5.0. *Computers, Materials & Continua, 83*(3), 3675–3712. DOI: 10.32604/cmc.2025.061411

Montgomery, D. C. (2007). *Introduction to Statistical Quality Control* (5th ed.). John Wiley & Sons.

Moor, M., Vaher, K., Riives, J., Kangru, T., & Otto, T. (2021). "Modern Robot Integrated Manufacturing Cell According to the Needs of Industry 4.0." IOP Conference Series: Materials Science and Engineering, Vol. 1140, Article 012012. DOI: DOI: 10.1088/1757-899X/1140/1/012034

Mutambik, I. (2025). IoT-Enabled Adaptive Traffic Management: A Multiagent Framework for Urban Mobility Optimisation. *Sensors (Basel)*, *25*(13), 4126. DOI: 10.3390/s25134126 PMID: 40648380

Narayanan, A., & Shmatikov, V. (2008). "Robust de-Anonymization of Large Sparse Datasets." In *2008 IEEE Symposium on Security and Privacy*, 111–25. DOI: 10.1109/SP.2008.33

Nawaz, S. A., Khan, D. M., & Qadri, S. (2022). Brain Tumor Classification Based on Hybrid Optimized Multi-Features Analysis Using Magnetic Resonance Imaging Dataset. *Applied Artificial Intelligence*, *36*(1), 2031824. DOI: 10.1080/08839514.2022.2031824

Nemali, K. S., & Van Iersel, M. W. (2006, November). An automated system for controlling drought stress and irrigation in potted plants. *Scientia Horticulturae*, *110*(3), 292–297. DOI: 10.1016/j.scienta.2006.07.009

Nisha Angeline, M., Selvakumar, P., Gupta, M. K., Nazar, N., Elumalai, D., & Bhattacharya, S. (2025). Emerging Materials for Energy Storage and Environmental Applications. In Muduli, K., Moharana, B., Ales, S., & Biswal, D. (Eds.), *Using Computational Intelligence for Sustainable Manufacturing of Advanced Materials* (pp. 53–72). IGI Global Scientific Publishing., DOI: 10.4018/979-8-3693-7974-5.ch003

Nunes, P., Figueiredo, R., & Brito, M. C. (2016). The use of parking lots to solar-charge electric vehicles. *Renewable & Sustainable Energy Reviews*, *66*, 679–693. DOI: 10.1016/j.rser.2016.08.015

O'Shaughnessy, S. A., & Evett, S. R. (2010, April). Canopy temperature based system effectively schedules and controls center pivot irrigation of cotton. *Agricultural Water Management*, *97*(9), 1310–1316. DOI: 10.1016/j.agwat.2010.03.012

Obermeyer, Z., Powers, B., Vogeli, C., & Mullainathan, S. (2019). Dissecting Racial Bias in an Algorithm Used to Manage the Health of Populations. *Science*, *366*(6464), 447–453. DOI: 10.1126/science.aax2342

Panneer Selvam, A., & Al-Humairi, S. N. S. (2024, December). Environmental Impact Evaluation Using Smart Real-Time Weather Monitoring Systems: A Systematic Review [Online]. *Innovative Infrastructure Solutions*, *10*(13), •••.

Pant, D., Bhatt, A., Khan, M., Nautiyal, O. P., & Adhikari, P. (2019). Automated IoT based smart water quality assessment system. In *Proceedings of the 8th International Conference on System Modeling and Advancement in Research Trends (SMART 2019)* DOI: 10.1109/SMART46866.2019.9117271

Patel, M. I., Suthar, S., & Thakar, J. (2019, May 17). Survey on image compression using machine learning and deep learning. In *Proceedings of the 2019 International Conference on Intelligent Computing and Control Systems (ICCS 2019)* (pp. 1103–1105). DOI: 10.1109/ICCS45141.2019.9065473

Paul, C., Ganesh, A., & Sunitha, C. (2018). An overview of IoT based smart homes. 2018 2nd International Conference on Inventive Systems and Control (ICISC), 43–46. DOI: 10.1109/ICISC.2018.8398858

Pearl, J., & Mackenzie, D. (2018). *The Book of Why: The New Science of Cause and Effect*. Basic Books.

People's Daily. (2019). "City Brain helps boost efficiency in Chinese cities," People's Daily Online, Apr. 24, 2019. https://en.people.cn/n3/2019/0424/c90000-9571706.html

Pérez-Chacón, J., Luna-Romera, J. M., Troncoso, A., & Martínez-Álvarez, F. (2010). Big data analytics for discovering electricity consumption patterns in smart cities. *Expert Systems with Applications*, *37*(12), 8333–8342.

Peyman, M., Copado, P. J., Tordecilla, R. D., Martins, L. C., Xhafa, F., & Juan, A. A. (2021). "Edge Computing and IoT Analytics for Agile Optimization in Intelligent Transportation Systems," Energies, vol. 14, no. 19, Article 6309, Oct. 2021. MDPI https://www.mdpi.com/1996-1073/14/19/6309

Philippi, D., Rothaus, K., & Castelli, M. (2023). A Vision Transformer Architecture for the Automated Segmentation of Retinal Lesions in Spectral-Domain Optical Coherence Tomography Images. *Scientific Reports*, *13*(1), 517. DOI: 10.1038/s41598-023-27616-1

Pirmagomedov, R., Bobrikova, E., Blinnikov, M., Amelyanovich, A., Glushakov, R., Loskutov, S., Koucheryavy, A., & Kirichek, R. (2018). IoT Based Earthquake Prediction Technology. In. Lecture Notes in Computer Science: Vol. 11118. *Internet of Things, Smart Spaces, and Next Generation Networks and Systems* (pp. 535–546). Springer. [Online], Available https://link.springer.com/chapter/10.1007/978-3-030-01168-0_48 DOI: 10.1007/978-3-030-01168-0_48

Ponce-Bobadilla, A. V., Schmitt, V., & Valderrama, D. (2024). Practical Guide to SHAP Analysis: Explaining Supervised Machine-Learning Model Predictions in Drug Development. *Clinical and Translational Science*, *17*(11), e70056. DOI: 10.1111/cts.70056

Pramanik, M., Khanna, M., Singh, M., Singh, D. K., Sudhishri, S., Bhatia, A., & Ranjan, R. (2022, February). Automation of soil moisture sensor-based basin irrigation system. *Smart Agricultural Technology*, *148*(2), 04022001. DOI: 10.1016/j.atech.2021.100032

Rajakaruna, S., Shahnia, F., & Ghosh, A. (2015). Plug. In *Electric Vehicles in Smart Grids: Energy Management*. Springer Singapore.

Rajak, P., Ganguly, A., Adhikary, S., & Bhattacharya, S. (2023). Internet of Things and smart sensors in agriculture: Scopes and challenges. *Journal of Agriculture and Food Research*, *14*, 100776. DOI: 10.1016/j.jafr.2023.100776

Rajkomar, A., Dean, J., & Kohane, I. (2019). Machine Learning in Medicine. *The New England Journal of Medicine*, *380*(14), 1347–1358. DOI: 10.1056/NEJMra1814259

Rao, K. V., Sivakumar, V., & Swathi, R. (2025). Generative AI and secure IoMT systems for enhanced real-time health monitoring and security. In Ouaissa, M., Ouaissa, M., Imad, M., Qurashi, J., & Farooq, M. (Eds.), *Utilizing AI of Medical Things for Healthcare Security and Sustainability* (pp. 469–502). IGI Global., DOI: 10.4018/979-8-3373-0690-2.ch015

Rashid, Z., Melia-Segui, J., Pous, R., & Peig, E. (2016). Using augmented reality and Internet of Things to improve accessibility of people with motor disabilities in the context of smart cities. *Future Generation Computer Systems*.

Rath, M., & Tomar, A. (2021). Chapter 7: Smart grid modernization using Internet of Things technology. In: Advances in Smart Grid Power System, 191-212.

Rath, M., & Pati, B. (2019). Security assertion of IoT devices using cloud of things perception. *International Journal of Interdisciplinary Telecommunications and Networking*, *11*(4), 17–31. DOI: 10.4018/IJITN.2019100102

Rathore, M. M., Ahmad, A., Paul, A., & Rho, S. (2016). Urban planning and building smart cities based on the Internet of Things using big data analytics. *Computer Networks*, *101*, 63–80. DOI: 10.1016/j.comnet.2015.12.023

Ray, A. K., & Bagwari, A. (2020). IoT based Smart home: Security Aspects and security architecture. 2020 IEEE 9th International Conference on Communication Systems and Network Technologies (CSNT), 218–222. DOI: 10.1109/CSNT48778.2020.9115737

Rekik, M., & Krichen, L. (2014). Active power filter based on wind turbine for electric power system quality improvement. In: 2014 IEEE 11th International Multi-Conference on Systems, Signals and Devices, SSD 2014, Barcelona, Spain. DOI: 10.1109/SSD.2014.6808768

Rekik, M., Abdelkafi, A., & Krichen, L. (2015). Synchronization of Wind Farm Power System to Utility Grid under Voltage and Frequency Variations. *International Journal of Renewable Energy Research*, *5*, 70–81.

Rekik, M., & Krichen, L. (2021). Double Layer Optimization Approach of Plug-in Electric Vehicle for Participation in Smart Grid Ancillary Services and Energy Markets. *International Journal of Renewable Energy Research*, *11*, 1107–1123.

Ribeiro, M., Singh, S., & Guestrin, C. (2016). ""Why Should i Trust You?": Explaining the Predictions of Any Classifier." In *Proceedings of the 22nd ACM SIGKDD International Conference on Knowledge Discovery and Data Mining*, 1135–44. DOI: 10.1145/2939672.2939778

Rojas, E., Carrascal, D., Lopez-Pajares, D., Alvarez-Horcajo, J., Carral, J. A., Arco, J. M., & Martinez-Yelmo, I. (2024). A Survey on AI-Empowered Softwarized Industrial IoT Networks. *Electronics (Basel)*, *13*(10), 1979. Advance online publication. DOI: 10.3390/electronics13101979

Ronneberger, O., Fischer, P., & Brox, T. (2015). "U-Net: Convolutional Networks for Biomedical Image Segmentation." In *Medical Image Computing and Computer-Assisted Intervention – MICCAI 2015*, 9351:234–41. DOI: 10.1007/978-3-319-24574-4_28

Rosenbaum, L. (2017). Data Sharing and the Ethics of Public Trust. *The New England Journal of Medicine*, *377*(5), 401–403. DOI: 10.1056/NEJMp1005897

Saeedi, S., Rezayi, S., Keshavarz, H., & Niakan Kalhori, S. R. (2023). MRI-Based Brain Tumor Detection Using Convolutional Deep Learning Methods and Chosen Machine Learning Techniques. *BMC Medical Informatics and Decision Making*, *23*(1), 16. DOI: 10.1186/s12911-023-02114-6 PMID: 36691030

Sahoo, S., & Lo, C. Y. (2022). Smart Manufacturing Powered by Recent Technological Advancements: A Review. *Journal of Manufacturing Systems*, *64*, 236–250. DOI: 10.1016/j.jmsy.2022.06.008

Sakib, M., & Mustajab, S. (2024). Enhanced multi-variate time series prediction through statistical-deep learning integration: The VAR-stacked LSTM model. *SN Computer Science*, *5*(5), 573. DOI: 10.1007/s42979-024-02950-x

Sakib, M., & Mustajab, S. (2025, July 4). CEEMDAN-TCN-Auto LSTM: A Triple-Force ensemble using cross-stitch networks for short-term load forecasting. *Arabian Journal for Science and Engineering*. Advance online publication. DOI: 10.1007/s13369-025-10398-1

Sakib, M., & Siddiqui, T. (2023). Multi-network-based ensemble deep learning model to forecast Ross River virus outbreak in Australia. *International Journal of Pattern Recognition and Artificial Intelligence*, *37*(10), 2352015. Advance online publication. DOI: 10.1142/S0218001423520158

Sakthivel, M., Jaganathan, A., Mohanraj, M., & Selvakumar, P. Mohit & Manjunath T, (2025). "Brand Loyalty through AI and Personalisation AI-Generated Content for Financial Marketing." In A J Nair, S Manohar, Y B Limbu, & B A Huhmann (Eds.), Intersecting Natural Language Processing and FinTech Innovations in Service Marketing (pp. 77–96). IGI Global Scientific Publishing. DOI: 10.4018/979-8-3693-9944-6.ch006

Sakurada, M., & Yairi, T. (2014). "Anomaly Detection Using Autoencoders with Nonlinear Dimensionality Reduction." In *Workshop on Machine Learning for Sensory Data Analysis (MLSDA)*, 4–11. DOI: 10.1145/2689746.2689747

Salah, W. A., Albreem, M. A., Alsayid, B., Zneid, B. A., Alkhasawneh, M., Al-Mofleh, A., Sneineh, A. A., & Abu Al-Aish, A. (2019). Electric vehicle technology impacts on energy. *International Journal of Power Electronics and Drive Systems*, *10*(1), 1–19. DOI: 10.11591/ijpeds.v10.i1.pp1-9

Sánchez-Moreno, L., Perez-Peña, A., Duran-Lopez, L., & Dominguez-Morales, J. P. (2025). Ensemble-Based Convolutional Neural Networks for Brain Tumor Classification in MRI: Enhancing Accuracy and Interpretability Using Explainable AI. *Computers in Biology and Medicine*, *195*, 110555. DOI: 10.1016/j.compbiomed.2025.110555 PMID: 40554976

Sarasketa-Zabala, E., Martinez-Laserna, E., Berecibar, M., Gandiaga, I., Rodriguez-Martinez, L., & Villarreal, I. (2016). Realistic lifetime prediction approach for Li-ion batteries. *Applied Energy*, *162*, 839–852. DOI: 10.1016/j.apenergy.2015.10.115

Schwab, P.. (2022). Counterfactuals and Causal Inference for Clinical Prediction Models. *Nature Machine Intelligence*, *4*, 813–824.

Sendak, M. P., D'Arcy, J., & Kashyap, S.. (2020). A Path for Translation of Machine Learning Products into Healthcare Delivery. *EMJ Innovations*, *4*(1), 56–64.

Sharma, S., Shrivastava, R. R., Sahu, A., Mishra, S., Srivastava, K., & Mohapatra, H. (2025). Cloud-driven public safety in smart cities. Enhancing Safety, Security, and Service in Organizations, 235-262. DOI: 10.4018/979-8-3373-0482-3.ch008

Sharma, A. K., Nandal, A., Dhaka, A., Zhou, L., Alhudhaif, A., Alenezi, F., & Polat, K. (2023). Brain Tumor Classification Using the Modified ResNet50 Model Based on Transfer Learning. *Biomedical Signal Processing and Control*, *86*, 105299. DOI: 10.1016/j.bspc.2023.105299

Sharma, V., & Shah, A. (2025, January). A Survey on Smart Traffic Monitoring System using AI. *International Journal of Innovative Research in Computer and Communication Engineering*, *13*(1). Advance online publication. DOI: 10.15680/IJIRCCE.2025.1301129

Sheena, K. S., & Nair, M. S. (2024). GenCoder: A novel convolutional neural network based autoencoder for genomic sequence data compression. *IEEE/ACM Transactions on Computational Biology and Bioinformatics*, *21*(3), 405–415. DOI: 10.1109/TCBB.2024.3366240 PMID: 38358865

Shickel, B., Tighe, P. J., Bihorac, A., & Rashidi, P. (2017). Deep EHR: A Survey of Recent Advances in Deep Learning Techniques for Electronic Health Record (EHR) Analysis. *IEEE Journal of Biomedical and Health Informatics*, *22*(5), 1589–1604. DOI: 10.1109/JBHI.2017.2767063

Shirazi, E., Zakariazadeh, A., & Jadid, S. (2015). Optimal joint scheduling of electrical and thermal appliances in a smart home environment. *Energy Conversion and Management*, *106*, 181–193. DOI: 10.1016/j.enconman.2015.09.017

Shi, Y., Tuan, H. D., Savkin, A. V., Duong, T. Q., & Poor, H. V. (2019). Model Predictive Control for Smart Grids with Multiple Electric-Vehicle Charging Stations. *IEEE Transactions on Smart Grid*, *10*(2), 2127–2136. DOI: 10.1109/TSG.2017.2789333

Shortliffe, E. H. (1976). *Computer-Based Medical Consultations: MYCIN*. Elsevier.

Sidid, S., & Gaur, S. (2017). *Smart network_grid building automation based on internet of things. In: 2017 Innovations in Power and Advanced Computing Technologies (i-PACT), 1-4.*

Simbeye, D. S., Mkiramweni, M. E., Karaman, B., & Taskin, S. (2022). Plant water stress monitoring and control system. *Agricultural Systems*, *188*(Jul), 103080. DOI: 10.1016/j.atech.2022.100066

Singh, D., Tripathi, G., & Jara, A. J. (2014). A survey of Internet-of-Things: Future vision, architecture, challenges and services. 2014 IEEE World Forum on Internet of Things (WF-IoT), 287–292. DOI: 10.1109/WF-IoT.2014.6803174

Singhal, K., Tu, T., & Gottweis, J.. (2023). Large Language Models Encode Clinical Knowledge. *Nature*, *620*(7972), 172–180. DOI: 10.1038/s41586-023-06291-2

Sivakumar, V., Reddy, A. H., & Varma, A. N. (2025). GRU in anomaly detection for IoT: A comparative study. In 2025 IEEE International Students' Conference on Electrical, Electronics and Computer Science (SCEECS) (pp. 1–7). IEEE. DOI: 10.1109/SCEECS64059.2025.10940416

Sivakumar, V., Sushruth, T. L., Suthersen, A. R. J., & Gugan, S. K. (2025). Weed detection in farmlands using RCNN. In 2025 3rd International Conference on Intelligent Data Communication Technologies and Internet of Things (IDCIoT) (pp. 1485–1491). IEEE. DOI: 10.1109/IDCIOT64235.2025.10915039

Sivakumar, V., Swathi, R., & Yuvaraj, V. (2021). An IoT-based energy meter for energy level monitoring, predicting, and optimization. In Challenges and Opportunities for the Convergence of IoT, Big Data, and Cloud Computing (Chapter 4, pp. 48–65). IGI Global.

Skodras, A., Christopoulos, C., & Ebrahimi, T. (2001). The JPEG 2000 still image compression standard. *IEEE Signal Processing Magazine*, *18*(5), 36–58. DOI: 10.1109/79.952804

Smaoui, M., Rekik, M., & Krichen, L. (2025). Optimal design of an off-grid electrical system in remote areas with different renewable energy scenarios. *Energy*, *318*, 134821. DOI: 10.1016/j.energy.2025.134821

Smith, B., Barlow, G. J., Xie, X., & Rubinstein, M. H. (2013). "SURTRAC: Scalable Urban Traffic Control – Performance Evaluation of a Multi-Agent Adaptive Signal Control System," in Transportation Research Board 92nd Annual Meeting, Paper 13-0315, 2013. https://www.ri.cmu.edu/pub_files/2013/1/13-0315.pdf

Soenksen, L. R., Ma, Y., Zeng, C., Boussioux, L., Carballo, K. V., Na, L., Wiberg, H. M., Li, M. L., Fuentes, I., & Bertsimas, D. (2022). Integrated Multimodal Artificial Intelligence Framework for Healthcare Applications. *NPJ Digital Medicine*, *5*(149), 149. Advance online publication. DOI: 10.1038/s41746-022-00689-4

Soliman, M., Abiodun, T., Hamouda, T., Zhou, J., & Lung, C.-H. (2013). Smart Home: Integrating Internet of Things with Web Services and Cloud Computing. 2013 IEEE 5th International Conference on Cloud Computing Technology and Science, 317–320. DOI: 10.1109/CloudCom.2013.155

Somayya, M., Ramaswamy, R., & Tripathi, S. (2015). Internet of Things (IoT): A Literature Review. *Journal of Computer and Communications*, *3*(5), 164–173. DOI: 10.4236/jcc.2015.35021

Sood, S. K. (2015). "Fog aided IoT-enabled patient health monitoring in smart homes," *Proceedings of the International Conference on Intelligent Computing and Internet of Things (ICIT)*, Harbin, China, pp. 41–44, doi: DOI: 10.1109/ICIT.2015.7293596

Stojkoska, B. L., & Trivodaliev, K. V. (2017). A review of Internet of Things for smart home: Challenges and solutions. *Journal of Cleaner Production*, *140*, 1454–1464. DOI: 10.1016/j.jclepro.2016.10.006

Subba, A. B., & Sunaniya, A. K. (2025). Computationally Optimized Brain Tumor Classification Using Attention Based GoogLeNet-Style CNN. *Expert Systems with Applications*, *260*, 125443. DOI: 10.1016/j.eswa.2024.125443

Subramanya, A. (2001). Image compression technique. *IEEE Potentials*, *20*(1), 19–23. DOI: 10.1109/45.913206

Sujitha, B., Parvathy, V. S., Lydia, E. L., Rani, P., Polkowski, Z., & Shankar, K. (2021). Optimal deep learning based image compression technique for data transmission on industrial Internet of things applications. *Transactions on Emerging Telecommunications Technologies*, *32*(7), e3976. DOI: 10.1002/ett.3976

Sun, Q., Yu, W., Kochurov, N., Hao, Q., & Hu, F. (2013). A Multi-Agent-Based Intelligent Sensor and Actuator Network Design for Smart House and Home Automation. *Journal of Sensor and Actuator Networks*, *2*(3), 557–588. DOI: 10.3390/jsan2030557

Sun, S., He, T., & Chen, Z. (2021). Semantic structured image coding framework for multiple intelligent applications. *IEEE Transactions on Circuits and Systems for Video Technology*, *31*(9), 3631–3642. DOI: 10.1109/TCSVT.2020.3042517

Taiwo, O., & Ezugwu, A. E. (2021). Internet of Things-Based Intelligent Smart Home Control System. *Security and Communication Networks*, *2021*, 1–17. DOI: 10.1155/2021/9928254

Taj, S., Asad, U., Azhar, M., & Kausar, S. (2019). Interoperability in IOT Based Smart Home: A Review. *Review of Computer Engineering Studies*, *5*(3), 50–55. DOI: 10.18280/rces.050302

Talari, S., Shafie-khah, M., Siano, P., Loia, V., Tommaetti, A., & Catalao, J. P. S. (2017). A review of smart cities based on the Internet of Things concept. *Energies*, *10*(4), 421. DOI: 10.3390/en10040421

Talukder, M. A., Islam, M. M., Uddin, M. A., Akhter, A., Md, A. J. P., Aryal, S., Muhammad, A. A. A., Hasan, K. F., & Moni, M. A. (2023). An Efficient Deep Learning Model to Categorize Brain Tumor Using Reconstruction and Fine-Tuning. *Expert Systems with Applications*, *230*, 120534. DOI: 10.1016/j.eswa.2023.120534

Tan, M., & Le, Q. (2019). "EfficientNet: Rethinking Model Scaling for Convolutional Neural Networks." In *Proceedings of the 36th International Conference on Machine Learning*, edited by Kamalika Chaudhuri and Ruslan Salakhutdinov, 97:6105–14. Proceedings of Machine Learning Research. PMLR. https://proceedings.mlr.press/v97/tan19a.html.

Taspınar, Y. S. (2023). Light weight convolutional neural network and low-dimensional images transformation approach for classification of thermal images. *Case Studies in Thermal Engineering*, *41*, 102670. DOI: 10.1016/j.csite.2022.102670

Tehsin, S., Nasir, I. M., Damaševičius, R., & Maskeliūnas, R. (2024). DaSAM: Disease and Spatial Attention Module-Based Explainable Model for Brain Tumor Detection. *Big Data and Cognitive Computing*, *8*(9), 97. DOI: 10.3390/bdcc8090097

Thakur, T., Srivastava, L., Panda, A., & Mohapatra, H. (2025). Smart parking in foggy condition real-time efficiency with PSO, ACO, and IoT sensors. Revolutionizing Urban Development and Governance With Emerging Technologies, 233-260. DOI: 10.4018/979-8-3373-1375-7.ch008

Thirunagalingam, A., & Selvakumar, P. (2025). Application of Automation and Robotics in the Oil and Gas Industry. In Abdullayev, V., & Khang, A. (Eds.), *Revolutionizing AI and Robotics in the Oil and Gas Industry* (pp. 229–244). IGI Global Scientific Publishing., DOI: 10.4018/979-8-3693-8156-4.ch011

Tison, G. H., Sanchez, J. M., Ben Ballinger, M. S., Olgin, J. E., Pletcher, M. J., & Vittinghoff, E.. (2018). Passive Detection of Atrial Fibrillation Using a Commercially Available Smartwatch. *JAMA Cardiology*, *3*(5), 409–416. DOI: 10.1001/jamacardio.2018.0136

Toderici, G. (2017). Full resolution image compression with recurrent neural networks. In *Proceedings of the IEEE Conference on Computer Vision and Pattern Recognition (CVPR)* (pp. 5435–5443) DOI: 10.1109/CVPR.2017.577

Topol, E. J. (2019). *Deep Medicine: How Artificial Intelligence Can Make Healthcare Human Again*. Basic Books.

Tsai, Y. H. H., Bai, S., & Yamada, M.. (2019). "Multimodal Transformer for Unaligned Multimodal Language Sequences." In *Proceedings of the Annual Meeting of the Association for Computational Linguistics*. DOI: 10.18653/v1/P19-1656

Tuballa, M. L., & Abundo, M. L. (2016). A review of the development of Smart Grid technologies. *Renewable & Sustainable Energy Reviews*, *59*, 710–725. DOI: 10.1016/j.rser.2016.01.011

U.S. Food and Drug Administration. (2019). "Proposed Regulatory Framework for Modifications to Artificial Intelligence/Machine Learning-Based Software as a Medical Device." FDA.gov.

Ullah, I., & Kim, D. (2017). An Improved Optimization Function for Maximizing User Comfort with Minimum Energy Consumption in Smart Homes. *Energies*, *10*(11), 1818. DOI: 10.3390/en10111818

Uma, J., Shobana, M., Abhiniti, G., Rahul, R., Kavinkumar, S. P., & Roshan, S. H. (2024). IoT-based smart farmland using deep learning. 2024 2nd International Conference on Artificial Intelligence and Machine Learning Applications (AIMLA), 1–6. DOI: 10.1109/AIMLA59606.2024.10531335

Upadhyaya, A. N., Debnath, P., Rani, P., Selvakumar, P., Manjunath, T. C., & Bhattacharya, S. (2025). AI-Based Disaster Prediction and Early Warning Systems. In Mobo, F. (Ed.), *AI and Emerging Technologies for Emergency Response and Smart Cities* (pp. 55–74). IGI Global Scientific Publishing., DOI: 10.4018/979-8-3693-9770-1.ch003

Venu, S., Kumar, R. G., Kumar, M. K., Prasad, T. G., Suresh, B., & Neelima, P. (2023). An Intelligent and Service Based Smart Agriculture Recommendation System. *International Journal of Intelligent Systems and Applications in Engineering*, *12*(7s), 153–158. https://www.ijisae.org/index.php/IJISAE/article/view/4051

Verma, R., Govindaraj, J., Chhabra, S., & Gupta, G. (2015). "DF 2.0: An automated, privacy preserving, and efficient digital forensic framework that leverages machine learning for evidence prediction and privacy evaluation," in *Proceedings of the 2015 Conference on Systematic Approaches to Digital Forensic Engineering (SADFE)*, pp. 1–10.

Viani, F., Robol, F., Polo, A., Rocca, P., Oliveri, G., & Massa, A. (2013). Wireless Architectures for Heterogeneous Sensing in Smart Home Applications: Concepts and Real Implementation. *Proceedings of the IEEE*, *101*(11), 2381–2396. DOI: 10.1109/JPROC.2013.2266858

Wahlang, I., Maji, A. K., Saha, G., Chakrabarti, P., Jasinski, M., Leonowicz, Z., & Jasinska, E. (2022). Brain Magnetic Resonance Imaging Classification Using Deep Learning Architectures with Gender and Age. *Sensors (Basel)*, *22*(5), 1766. DOI: 10.3390/s22051766 PMID: 35270913

Wang, F., Shen, L., Teng, Q., & Tian, Z. (2024). DSCIC: Deep screen content image compression. *IEEE Transactions on Circuits and Systems for Video Technology*, *34*(11), 11965–11979. DOI: 10.1109/TCSVT.2024.3419575

Wang, J., Lu, S.-Y., Wang, S.-H., & Zhang, Y.-D. (2024). RanMerFormer: Randomized Vision Transformer with Token Merging for Brain Tumor Classification. *Neurocomputing*, *573*, 127216. DOI: 10.1016/j.neucom.2023.127216

Wang, M., Zhang, G., Zhang, C., Zhang, J., & Li, C. (2013). An IoT-based appliance control system for smart homes. *2013 Fourth International Conference on Intelligent Control and Information Processing (ICICIP)*, 744–747. DOI: 10.1109/ICICIP.2013.6568171

Wang, S., Wang, Z., Wang, S., & Ye, Y. (2022). Deep image compression towards machine vision: A unified optimization framework. *IEEE Transactions on Circuits and Systems for Video Technology*, *33*(6), 2979–2989. DOI: 10.1109/TCSVT.2022.3230843

Wang, X., Yang, W., Wheaton, A., Cooley, N., & Moran, B. (2010, November). Efficient registration of optical and IR images for automatic plant water stress assessment. *Computers and Electronics in Agriculture*, *74*(2), 230–237. DOI: 10.1016/j.compag.2010.08.004

Weber, G. M.. (2020). Regulating AI in Health Care: An Overview of Proposed U.S. and EU Approaches. *Health Affairs*, *39*(11), 2024–2031.

Williams, V., Terence, S. J., & Immaculate, J. (2019). Survey on Internet of Things based Smart Home. *2019 International Conference on Intelligent Sustainable Systems (ICISS)*. DOI: 10.1109/ISS1.2019.8908112

Wu, Y., Dai, H. N., Wang, H., Xiong, Z., & Guo, S. (2022). A Survey of Intelligent Network Slicing Management for Industrial IoT: Integrated Approaches for Smart Transportation, Smart Energy, and Smart Factory. *IEEE Communications Surveys and Tutorials*, *24*(2), 1175–1211. DOI: 10.1109/COMST.2022.3158270

Xie, Y., Cheng, K. L., & Chen, Q. (2021). Enhanced invertible encoding for learned image compression. In *Proceedings of the 29th ACM International Conference on Multimedia* (pp. 162–170). DOI: 10.1145/3474085.3475213

Xiong, Z., Sheng, H., Rong, W., & Cooper, D. E. (2012). Intelligent transportation systems for smart cities: A progress review. *Science China. Information Sciences*, *55*(12), 2908–2914. DOI: 10.1007/s11432-012-4725-1

Xu, C., Tao, D., & Xu, C. (2013). "A Survey on Multi-View Learning." *arXiv Preprint arXiv:1304.5634*.

Xu, J., Hu, Z., Miao, J., Li, K., & Liu, L. (2024). Machine Learning for Predicting Hemodynamic Deterioration of Patients with Intermediate-Risk Pulmonary Embolism in the Intensive Care Unit. *Shock (Augusta, Ga.)*, *61*(1), 51–58. DOI: 10.1097/SHK.0000000000002261

Xu, R., Brown, L., Ahuja, V., & Li, N. (2020). Towards Natural Language Processing for Symptom Checking: A Digital Dispatcher Study. *Translational International Medical Applications Science*, *12*(1), 109–114.

Yang, Q., Liu, Y., Chen, T., & Tong, Y. (2019). Federated Machine Learning: Concept and Applications. *ACM Transactions on Intelligent Systems and Technology*, *10*(2), 12. DOI: 10.1145/3298981

Yang, T., Zhang, W., & Wang, W.. (2021). "Large-scale Robust Deep AUC Maximization: A New Surrogate Loss and Empirical Studies on Medical Image Classification." In *Proceedings of the IEEE/CVF Conference on Computer Vision and Pattern Recognition*.

Yao, W., Zhao, J., Wen, F., Dong, Z., Xue, Y., Xu, Y., & Meng, K. (2014). A multi-objective collaborative planning strategy for integrated power distribution and electric vehicle charging systems. *IEEE Transactions on Power Systems*, *29*(4), 1811–1821. DOI: 10.1109/TPWRS.2013.2296615

Ye, J., Woods, D., Jordan, N., & Starren, J. (2024). The Role of Artificial Intelligence for the Application of Integrating Electronic Health Records and Patient-Generated Data in Clinical Decision Support. *AMIA Joint Summits on Translational Science Proceedings AMIA Summit on Translational Science*, •••, 459–467.

Yuehong, Y. I. N., Zeng, Y., Chen, X., & Fan, Y. (2016). The internet of things in healthcare: An overview. *Journal of Industrial Information Integration*, *1*, 3–13.

Zhang, D. D., Zhang, S. M., Liu, X. X., & Zhao, Y. W. (2023, April). Assessment for crop water stress with hyperspectral imaging and deep learning technologies. *Precision Agriculture*, *24*(2), 223–239. DOI: 10.1016/j.compag.2021.106019

Zhang, H. (2022). ResNest: Split-attention networks. In *Proceedings of the IEEE/CVF Conference on Computer Vision and Pattern Recognition Workshops (CVPRW)* (pp. 2735–2745).

Zhang, J., Hua, X., Huang, J., Shen, X., Chen, J., Zhou, Q., Fu, Z., & Zhao, Y. (2019). City Brain: Practice of Large-Scale Artificial Intelligence in the Real World. *IET Smart Cities*, *1*(1), 78–84. DOI: 10.1049/iet-smc.2019.0034

Zhang, Y., Zou, W., Chen, X., Yang, C., & Cao, J. (2014). The security for power internet of things: framework, policies, and countermeasures. In: *2014 International Conference on Cyber-Enabled Distributed Computing and Knowledge Discovery*, Shanghai, 2014, pp. 139-142. DOI: 10.1109/CyberC.2014.32

Zhao, L., Jia, L., & Greenspan, H. (2018). Multi-Scale Coarse-Fine Fusion for Volumetric Segmentation of Medical Images. *Medical Image Analysis*, *47*, 20–30. DOI: 10.1016/j.media.2018.08.002

Zhenga, Y., Niua, S., Shanga, Y., Shaoc, Z., & Jian, L. (2019). Integrating plug-in electric vehicles into power grids: A comprehensive review on power interaction mode, scheduling methodology and mathematical foundation. *Renewable & Sustainable Energy Reviews*, *112*, 424–439. DOI: 10.1016/j.rser.2019.05.059

Zitnik, M., Agrawal, M., & Leskovec, J. (2018). Modeling Polypharmacy Side Effects with Graph Convolutional Networks. *Bioinformatics (Oxford, England)*, *34*(13), i457–i466. DOI: 10.1093/bioinformatics/bty294

About the Contributors

Muneer Khan is an accomplished professional innovator in the field of Electrical Engineering, with a strong background in both academia and industry. He holds a Master of Science in Electrical Engineering from Columbia University, specializing in intelligent and connected devices, sensors, Embedded AI, semiconductor physics, and integrated photonics. Khan has extensive experience in research and applied science, having worked as a Research Assistant at Columbia University's Laboratory of Unconventional Electronics. He has completed internships in France and India. His work spans advanced technologies, including PCB design, machine learning algorithms, hardware inspection, and robotic sensors. In addition to his academic and research roles, Khan is a successful entrepreneur, having founded Cadre Technologies Services. His venture focuses on AI, machine learning, and assistive technologies. He has been recognized for his contributions to science and technology with numerous awards, including the Distinguished Young Scientist Award from the Government of India, the Engineering for Humanity Award in New York for his outstanding innovation for blind individuals, and multiple accolades from academic institutions. He won the First Position at the Wearable Innovation Forum 2025 at Cambridge University, UK, for his groundbreaking work in wearable assistive technology. Khan's work has led to several patents, publications, and major funding, including the Engineering for Humanity Award 2025 for his startup. His technical skills span programming, circuit design, and advanced testing, and he continues to lead innovative projects and product developments worldwide.

Shantanu Awasthi has an extensive academic background, holding a Ph.D. in Mathematics from North Dakota State University, an M.S. in Mathematics from Virginia State University, and a B.Tech in Electronics and Communication Engineering from Maharaja Agrasen Institute of Technology. His professional experience spans several academic and industry roles, including positions as an Assistant Professor of Data Analytics at Missouri Southern State University and

Data Science roles at various institutions, such as Sense 360 and the University of North Dakota. His research interests focus on stochastic processes, machine learning, and deep learning, with publications in journals like the Journal of Safety Research and the Journal of Stochastic Analysis. In addition to his research, Shantanu has presented his work at notable conferences, including the North American Meetings of the Regional Science Association International and the SIAM Northern States Section Student Chapter Conference. His teaching experience covers data science, statistics, and mathematics courses at multiple universities. He has also earned several honors, including a Graduate Student Travel Grant and a Department Research Award. Furthermore, he holds certifications in actuarial sciences, SAS programming, and deep learning.

Mohd Sakib is an Assistant Professor in the School of Computer Science at UPES University, Dehradun, India. He earned his Ph.D. in Computer Science from Aligarh Muslim University, specializing in ensemble learning, deep learning, and time series analysis. Before joining UPES, he served as an Assistant Professor at Madanapalle Institute of Science & Technology (MITS), Andhra Pradesh, where he coordinated student holistic activities, NBA documentation, and national initiatives such as AICTE PARAKH. He has taught and supervised projects in areas including artificial intelligence, deep learning, and data science, at both undergraduate and postgraduate levels. His research focuses on developing intelligent AI-driven models for forecasting, anomaly detection, and smart energy systems. He has published widely in reputed Q1, SCI/SCIE and Scopus-indexed journals and presented his work at IEEE international conferences. He actively contributes as a reviewer and technical committee member for journals and conferences, and serves on the editorial board of IGI Global. He is also a Life Member of the Indian Society for Technical Education (ISTE) and regularly delivers invited talks and organizes faculty development programs.

Mohammad Vaseem khan (B.Tech in ECE), M.Tech (Electronics) from University of Mumbai-Maharashtra, Gate-2004,2005,2014 Qualified), is a distinguished academician and researcher with over 16 years of experience in Electronics and Communication Engineering. Currently serving as Associate Professor at Bansal Institute of Engineering and Technology, Lucknow(UP,India), he has previously held key position including Head of Department and Assistant Controller of Examination in reputed institutions and Dean Academics at Rajkiya Engineering college Kannauj(UP,India). With expertise in signal processing, basic Electronics, signal systems, Electronics devices and circuits, robotics and IoT. He has significantly contributed to the field through his research on signal processing and IoT. His work has been published in prestigious international journals like IRJET and IJRASET. Beyond academia, his leadership in organizing technical events, state level competitions. A passionate educator and lifelong learner, He continue to inspire students and professionals with his initiative teaching methods.

***

Pratham Aggarwal is a student researcher, passionate for emerging technologies and innovations to derive meaningful products or services with the help of it!

Atul Anand is currently pursuing an M.Tech degree in computer science and engineering at IIT Patna. He earned his Bachelor of Engineering degree from Visvesvaraya Technological University in Karnataka, India. Additionally, he has worked as a Project Fellow in the Department of Computer Science and Engineering at IIT Kanpur. His research interests include the Internet of Things (IoT), Federated Learning, and Post-Quantum Security.

Nilesh Anute is an Associate Professor at Sri Balaji University, Pune, where he plays a pivotal role in shaping future leaders through his academic and research expertise. With a strong background in management education, Dr. Anute is known for his dynamic teaching style, deep subject knowledge, and dedication to student success. His contributions to research, curriculum development, and institutional initiatives reflect his commitment to academic excellence and innovation in higher education.

Jagjot Bhardwaj is a software engineer focused on BI

Dhruv Dhayal is actively engaged in emerging technologies at the intersection of artificial intelligence and intelligent systems. His work reflects a strong commitment to innovation, interdisciplinary research, and ethical tech development. With a forward-looking approach, he aims to contribute to scalable, real-world solutions that drive meaningful impact in both academia and industry.

Rajeshkumar Golani is a software engineer with 12+ years of experience in global-scale product engineering, focused on mass-market systems and AI-powered innovations. He has worked at Meta, Microsoft, and Amazon, building user-facing features and infrastructure that impact hundreds of millions. At Meta, he led the integration of Meta AI into Facebook Feed, enhancing engagement for over 250M users. He also drove features for private sharing and meaningful interactions across Facebook, Instagram, and WhatsApp. At Microsoft, he helped scale Azure's compute infrastructure to support growing global demand. He holds an MS in Computer Science from Stony Brook University and a BE in Computer Engineering from L.D.

College of Engineering. His peer-reviewed research on mobile network transparency and ISP traffic differentiation has received over 90 citations. He brings a strong combination of technical depth, product focus, and large-scale system expertise to the peer review process. He is also IEEE Senior Member and Fellow Member at Institue of Analytics, UK.

M. Kanishka is currently pursuing an integrated MTech in Computer Science with a specialization in Data Science at VIT Vellore. As a fourth-year student, Kanishka is building a strong foundation in data science principles, aiming to contribute to the field through innovative research and practical applications.

Siwar Khemakhem is an assistant professor at the Faculty of Sciences of Gafsa and a member of the Electrical Systems and Renewable Energies Laboratory (LSEER) at the National Engineering School of Sfax, Tunisia. Their research interests include smart grids, electric vehicles, renewable energy and energy management.

Lotfi Krichen is a professor in the Department of Electrical Engineering at the National Engineering School of Sfax (ENIS) and has been the head of the Electrical Systems and Renewable Energies Laboratory (LSEER) at ENIS since October 2018. Their research interests include electric machine control, renewable energies, storage systems, smart grids, and electric vehicles.

T.C. Manjunath was born on 6th February 1967. He did his BE from RV College of Engg. in 1989, ME in Robotics & Automation from LD College of Engg. in 1992 with First Rank and PhD in Systems & Control Engg. from IIT Bombay in the year 2007. He is currently working as the Dean Research R D and as Professor in the Dept. of Computer Science & Engineering with specialization in IoT, Cyber Security & Blockchain Technology in Rajarajeswari College of Engineering, Bengaluru, Karnataka. He has completed B.E. Degree from R.V. College of Engg., M.E. Govt.'s LD College of Engg., Gujarat Univ., Ahmedabad in First Rank and Ph.D. in Systems and Control Engineering from the Indian Institute of Technology Bombay IIT Bombay He has got a teaching academic, research and administrative experience of more than 35+ years 10 years as Principal, 8 years as HoD rest as faculty and has published more than 1200+ research papers in various national & international journals and conferences, with 175+ patents, 25 book-chapters, 25 awards and also published 25 text books in various field of science, engineering and technology, completed projects worth Rs. 9 Crores till date apart from guiding 10 PhD scholars. He is a member of 30 Professional Society Chapters and is a fellow of IE, IETE, AMIEE and a chartered engineer. He got the best research scholar award in 2006 in IIT Bombay amongst all the research scholars of the institute. He has got 10 awards in academics and research activities.

Hitesh Mohapatra received the B.E. degree in Information Technology from Biju Patnaik University of Technology (BPUT), Odisha in 2006, and the MTech. Degree in CSE from Odisha University of Technology and Research (OUTR) University, Odisha in 2009. He received his Ph.D. in Computer Science & Engineering in 2021 from Veer Surendra Sai University of Technology (VSSUT), Burla, India. He has contributed 50+ SCI and Scopus indexed research articles. He has authored two books on Software Engineering and C Programming respectively. He has 15 years of teaching experience both in industry and academia. He has served the research community with various capacities like session chair, technical chair, keynote speaker etc. His research interests include wireless sensor network, smart city, smart grid and smart water. Currently he is working as Associate Professor at School of Computer Engineering, KIIT (Deemed to be) University, Bhubaneswar, Odisha, India.

Mohammed Mubasheer is currently pursuing an integrated MSc in Mathematics at VIT Vellore. As a fourth-year student, he is focused on advanced mathematical theories and their practical applications. Mubasheer aims to contribute to the field through innovative research and practical applications.

SelvakumarP. is currently an Associate Professor in the Department of Science and Humanities at Nehru Institute of Technology, Coimbatore, Tamil Nadu, India. Prior to joining Nehru Institute of Technology, he held positions as an Assistant Professor in the Department of Science and Humanities various engineering colleges in Coimbatore, Tamil Nadu, India, spanning from 2011 to 2023. Dr.P. Selvakumar's academic and research interests lie primarily in the field of phytochemistry. He has made significant contributions to his field, with an impressive record that includes 30 journal publications, 12 patents, and 20 conference papers. Additionally, he has organized 5 seminars and conferences, further demonstrating his commitment to academic and scientific exchange. Throughout his career, Dr. P.Selvakumar has been recognized for his dedication and excellence in teaching, having been honored with several awards, including the Best Teacher Awards. He holds a prominent position in the academic community, contributing actively to research.

Ganesh Pathak is an Associate Professor at Sri Balaji University, Pune (SBUP), where he brings extensive academic and industry experience to his teaching and research. Known for his commitment to excellence in higher education, Dr. Pathak has contributed significantly to the fields of management and business studies through his scholarly work, mentorship, and active participation in academic initiatives.

His interdisciplinary approach and dedication to student development make him a respected figure within the university and the broader academic community.

Swathi R, is currently working as Assistant Professor in the Department of Computer Applications at Sree Abiraami Arts and Science College for Women, Gudiyatham, Affiliated to Thiruvalluvar University, Vellore, Tamil Nadu, India. More than 9 years of teaching experiences. UGC NET Qualified. Her research interests include Networks, Artificial intelligence (AI), Machine learning (ML). She has also organized several international level conferences and national level workshops. She has published more than 10 International journals & more than 10 papers are presented in the International & National conferences.4 Book Chapters has published by IGI Global Publisher.

Nithin Subba Rao a distinguished manufacturing excellence leader with over 15 years of experience in the automotive industry, Nithin Subba Rao has established himself as a transformational force in operational optimization and cost reduction. He holds a Master's degree in Mechanical Engineering from Michigan Technological University and an MBA in Operations Management, combining technical expertise with strategic business acumen. As a Six Sigma Black Belt with 10+ approved patents in manufacturing process improvement, Nithin Subba Rao has demonstrated exceptional capability in driving operational transformation and delivering measurable results. His track record includes delivering over $80 million in combined cost savings and restructuring benefits while implementing Industry 4.0 technologies across multiple manufacturing facilities. Nithin Subba Rao's expertise spans digital transformation, lean manufacturing, ERP integration, and the development of sustainable continuous improvement cultures that drive long-term competitive advantage. His personal philosophy of continuous improvement – symbolized by naming his daughter “Aarohi,” meaning continuous improvement / ascending – reflects his deep commitment to daily progress and systematic excellence. He firmly believes that consistency beats intensity, working toward his goals one step at a time with a target of 1% daily improvement, a principle that has delivered miraculous results in both his personal and professional life

Mouna Rekik was born in Sfax, Tunisia, in 1988. She received her BS, Master's, Ph.D., and HDR degrees in 2011, 2012, 2015, and 2022, respectively, all in electrical engineering from the National Engineering School of Sfax, University of Sfax, Tunisia. She is an Associate Professor in the Department of Electrical Engineering at the National Engineering School of Sfax (ENIS), Tunisia. She is a member of the Electrical Systems and Renewable Energy Laboratory at the University of Sfax and an IEEE member. Her main research interests include power system stability, microgrids, renewable energy systems, plug-in hybrid electric vehicles, and smart grid technologies.

Madhu S has 31+ years of experience in academia and industry. A prolific author and researcher, he has authored 85+ books, holds 26 design patent, 5 utility patent published, published 25+ articles, and has won multiple awards. He serves as a consultant, keynote speaker, jury member, resource person, and academic mentor, guiding hundreds of students and contributing widely to research and innovation. BoS and BoE member for various universities.

Sandeep Sarkale is an experienced academician and researcher with a strong background in higher education, specializing in Production & Materials Management and Operations Management. With extensive expertise in academic and administrative activities under various government bodies related to technical and professional education, he has contributed significantly to institutional development, his research aptitude extends to healthcare and academic practices, demonstrating a commitment to advancing knowledge and improving educational methodologies. Experienced academician and researcher in Operations & Supply Chain Management with over a decade of teaching and administrative experience. Demonstrated expertise in Production Management, Six Sigma, Inventory Management, and Total Quality Management. Holds a Ph.D. from SRTMU, Nanded. Actively published in Scopus, Web of Science, and UGC Care journals. Passionate about contributing to peer review and academic quality assurance.

Sivakumar V. has received Ph.D. degree in Anna University, Chennai. Currently he working as an Assistant Professor Senior Grade in Vellore Institute of Technology (VIT), Vellore, India. Previously worked as Associate Professor in Dayananda Sagar Academy of Technology and Management, Bangalore. More than 15 years of teaching experiences along with 1 year of Industrial experience. His research interests include Networks, IOT, AI & ML, Data science and Cloud Computing. Act as a Reviewer & Technical Committee Member in many IEEE International Conferences. Act as an International Advisory Board Member, International Committee Member & Reviewer in the International Conferences. He has filed 3 patents and 2 patent has granted. He has published more than 30 International journals & more than 30 papers are presented in the International & National conferences. 4 Book Chapters has published by IGI Global Publisher. One Book Series has published by Springer. He has reviewed Papers in the IEEE Access (More than 52 Papers), IEEE Transactions on Mobile Computing, Elsevier - Results in Engineering, Springer - Cluster Computing, Springer Nature - SN Applied Sciences, IET Circuits Devices & Systems, TAYLOR & FRANCIS - Desalination and Water Treatment Journals.

Index

A

B

C

D

E

F

G

H

I

www.ingramcontent.com/pod-product-compliance
Lightning Source LLC
LaVergne TN
LVHW060004031026
849161LV00025B/1372